Accessible Technology and the Developing World

Accessible Technology and the Developing World

MICHAEL ASHLEY STEIN AND
JONATHAN LAZAR

OXFORD

UNIVERSITY PRESS

OXFORD

UNIVERSITY PRESS

Great Clarendon Street, Oxford, OX2 6DP,
United Kingdom

Oxford University Press is a department of the University of Oxford.
It furthers the University's objective of excellence in research, scholarship,
and education by publishing worldwide. Oxford is a registered trade mark of
Oxford University Press in the UK and in certain other countries

First Edition published in 2021

Impression: 3

Published in the United States of America by Oxford University Press
198 Madison Avenue, New York, NY 10016, United States of America

British Library Cataloguing in Publication Data

Data available

Library of Congress Control Number: 2021934835

ISBN 978–0–19–884641–3

DOI: 10.1093/oso/9780198846413.001.0001

Printed and bound by
CPI Group (UK) Ltd, Croydon, CR0 4YY

Foreword
Judith Heumann

Information and communications technology (ICT) has dramatically changed the world we live in, and will continue to do so even more markedly in the future, with new innovations appearing on a daily basis. What was once inconceivable is now not only possible, but the "new normal." At the same time, the digital divide that separates those with and without resources, based on their access to ICT, continues to grow and to have practical social and economic implications. This new measure of inequality has a profound impact on persons with disabilities (PWDs), and especially those living in low resourced settings.

Two stories are commonly told about the origins of these two digital divides. First, that ICT innovation begins in the Global North and is then exported to the Global South through development and cooperation programs. Second, that making ICT accessible to PWDs is expensive, and therefore a secondary consideration. The chapter contributions in *Accessible Technology and the Developing World*, edited by Michael Ashley Stein and Jonathan Lazar, challenge both of those received wisdoms by providing ample examples of accessible ICT innovations arising in the developing world. Ultimately, ICT that is designed without the needs and abilities of PWDs creates additional barriers to their social inclusion, while accessible ICT enables and empowers all users, with and without disabilities.

In 2006, the United Nations General Assembly adopted, by consensus, the Convention on the Rights of Persons with Disabilities (CRPD), a paradigm-shifting human rights treaty aimed at ensuring the equal place and inclusion of PWDs in their societies. The CRPD has now been ratified by 181 States, making it a nearly universal mandate. Among the Convention's 50 articles, Article 9 requires accessibility for PWDs to all areas of life, including ICT; Article 21 assures equal access by PWDs to information; and Article 32 obligates States Parties to make their international development and other programming accessible to and inclusive of PWDs.

I know from personal experience the importance of accessibility in the built environment and with ICT for disabled people. The world has changed dramatically since I had polio in 1949, when television was just beginning to be made more widely available in the United States. Then, accessibility of any form was not seen as essential for disabled people. Buses and trains were not built accessibly, nor were ramps put on buildings or bathrooms made accessible, despite returning disabled veterans and other disabled people speaking out. I mention this, because we have learned that including access in original design and implementation is much more cost effective. The failure to include access features in ICT means that we are still not seen as an essential part of the customer base, with disposable income as consumers, nor as an essential part of the design, development, and implementation phases.

Over the years, I have benefited from ICT innovations that facilitate my access to print materials, make it easier for me to use my computer, read, and ensure that I can be a valued employee as well as readily communicate with friends, family, and colleagues across the world. Through my work at the United States Department of Education, World Bank, United States Department of State, and at various disability-run civil society organizations, I have witnessed the potential of accessible ICT to empower PWDs, ensure access and inclusion, and challenge stereotypes and assumptions of the role we can play in our families, communities, and societies. I have also seen how much we have had to work to get many of these entities to step up and commit to making fundamental changes that will ensure accessibility from the outset.

I also understand the gaping inequalities that result when ICT is not accessible. We know that accessible ICT has an important role to play in ensuring equality of access for disabled people in education, employment, healthcare, political partici-pation, and all areas of life. The CRPD has been a welcome and important tool in pushing for the changes needed—a tool developed by United Nations Member States in partnership with the international disability community—and we as disabled people are gaining influence. However, it is essential to say that we are not yet where we need to be. ICT is a perfect example of how we are seeing tremendous efforts made to pass and implement laws, rules, and standards advancing accessibility, but our all-too-frequent absence from leadership and decision-making roles still results in disabled people not having the influence needed to effect meaningful and sustained changes.

We must not be an afterthought, but be seen as an integral part of the design of all ICT. Disability rights organizations are slowly having more influence. However, employers in the ICT market still do not regularly demand that new hires under-stand accessibility, necessitating remedial training. Too many leaders in govern-ment, academia, and business perceive us to be an isolated group of people asking for charity, when in reality we are nothing of the sort. We can and should be full participants in the communities advancing progress in every country—progress from which all members of society can benefit. Critical to changing approaches is changing understandings of accessible ICT and the important role that disabled people can and must play in its development. I welcome *Accessible Technology and the Developing World* as an important contribution to expanding understand-ing of these issues, and in situating disabled people—especially from the Global South—at the center of this work.

Contents

LEGAL FRAMEWORKS, DESIGN APPROACHES, AND APPLICATIONS

Introduction

Nearly all research on information and communications technology (ICT) accessibility and innovation for persons with disabilities (PWDs)—whether from the legal, technical, or development fields—has focused on developed countries and the contributions they make (or should make) to developing countries, with very little being written about initiatives arising from resource-constrained states and their significant global contributions.

Several reasons commonly explain this asymmetry. Developed countries are more likely to have well-established technical telecommunications infrastructures; promulgate institutional mechanisms including agencies, law, and policies to require public compliance; possess sufficient resources to implement accessibility standards and modalities; prioritize the education of service providers and individuals with disabilities about ICT use; and provide socioeconomic environments conducive to fomenting innovative concepts and solutions. On the other hand, developing countries often have technology and service infrastructures that differ from more developed states. For instance, in developing countries, where wired telecommunications infrastructures are often lacking, wireless infrastructure and phones more commonly serve as the primary methods of communication. And, when technology-inclusive mandates do exist for people with disabilities in these nations, they are focused on the more basic goal of getting assistive technologies to users, not on ensuring equal access to all digital information.

Consequently, digital accessibility is often misperceived by ICT companies and developers. They wrongly assume that digital accessibility is expensive, when it's not. They likewise misapprehend that accessibility can simply be added at the end of technology development, but such changes cannot always be incorporated into the established technology, and those modifications inevitably increase the overall costs. ICT companies and developers also wrongly assume that individuals with disabilities won't use their technologies, hence there is no need to make it accessible. This is exemplified by the infamous and circular justification that "no blind people use our product!" which in turn is used to rationalize leaving those products inaccessible to the visually impaired. Along the same lines, ICT companies and developers often and erroneously presume that PWDs cannot be involved in product development because, allegedly, unlike ICT experts, they "don't know what they need." Each of these misperceptions is inaccurate. Yet, because digital accessibility continues to be viewed through a stereotyping lens, barriers to accessible and effective ICT use by PWDs remain in place and continue to be created.

Accessible Technology and the Developing World. Michael Ashley Stein and Jonathan Lazar, Oxford University Press.
© Oxford University Press 2021. DOI: 10.1093/oso/9780198846413.003.0001

Concurrently, a series of misperceptions pervade notions regarding digital accessibility in developing areas of the world. One erroneous assumption is that people who live in resource-constrained locations are not interested in ICT and its related content. Another equally wrong view is that innovation in the area of digital accessibility occurs exclusively in the developed part of the globe and is then transferred, beneficently, to the developing parts of the world. In this, as in many ways, people with disabilities and people who live in developing areas face parallel challenges: the paternalism of people without disabilities deciding "what's best" for PWDs (ableism), and the arrogance of elites in privileged societies determining "what's best" for individuals living in the developing world (colonialism). By contrast, the lives and decisions of people with disabilities, and the ICT developed for them, should always be determined by and with those individuals, not by external actors, however well intended they believe themselves to be. The mantra of the global disability rights movement, "Nothing About Us Without Us," can and should be equally applied to recipients of ICT development schemes.

This book focuses on ICT accessibility in the developing world, meaning the intersectionality of technology accessibility and development in resource-constrained places. Although technologists often use the term "development" to mean the building of software or hardware, here we use the term to mean the idea and process of encouraging economic and infrastructure growth in low-resource settings. Notably, there is a great deal published about ICT accessibility for people with disabilities in developed areas of the world (Global North), but there is limited work about ICT accessibility in developing parts of the world (Global South). Moreover, the innovations being created in developing states often get little or no attention. We often give the example of the government of Ecuador deploying a network of fieldworkers using global positioning system technology to identify and be able to subsequently assist PWDs across that country, including those living in hard-to-reach terrain, in the event of natural disaster. In comparison, many localities in the United States have established emergency evacuation notices in case of natural disaster, yet those notifications are frequently inaccessible to users with disabilities. This is only one example of the underexplored phenomena of developing states, precisely because they lack resources, advancing creative and cost-effective uses of ICT to include and enable PWDs.

Equally, it is important to ensure that when Global South countries evolve their technology infrastructures, barriers are not erected to equal access for people with disabilities. It is also crucial to not just apply techniques from the Global North, which most often will not transfer well to the developing world. In fact, developed nations have much to learn from developing countries about potential technical advances and solutions. For instance, it seems quixotic and inappropriate to premise dissemination of information regarding basic services, such as availability of vaccinations or locations of clean water and sanitation, on the possibility that people in Botswana will have iPads with internet access (even if some do). However, a radio transmission or a mobile phone app might provide this same information in a more immediately practical manner, although access to those technologies themselves may require certain disability-specific accommodations (e.g., captioning for the deaf and hard-of-hearing). In this respect, it is vital not to strike a colonial attitude but to acknowledge that, although many technological transfers will flow from the North to the South with needed adaptions to the local context in a developing country, many innovations and creative solutions will also be generated and disseminated from the Global South to the Global North.

*　　*　　*　　*　　*

Three international legal and policy initiatives that are at the forefront of ensuring the accessibility and equal availability of ICT are described in detail in Schulze's Chapter 1, as well as frequently referenced throughout the volume. The United Nations (UN) Convention on the Rights of Persons with Disabilities (CRPD), the first disability-specific international human rights treaty, contains several relevant provisions. Article 9 requires States Parties to ensure accessibility to PWDs to ICT on an equal basis with others; Article 21 obligates States Parties to take appropriate measures for ensuring freedom of expression as well as equal access to information for PWDs; and Article 32 mandates that all international cooperation, including the transfers of ICT, be "inclusive of and accessible to PWDs." The Marrakesh Treaty to Facilitate Access to Published Works for Persons Who Are Blind, Visually Impaired or Otherwise Print Disabled (Marrakesh Treaty), which forms the basis of Harpur and Stein's chapter, is an international copyright treaty administered by the World Intellectual Property Organization. The Marrakesh Treaty mandates copyright exceptions that enable access to printed works by individuals with print disabilities. The 17 Sustainable Development Goals are aimed at reducing global poverty and inequality, reference disability 11 times, and include requirements for inclusive infrastructure in Goal 9 and sustainable cities in Goal 11.

Hence, governments, development agencies, bilateral donors, non-governmental organizations, disability advocacy groups, and technical communities all need to focus their attention and resources on how to make ICT accessible to the 800 million PWDs living in the developing world. As many technology infrastructures (both network and organizational) in the Global South are currently being built, this is an especially important time to focus on the topic of ICT accessibility, in order to find ways of ensuring that barriers are not needlessly put in place. The quality of life for 800 million individuals, as well as billions of dollars of investment in development aid and loans, are at stake.

*   *   *   *   *

Skeptics often say that "accessibility is a barrier to innovation," missing the point that accessibility *is* a form of innovation. Numerous mainstream technologies, such as voice recognition, e-books, eye tracking, and speech-to-text, started out as assistive ICT, but have been broadly adopted across the world. Indeed, accessibility is a hotbed of innovation, and accessibility innovations, when originating in developing areas of the world, are even more creative and more innovative. Put another way, when accessibility innovation drives creative solutions in resource-limited countries with limited resources and less infrastructure, it can result in more imaginative, less costly innovations. This begs the question of why there is only limited research on and acknowledgment of ICT accessibility in the Global South. We propose a few potential theories in the next few paragraphs. Each discusses chapters in the book that aim to improve the understanding of these topics.

The technical infrastructures in developing areas are often very different from the technical infrastructures in the developed world

While most developed countries already have an established wired communications infrastructure, to which wireless was added, many developing areas are establishing their telecommunications infrastructure as a wireless-first infrastructure. Depending on the geographic area, the electrical infrastructure (the power grid) may also be a challenge to adoption of assistive and accessible technology. Accessibility innovations transferred to the Global South often miss these important factors, leading to inappropriate expectations of the tech infrastructure and therefore adoption in the developing world. In Chapter 10, Landers describes how individuals in developing areas often develop less resource-intensive accessibility solutions, yet those innovations may fail to meet the requirements for patent protection.

Developing areas may not have the existing educational and advocacy infrastructure of developed areas

In much of the Global South, awareness of and support for assistive technology and accessibility comes out of universities or long-existing advocacy groups. Yet assumptions about how to increase adoption of assistive technologies and accessibility in developing areas often incorrectly assume the existence of such an educational and advocacy infrastructure. Chapters 14 and 16 by Pavithran and Ojok, respectively, discuss the relationship between education and educational institutions, and digital accessibility in low-resource settings. In Chapter 7, Beaumon describes many of the organizations that currently play a role, and the types of partnerships that need to occur for improved accessibility in developing areas.

The legal frameworks for disability rights that exist in developed countries often do not exist in developing countries

Most countries around the world have ratified the CRPD (as of August 2020, 181 UN Member States have ratifications or accessions to the treaty, with the United States a noticeable absence). Schulze in her chapter describes the international legal framework for inclusive and accessible development. However, often the specific national laws or regional laws related to disability, when implemented, do not provide legal requirements or oversight for the areas of digital accessibility. For example, when considering the utilization of the Web Content Accessibility Guidelines (WCAG) in government laws and policies, the difference between developed areas with laws that specifically address or discuss the WCAG, and developing areas that do not have laws that specifically address or discuss the WCAG, becomes very clear. Almost all of Africa, most of Central Asia and Southeast Asia (with China, India, and Japan as notable exceptions), and parts of Central and South America, do not have laws or policies that specify WCAG for web content. Chapter 11, by Khan, Sheikh, and Akram, describes the impact of the legal framework for accessibility in Pakistan, while Harpur and Stein (Chapter 9) describe the role of the CRPD and the Marrakesh Treaty in improving access to accessible digital books.

Design methods used for assistive technology and accessible technologies are often designed with a Global North mindset

The biggest barriers to acceptance and adoption of accessible ICT and assistive ICT, especially in developing areas, are often culture, economics, and telecommunications (and electrical) infrastructure. Yet most design methods do not hold these as a central concern, assuming that the telecommunications and power infrastructure are sufficient to have a device that always has access to WiFi and power outlets. Other aspects that need to be addressed as a core part of the design methods are the involvement of and management by local residents (not outsiders), literacy levels, and cultural perceptions of disability. Chapter 12, by Chavarria, Mugeere, Schönenberger, Hurst, and Rivas Velarde, describes design methods that are more appropriate for the Global South; Swaminathan and Pal (Chapter 13) discuss the concept of "ludic design"—designing technologies for exploration and fun in the Global South; and in Chapter 8 Giannoumis and Skjerve examine the relationship between intersectional identity and accessibility.

Those without experience in development often assume that development only occurs in countries that are of low socioeconomic status

If you ask someone without experience about developing countries, they will often point to countries with low average gross domestic product. Yet describing developing countries in this fashion misses the point that there are many high-income countries which contain areas that are still developing, with low access to water and sanitation, electricity, and internet access. Brewer and Abou-Zahra, in Chapter 6, mention such areas within the United States, including locales where indigenous populations live, which can also be considered "developing." Similarly, there are many areas in countries that could be considered developing countries that actually are very high in terms of economic development and ICT use. McClain-Nhlapo and Raja (Chapter 2) offer extensive and far-reaching evidence from the World Bank on the relationship between disability-inclusive economic development programming and what rationales motivate the funding of accessible digital technology.

Data needs to be collected about technology accessibility in developing areas

Data is often collected on ICT accessibility in North America and Europe, ignoring the need for understanding accessibility elsewhere. Frequently, when one reads studies about accessibility in developing areas, the focus is on something one-dimensional, such as website accessibility for government agencies. Yet understanding the infrastructure for accessible technology is a complex undertaking, involving laws, policies, advocacy groups, and educational institutions, among others. This book includes three chapters from the Global Initiative for Inclusive ICT (Chapters 3, 4, and 5), written, respectively, by Gallegos, Gould, and Leblois; Cesa Bianchi; and Narasimhan. The Global Initiative for Inclusive Information and Communication Technologies (ICT) coordinates the Digital

Accessibility Rights Evaluation Index, the most comprehensive data collection effort in the world related to digital accessibility. While the Digital Accessibility Rights Evaluation Index is not well known in many accessibility circles, it is comprehensive, multi-faceted, and we believe that it is important for all digital accessibility researchers and practitioners to be familiar with it. In addition, Brewer and Abou-Zahra, in chapter 6, describe the role of technical standards for improving digital accessibility in developing areas.

Individuals often miss the connection between accessible technology and economic independence

The infrastructure of daily living, including public transportation, banking, shopping, and government, is often assumed to be accessible or at least partially accessible. Yet, in many developing areas, the infrastructure has barriers that stop those with motor or perceptual impairments from even accessing them. Two chapters in the book describe innovative uses of technology to improve access to independent living and economic independence in India. In Chapter 17 Kushalnagar writes about the use of accessible mobile banking, and Kameswaran and Pal (Chapter 15) write about the use of accessible ride-sharing.

*   *   *   *   *

This book brings together a unique combination of authors. The chapter contributors come from non-governmental organizations that are part of the public zeitgeist (the World Wide Web Consortium); significant UN entities (the World Bank and Global Initiative for Inclusive ICT); universities in the developing world (Pakistan and Uganda); universities in the developed world (the United States and Norway); and Global North industrial labs innovating in the Global South (Microsoft Research, India). We also included a diverse set of disciplinary backgrounds (technology, law, development, and education). Our sincere hope is that this book brings increased attention to studying and enabling ICT accessibility in developing areas.

*   *   *   *   *

We thank the Radcliffe Institute for Advanced Study at Harvard University for sponsoring an Exploratory Seminar in June 2018 that brought together several of the volume's contributors as well as related experts. In doing so, Radcliffe provided an opportunity to investigate and think through many of the cross-cutting themes of the book, and we are grateful. We also thank the Weatherhead Center for International Affairs for support; Isabel Ruehl for research assistance; and our terrific editors at Oxford University Press for their encouragement and support. Michael thanks his colleagues at the Harvard Law School Project on Disability (www.hpod.org); Jonathan appreciates the support of the College of Information Studies (ischool.umd.edu), Trace Research and Development Center (trace.umd. edu), and the Human-Computer Interaction Lab (hcil.umd.edu), all at the University of Maryland.

GLOBAL AND REGIONAL PERSPECTIVES

1

Development for All

How Human Rights Break Down Barriers to Technology

Marianne Schulze
Independent Legal Consultant and Human Rights Expert

1.1 Introduction

Persons with disabilities (PWDs) tend to be overlooked and left out. This is also true of policies around "development" aid—financial and technical support provided to countries perceived as needing support in governance and financial matters. The Convention on the Rights of Persons with Disabilities (CRPD) addresses the disparities for PWDs in development efforts by way of a stand-alone provision: Article 32.

The understanding of disability has been transformed by the CRPD by placing the focus on the social factors contributing to disability. It is less the perceived impairment, and more the attitudes and lack of inclusion in planning, which contributes to PWDs being left out, overlooked, and ultimately left behind. The concept of "development" is simultaneously undergoing a fundamental shift, as it is increasingly understood to be based on a flawed premise: the reasons for countries with low economic means are manifold, and they often have a direct link to historical exploitation of one sort or another. And framing such contexts as "underdeveloped" goes against the aspiration of making countries equal, because the underlying assumption is that of being an underdog in need of "Western benevolence" (Hickel, 2017, p. 25) rather than an aspiring equal partner.

Against this backdrop, this chapter discusses the concepts of disability, accessibility, and "development," and contextualizes them within international human rights standards. In doing so, the negotiations of the provision on "inclusive development" in the CRPD is outlined and connected to milestones at the international level in the decade following the conclusion of the treaty.

Technology plays a pivotal role in enabling accessibility. For that to happen, technology itself needs to be designed accessibly and also economically accessible (affordable) to users. Furthermore, users have to build the capacity to use and benefit from the technology. These and other aspects of technology, as seen from a human rights perspective, round out this text.

Accessible Technology and the Developing World. Michael Ashley Stein and Jonathan Lazar, Oxford University Press.
© Oxford University Press 2021. DOI: 10.1093/oso/9780198846413.003.0002

1.2 What Is Disability?

The reasons for the exclusion of PWDs are manifold, largely social, and certainly socially constructed: stigma, stereotypes, fear of doing harm, or being unhelpful. Frequently there is also a lack of "space" for day-to-day interaction with individuals with disabilities as a result of this very marginalization. Newborns with obvious impairments being left to die; family members with disabilities hidden; persons with mental health impairments being held in cages or chained—frequently without clothes, sanitation, shelter, let alone medication—possibly in full public view; persecution because of assumptions of sorcery and witchcraft: the violence and denial of basic needs and therewith human rights for PWDs is long.

Human rights experts frame such exclusion as both a legal and a social problem:

> Both de jure and de facto discrimination against PWDs have a long history and take various forms. They range from invidious discrimination, such as the denial of educational opportunities, to more "subtle" forms of discrimination such as segregation and isolation achieved through the imposition of physical and social barriers.
>
> (Covenant on Economic, Social and Cultural Rights, 1994, p. 15)

Being overlooked and unaccounted for is a defining trait of the treatment of PWDs. When their existence is acknowledged, the perceived otherness combined with helplessness frequently results in scales of paternalistic approaches. In parts also sparked by faith, a culture of pity and mercy fuels alms and other forms of welfarism intended to "do good." PWDs are often spoken-for rather than asked, assumptions about their needs and wishes are ubiquitous, and their capacity tends to be underestimated. Where a certain level of engagement is made possible—including special schools and sheltered workshops—a culture of over-protection unfolds. For most PWDs, this curtails a sense of making their own choices, including their right to make and learn from mistakes, an essential part of experiencing dignity (Deegan, 1996).

Dignity and a life in dignity is the central goal of human rights. The aspiration of the 1948 Universal Declaration is that "all human beings are born free and equal," (Article 1), yet the clause prohibiting discrimination on various grounds (Article 2) mentions neither disability nor impairment. This "omission" in the Universal Declaration of Human Rights is one of many legal markers that reinforce a social pattern of excluding and largely shunning PWDs.

Concluded in 2006, the CRPD seeks both to address the social factors contributing to the exclusion of PWDs and to remedy the legal gaps caused by that very exclusion. A human rights based understanding of impairment and disability is focused not so much on establishing and recording perceived deficits but rather on environmental and attitudinal factors that contribute to the marginalization and exclusion. Given that disability is in many ways a social construct, the CRPD recognizes its "evolving nature" (Preamble [e] CRPD). Acknowledging that there is no finite way of describing impairments and disability medically or socially, the CRPD enshrines an open-ended description of PWDs as a "non-definition": "PWDs include those who have long-term physical, mental, intellectual or sensory impairments, which in interaction with various barriers may hinder their full and effective participation in society on an equal basis with others" (Article 1 CRPD).

In addition to the explicit mention of "mental impairment," the emphasis placed on other barriers, particularly those of an "attitudinal" nature, which "in interaction with environmental barriers" negate the inclusion of and accessibility for PWDs, should be noted (Preamble [e] CRPD). The importance of such external factors, among others, is evidenced in data on the likelihood of persons with mental impairment being subject to violence and abuse as well as on the frequency with which their human rights are violated (WHO, 2010).

The shift of focus to the "others"—as a contributor to dis-abling environments—necessitates a clear move away from a purely medical view, a deficit-only description of impairment and PWDs, respectively. The bio-psycho-social model (WHO, 2001) receives new impetus in the CRPD's aim to fix attitudes rather than perceived impairments and "deficits," respectively.

Disability as a construct, then, has many aspects and is "a combination of various individual, institutional and societal factors that define the environment within which a person with impairment exists" (Trani, 2011). This environment is full of stereotypes, prejudices, and harmful practices, "including those based on sex and age, in all areas of life" (Article 8 CRPD) as well as hate crimes and "multiple or aggravated forms of discrimination on the basis of race, colour, sex, language, religion, political or other opinion, national, ethnic, indigenous or social origin, property, birth, age or other status," (Preamble [p] CRPD), all of which need to be addressed (Article 4 [1] [b] CRPD).

1.3 Accessibility

The most widely recognized dimension of accessibility is the physical or architectural one: ensuring that mobility-impaired persons—as well as elderly people and those using prams, rollers, and other mobility devices, or persons lugging suitcases—have access to all premises and public spaces. Secondly, redressing the more "subtle" forms of discrimination and, importantly, the manifold ways in which the agency of PWDs is structurally impaired requires the dismantling of attitudinal barriers, enabling social accessibility.

A third dimension, communication (Article 2 CRPD), ensures access to information, including means and modes of communication other than verbal, highlighting augmentative and other devices. Orientation in buildings, including hospitals, also falls within this category. The paths to communication, also for non- and semi-verbal persons, are manifold, as the CRPD clearly states. Due to its essential role, one aspect of communication is specifically mentioned as a fourth dimension: intellectual accessibility through easy-to-understand formats. All of these dimensions of accessibility need to be reflected in the design and handling of technology. Note that the broad understanding of accessibility in the CRPD also covers personal assistants as well as refuge spaces.

A human rights based approach highlights that PWDs have largely been deprived of adequate opportunities to enable them to earn sufficient income: economic accessibility or affordability, respectively, is therefore of great importance.[1] Finally, institutional accessibility shall ensure that structural barriers, which limit the opportunities for PWDs to participate on an equal basis with others, are removed: in planning, programming, and so forth. Accessibility is key to inclusion, particularly in design but also in social terms. Addressing the factors that create exclusion and making sure everyone can feel comfortable in their own skin is thus key.[2]

Technology can help enable that: initially by assisting accountability on births and early identification and referral but even more so for PWDs in obtaining the knowledge and skills that empower them to live equally in the communities of their choosing, comfortable with who they are.

How do all these factors—disability, exclusion, accessibility—play out in the context of "development," that is, situations where institutional shortcomings and economic resource constraints are dominant?

1.4 "Development"

The term "development" is used by the United Nations (UN) to designate those countries who are in need of various forms of aid to progress to a certain level of—largely economic—success that takes them to the threshold of a developed

---

[1] The foundations for Availability, Acceptability, Adequacy, and Quality were laid out in General Comment 14, CESCR Committee, Right to Health, E/C.12/2000/4, Para. 12.

[2] In many communities this is still embodied in the right to life, or at least the right to live and thrive.

country. The oversimplification of "development" and the perception of "developing" countries as opposed to "developed" countries is highly problematic and charged with deep historical injustices which continue to dominate to this day, including colonialism and therewith racism (Escobar, 1995). The division of the globe into a—supposedly prosperous and highly developed—North and a—supposedly poor and less developed—South[3] is a stark and troubling reflection of how deeply etched the sense of supremacy of colonialists is in the consciousness of many people in formerly colonial countries and their allies. The dichotomy of "us, the (incredibly helpful and know-it-all) developers" versus "them, the beneficiaries (mostly clueless)" has entrenched the division of political power and—as some argue—increased the level of disempowerment in the name of empowering (Crewe, 1998).[4]

The term "development" indicates a potential for growth; in the context of the UN, it has dominantly been interpreted as an index for economic growth. For the purposes of this discussion, it should be noted that growth is not only an economic vector that reflects select parts of economic expansion and that there are significant achievements amid contexts which might seem poorly resourced to an outsider. The loaded nature of terms and its consequences for the description of relationships and situations should be noted also, particularly in how it may reflect on the distribution of power. This is particularly true for PWDs, who tend to be portrayed as powerless. Reflecting this awareness, the term "resource-constrained contexts" is being used.

Obviously, human rights and development can hardly be discussed as separate planet systems. However, history and politics have wedged a split between the two that has left a distinct mark on inter-relation and reinforcement. The former High Commissioner for Human Rights, Mary Robinson, sees different cultures at play: "Unlike development, human rights is not a pragmatic tradition" (Robinson, 2005, p. 34). The extent of the gap and the puzzlement over how to bridge it are well captured by the observation of the former President of the World Bank, James D. Wolfensohn: "We're an institution which is owned and, to some extent, dominated by the very same people who wrote the Universal Declaration of Human Rights and the subsequent human rights instruments" (Wolfensohn, 2005, p. 21).

"While the majority of development policies and frameworks incorporate human rights concerns, many do so only implicitly" (McInerney-Lankford, 2009, p. 21), and, as a result, converging principles such as participation, inclusion, and particularly equality, which are both well established in development discourse

---

[3] As one example of many "poorest people in the poorest countries," Nigel Crisp, *Turning the world upside down*, 6.

[4] See, also, the observation by Stephen P. Marks on "worthwhile development" versus "undesirable maldevelopment" (Marks, 2018).

and are human rights (McInerney-Lankford, 2009, p. 53), are not used as mutually reinforcing. The power issues at play are largely put aside under the premise of states' "sovereignty" and the corresponding claim that human rights equate to internal matters only, which other countries have no business wading into. All the while, as Wolfensohn observes, countries violate human rights obligations, to which they—voluntarily—signed up. As a result, they trample on one of the core tenets of both rights and development: "do no harm."

In the midst between development debates and human rights obligations, the idea of an actual "right to development" sprung up. Originally, the debate took hold as part of the discussions of a New International Economic Order. As part of the UN Conference on Trade and Development, a number of countries put forward a series of proposals to promote trade on their terms in the 1970s. Subsequently, a Declaration on the Right to Development was agreed (UN, 1986). However, the notion of such a right is still questioned by many, particularly those countries that contribute significant economic support to "development" efforts. Thus, the discussions of a pertinent provision in the CRPD reflected the conflicted understanding of what a human rights based approach to development[5] could and should mean (Article 32 [2] CRPD).

The CRPD thus enshrines the principle of "progressive realization," which was coined as an obligation of states to fulfill their human rights obligations in line with the resources available to them. The intention at the time[6] was to have countries with low economic resources sign up to human rights with the promise that the level of implementation would be measured against their fiscal capacity. The phrase "to the maximum of available resources"—per Article 2 Covenant on Economic, Social and Cultural Rights—is "a difficult phrase—two warring adjectives describing an undefined noun. 'Maximum' stands for idealism; 'available' stands for reality," as Robertson[7] pointedly observes.

1.5 "Development" in Core Human Rights Treaties

To appreciate the breakthrough nature of Article 32 CRPD—the stand-alone provision on inclusive development—it is worthwhile to explore the way in which "development" is framed and represented in the preceding core human rights treaties of the UN,[8] among them:

---

[5] Chapman, 27, points to the 1995 Copenhagen Declaration on Social Development and Programme of Action as an indicator of using "vocabulary of development rather than rights."

[6] Article 2 CESCR; an extensive debate was undertaken by the Committee in General Comment 3.

[7] Robert E. Robertson (1994), Measuring state compliance with the obligation to devote the "maximum available resources" to realizing economic, social and cultural rights. *Human Rights Quarterly*, 16, 694, as cited by Chapman, p. 31.

[8] Other international agreements, including at the regional level, will not be explored for this context.

- Covenant on Economic, Social and Cultural Rights
- Covenant on Civil and Political Rights
- Convention on the Elimination of All Forms of Discrimination Against Women
- Convention on the Rights of the Child

The Covenant on Economic, Social and Cultural Rights (CESCR) starts with the principle of "progressive realization": "Each State Party to the present Covenant undertakes to take steps, individually and through international assistance and cooperation, especially economic and technical to the maximum of its available resources, with a view to achieving progressively the full realization of the rights recognized in the present Covenant" (Article 2 CESCR). The right to an adequate standard of living is tied to the "essential importance of international cooperation" as is the right to be from hunger (Article 11 CESCR). The understanding of "international cooperation" as a development measure is also reflected in the treaty's provision on scientific advances, which also mentions "international contacts" as a means of cooperation (Article 15 [4] CESCR). The CESCR also specifically mentions the importance of "technical assistance" as a means for support in resource-constrained contexts.

It bears noting that the twin-treaty of the CESCR, namely the Covenant on Civil and Political Rights (CCPR), discusses the question of "development" and "cooperation" in the frame of the self-determination of peoples—rather than as a measure tied to the fulfillment of individual rights for individuals (Article 1 CCPR). The Convention on the Elimination of All Forms of Discrimination Against Women (CEDAW) significantly narrows the relevance of development and does not mention "international cooperation" per se; rather the focus is on the particular problems women in rural areas face, and the importance of ensuring that efforts for "development" enable the participation of women (Article 14 CEDAW).

The UN's most widely ratified human rights treaty, the 1989 Convention on the Rights of the Child (CRC), makes scattered references to international cooperation and—anticipating the change in the "development" debate—emphasizes the importance of information and its exchange (Article 22 CRC). With relation to mass media and its production, the use of international cooperation to ensure diversity of cultural, national, and international content is emphasized (Article 17 CRC). Reflecting the times, the provision on children with disabilities discusses the promotion of "preventive health care and medical, psychological and functional treatment of disabled children," "in the spirit of international cooperation" (Article 23 [4] CRC). The CRC also contains a provision on progressive realization (Article 24 CRC). As for rights-specific support, "international cooperation" is encouraged in realizing the right to education, which includes the elimination of "ignorance" (Article 28 [3] CRC).[9] The provision specifically refers to "the needs of developing countries." Apart from the UN's core human rights treaties, the World Programme Concerning Action on Disabled Persons, which distinguishes between technical and economic cooperation (Rule 21) and international cooperation (Rule 22), bears mentioning for the intersection of disability and "development."

Against this backdrop, the negotiators of the CRPD agreed to a stand-alone provision on development inclusive of and accessible to PWDs—Article 32.

1.6 Enshrining "Inclusive Development"

Since the adoption of the 1989 CRC, which is the first of the UN's core human rights treaties to specifically include an anti-discrimination clause for disability and a stand-alone provision for children with disabilities (Articles 2 + 23 CRC), the full and effective enjoyment of all human rights for persons—better: children—with disabilities has not seen any noteworthy improvements (Quinn, 2002). The agreement reached in the aftermath of the Cold War to strengthen human rights globally, the 1993 Vienna Declaration and Programme of Action, makes a clear reference to the importance of implementing human rights for PWDs (UN, 1993, p. 63). Also, a pertinent General Comment by the CESCR a year later did not yield the anticipated change toward more recognition and inclusion (CESCR, 1994). Thus, a more substantial clarification of the human rights of PWDs was called for.

Following a few salient attempts, the Mexican government made a foray into the area at the 2001 World Conference against Racism, Racial Discrimination, Xenophobia and Related Intolerance in Durban, followed by a proposal by then president Vicente Fox, which led to the negotiations of a Comprehensive and Integral International Convention on the Rights and Dignity of PWDs (UN, 2001). The negotiators were explicitly tasked with bridging the gap between human rights and "development," as the draft convention was to be "based on a holistic approach in the work done in the fields of social development, human rights and non-discrimination" (UN, 2001).

The Mexican government pushed hard to get the negotiations underway (Sabatello, 2013), and was sure to make good on the promise of a social development angle. While in many ways exceptional, the Ad Hoc Committee was not immune to the ongoing battles over the role and function of "development" and its interplay with rights at the national level. Accordingly, the "issue proved too

---

[9] The inclusion of children with disabilities was further developed in a pertinent General Comment in 2006, parallel to the conclusion of the CRPD negotiations.

divisive to merit mention in the Working Group text" (Lord, 2018) which followed the third negotiation round. Discussions were largely around whether there should be a provision at all—given that there was no precedent in human rights treaties—and how the question of sovereignty and potential demands of countries providing financial and technical aid should be resolved.

1.6.1 Parliamentary Assistance from Brussels

As the Seventh Ad Hoc Committee Meeting got underway on January 16, 2006, the main donor countries—represented by the European Union—were still starkly opposed to an elaborate mention, let alone a stand-alone provision on "development." Meanwhile, the European Union's Parliament adopted a resolution (EU, 2006) which recalled various provisions of the European Convention for the Protection of Human Rights and Fundamental Freedoms and, among other items, restated the fact that, due to their being disproportionately poor, there was a need to specifically include "PWDs" in poverty alleviation and—strikingly—"in all policies and activities" of the European Union's delegations. Finally, the resolution called on the Commission of the European Union to support a separate Article on international cooperation in the Convention as a "necessary foundation for collaborative actions in pursuit of inclusive development and to facilitate bilateral and multilateral exchange of expertise, between developing countries and between such countries and the European Union."

The resolution was shared with select delegations pushing for a stand-alone provision on "development." In the January 2006 facilitations on a potential development provision, the delegation of the European Union cringed as the resolution was mentioned. Within the European Union, the Parliament is not seen as a political heavy-weight. But at a multi-lateral level such as the UN, there was no arguing over the weightiness of the EU Parliament. The move sealed Article 32: International Cooperation.

The first stand-alone provision on "development" in a UN core human rights treaty reads as follows:

Article 32

International Cooperation

1. States Parties recognize the importance of international cooperation and its promotion, in support of national efforts for the realization of the purpose and objectives of the present Convention, and will undertake appropriate and effective measures in this regard, between and among States and, as appropriate, in partnership with relevant international and regional organizations and

civil society, in particular organizations of PWDs. Such measures could include, inter alia:

(a) Ensuring that international cooperation, including international development programs, is inclusive of and accessible to PWDs;
(b) Facilitating and supporting capacity-building, including through the exchange and sharing of information, experiences, training programs and best practices;
(c) Facilitating cooperation in research and access to scientific and technical knowledge;
(d) Providing, as appropriate, technical and economic assistance, including by facilitating access to and sharing of accessible and assistive technologies, and through the transfer of technologies.

2. The provisions of this article are without prejudice to the obligations of each State Party to fulfill its obligations under the present CRPD.

Note that "accessible and assistive technologies" are specifically mentioned—as well as a few other features which bear highlighting, such as the strong emphasis on the participation of not only PWDs as experts in their own right but also civil society at large as well as international organizations. This reinforces the general obligation (Article 3 [c] CRPD) to ensure participation (as per Article 4 [3] CRPD) and the requirement in the monitoring provision which immediately follows Article 32 (Article 33 [3] CRPD). This strong base for "inclusive development" means that all planning considerations and discussions have to involve experts in their own right: PWDs. Per (1) (a), no less than an international level of cooperation per se is to be inclusive of and accessible to PWDs. Planning, importantly, starts a lot sooner than most consultation processes allow and is certainly not limited to legislative proposals only, which are—if at all—the forum for participatory efforts.

1.6.2 New Meanings

Section 1.01 As the UN Secretary General noted soon after the adoption of the CRPD, "As a human rights instrument with an explicit social development dimension, the CRPD is both a human rights treaty and a development tool" (UN, 2008, p. 61), thus giving renewed impetus to the human rights based approach to "development," the procedural aspects of which have been highlighted elsewhere (OHCHR, 2006).

Also noteworthy is the holistic understanding of "cooperation" that underlies Article 32, which includes capacity-building by way of sharing information but also experience (Article 32 [1] [b] Convention). The provision on facilitating cooperation in research and scientific and technical knowledge (Article 32 [1] [c] CRPD) is particularly helpful when read in conjunction with the obligation that the *advances* of scientific research should be widely shared (Article 15 CESCR). The UN Declaration on the Use of Scientific and Technological Progress in the Interests of Peace and for the Benefit of Mankind clearly states that scientific and technological achievements shall benefit "the material and spiritual needs for all sectors of the population" (UN, 1975). In a detailed discussion of the obligations to share the benefits of scientific progress, the UN Special Rapporteur in the field of cultural rights, Farida Shaheed, emphasizes the need to eliminate both de jure and de facto barriers in access, explicitly mentioning PWDs and underscoring access to processes, also through consultative processes (UN, 2012a).

All in all, there is a clear departure from the notion that "development" or "international cooperation" should be largely financial or technical. Moreover, the understanding that there is a "North–South" divide and that "development" is monolithically in the Northern sphere (Crewe, 1998) is replaced by a multi-directional understanding which may go any which way and, importantly, may create paths that strengthen genuine eye-level give-and-take.

1.6.3 Monitoring

Accountability and transparency have been ongoing challenges for human rights obligations. It is therefore particularly poignant that the CRPD's provision on international cooperation is immediately followed by the first-ever requirement for a national monitoring framework in the core treaty (Schulze, 2013). Consequently, these national frameworks should explicitly be tasked with oversight of countries' Article 32 efforts.

Article 32 (2) addresses an inherent dilemma: Can the provision of aid have strings attached? Can the aid be made dependent on fulfilling human rights obligations? Importantly, an agreement was found on the inherent tension between the sovereignty of countries and tying together human rights obligations and the provision of international cooperation—or support. Article 32 (2) was agreed to by consensus and foresees that there be no ramifications for delivery. Furthermore, the provision on progressive realization—Article 4 (4)—equally reinforces the notion that some rights need to be realized as a matter of urgency.

First and foremost, this applies to freedom from discrimination: The CRPD includes a definition of discrimination, and includes the denial of reasonable accommodation in this clause (Article 2 CRPD). Reinforcing all this, the provision of reasonable accommodation has to be ensured (Article 5 [3] CRPD). Consequently, any budding new policies and projects need to abstain from creating new barriers or adding to existing ones. As a result, all new school buildings and health clinics designed and built should be fully accessible. The same applies to participation: There is no obvious reason to justify the exclusion of PWDs from consultation processes.

An area that requires particularly urgent attention is the lack of data. The CRPD enshrines a pertinent provision—Article 31; as has been emphasized elsewhere, data on disabling barriers is paramount (WHO, 2011).

Obviously, all provisions of the Convention are interlinked—or, as the Vienna Declaration would have it, interdependent, interrelated, indivisible and universal (UN, 1993, Para. 5)—and therefore mutually reinforcing. Two provisions stand out: Article 11 on Situations of Risk and the specific mention of poverty reduction programs in Article 28 (2) (b).

1.7 Technology

In addition to the explicit reference to technology in Article 32, the CRPD enshrines an obligation to "undertake or promote research and development of, and to promote the availability and use of new technologies, including information and communications technologies, mobility aids, devices and assistive technologies, suitable for PWDs, giving priority to technologies at an affordable cost" (Article 4 [1] [g] CRPD). The emphasis on economic accessibility, that is affordability, bears repeating, even though this provision is titled "General Obligations" to start with. The provision of accessible information is also directly tied to assistive technology and its availability (Article 4 [1] [h] CRPD).

Furthermore, access to new information and communications technologies shall be ensured and the design, development, and production of accessible information promoted (Article 9 [g] + [h] CRPD). The importance of accessible technology is also highlighted in the context of personal mobility and training of mobility skills as well as production of devices (Article 20 CRPD). Rehabilitation makes a reference to assistive technology, as does the provision on the right to political participation and elections (Article 26 + 29 CRPD). The latter has been highlighted elsewhere (Lord, 2018, Footnote 22); the application across the board may be emphasized by also explicitly mentioning the importance technology can play in access to justice (Article 13 CRPD).

The need for broad application has been reinforced by the Committee on the CESCR: "In supporting research and development for new products and services, States parties should aim at the fulfilment of Covenant rights, for instance by supporting the development of universally designed goods, services, equipment and facilities, to advance the inclusion of PWDs" (CESCR, 2017, Para. 24).

This followed on from the Committee's observation that technologies essential to sexual and reproductive health needed to be provided (CESCR, 2016, Para. 49 [g]).

Going forward, a broad understanding of accessibility, which bridges to universal design, should be embraced, as it also makes the case for accessibility beyond the realm of PWDs, thus reinforcing the principle of inclusion (World Bank, 2013). The potential is obvious from the shift that is embodied both in the CRPD and specifically in Article 32, as well as the policy changes that have been instigated since the treaty's adoption.

The principal importance of the obligation to ensure participation is particularly relevant in research, including who designs and participates as a researcher in projects. The impact on how accessibility is understood and how its potential is viewed changes dramatically. A game-changing capacity lies in budgeting based on anti-discrimination (including accessibility and inclusion), as well as procurement[10] founded on such an approach.

1.8 What Has Been Achieved in the Initial Decade

The unique setting of the Ad Hoc Committee and the exceptional achievement of a stand-alone provision on "development" in a human rights treaty meant that both the uniqueness of the CRPD and this historic provision were hardly known beyond the community of CRPD negotiators. In turn, that meant that the key human rights fora, including those within the UN system, maintained the same level of knowledge that had created the urgency around inclusive development. Following the adoption of the CRPD, it was therefore a matter of necessity to ensure that the treaty would be treated on a par with other UN core human rights treaties and to start the process of utilizing the "development" provision for the CRPD and for human rights discourse more broadly.

The piled up stereotypes and a prevailing welfarist approach to PWDs were thus reflected in the agenda itemization in various UN bodies, including the General Assembly's Third Committee (which deals with human rights), the Second Committee (which largely defines the development agenda), the Human Rights Council, the Commission on Social Development, the Statistics Commission, and the Annual Ministerial Review. The same applied to various UN agencies.

---

[10] See Chapter 3: Global Trends for Accessible Technologies in the Developing World for more on this topic.

1.8.1 Data Collection Challenges

Putting aside specificities, which apply to the exclusion of PWDs, the core challenge was the lack of an entry point to collect data on PWDs. The development efforts of the UN hinge on collecting—importantly: comparable—data on various aspects, which largely define resource-constrained settings. This trend was consolidated following the adoption of the Millennium Declaration (UN, 2000), which merged various development programs of the UN under one umbrella. The Millennium Development Goals (MDGs) set out various targets and indicators to be achieved by 2015, notably the aim of halving poverty globally—the effort was thwarted by various challenges, including, not least, the so-called financial crisis of 2008. The omission which prevented the collection of data about PWDs was the mention of neither "disabilities" nor "PWDs" in the Declaration. Conversely, there is mention of refugees in the Declaration (UN, 2000, Para. 26), and correspondingly data was collected.

The uphill battle was to demand the collection of data on "disabilities" amid the resource cutbacks of the financial crisis. Collecting data on "disabilities" is a challenge unto itself: given the cultural specificities of the concept and the various layers of welfarism and medicalization piled under the tendency to overlook and deny impairments, it is tricky to find common ground on data collection. What is more: data on "disabilities" is not merely about knowing how many PWDs there are, it is to a large extent an effort of understanding what factors contribute to the exclusion of PWDs; for example, in accessing technology. Pointing to the stand-alone provision on data collection in the CRPD (Article 31)[11]—another first of that treaty—was largely met with helpless shrugs within the UN system.

Thus started a process which essentially follows the twin-track approach of the CRPD: one track focused on PWDs and ensuring there are policies in place to ensure inclusion and accessibility, and a second track aimed at ensuring the inclusion in policies designed for the so-called mainstream. Starting with the 61st General Assembly (2007), efforts got underway to amend UN resolutions and policies, which used outdated terms and failed to recognize the data-gap and to ensure that "the inclusion of and accessibility to PWDs"—as Article 32 CRPD enshrines it—is specifically added to most any policy and resolution as a mainstreaming effort. Ahead of a stock-taking of the MDGs in 2010, a resolution was adopted to provide an entry point for the inclusion of PWDs in MDG related processes (UN, 2009a). Almost parallel to this, the Security Council made a

---

[11] A noteworthy development in this area was the Busan Declaration of the Fourth High Level Forum on Aid Effectiveness, 2011.

first-ever reference to PWDs (UN, 2009b) and placed PWDs on its list of issues for permanent consideration. A decade later, the Security Council adopted the first stand-alone resolution on PWDs (UN, 2019).

1.8.2 Human Rights Council

Ever since 2009, the Human Rights Council has held an annual debate on PWDs based on a report prepared by the Secretariat of the Office of the High Commissioner for Human Rights. In 2010 it notably focused on inclusive development (UN, 2010); in 2015 the Council discussed humanitarian emergencies (UN, 2015c). In lock-step with this, other UN fora as well as its agencies stepped up their efforts to make their policies more inclusive of, and accessible to, PWDs. The main resolution on budgetary priorities, which come under the title of the Quadrennial Comprehensive Policy Review, took up PWDs in 2012 (UN, 2012b). The opening discussion of the 68th General Assembly in September 2013 provided a High Level Meeting on Disability and Development with the overarching theme "The Way Forward: A Disability-Inclusive Development Agenda Towards 2015 and Beyond", which resulted in a Declaration (UN, 2013). A noteworthy process was undertaken by the UN High Commission for Refugees, when its Executive Committee—comprised of member states of the UN—adopted a conclusion on addressing the exclusion of PWDs (UNHCR, 2010). The debate was also sparked by pertinent reports of the Women's Refugee Committee (2008) and the International Committee of the Red Cross (Kett, 2007). The latter has since adopted a guidance note on PWDs (ICRC, 2017).

1.8.3 Development Programming

In 2015, the MDGs were replaced by a more refined development policy, the Sustainable Development Goals (SDGs), which specifically mention PWDs (UN, 2015a). The 2018 *Flagship Report on Disability and Development* provides an analysis of the impact of mentioning PWDs in the SDGs and concludes that energy access is an area of great underachievement. In redressing this, the report notes: "disability tends to be under the responsibility of a ministry or a department of health or social welfare, while assistive technology tends to be under the mandate of the ministry of health, and energy issues fall under the mandate of a ministry or a department of energy" (UN, 2018, p. 183).

It should be noted that the realm of humanitarian assistance—which continues to be treated as a distinct area—has also been impacted by discussion around the inclusion of and accessibility to PWDs (UN, 2011). There have also been setbacks, some of them due to the dynamics within the disability community. An important example is the UN Standard Minimum Rules for the Treatment of Prisoners (SMR), which were overhauled after five decades with major contributions from civil society. Imprisonment is historically very entwined with issues of health, including mental health. Efforts to broaden this particular aspect in what have come to be known as the Nelson Mandela Rules (UN, 2015b) failed largely due to outlier positions within the representation of PWDs.

The broader need to build bridges and alliances, particularly outside the realms of the CRPD, is also apparent in ensuring the accessibility of technology. A holistic, yet meaningful, accessibility paradigm needs to be pushed, in alliance with a broad range of stakeholders. While accessibility is key (and often life-saving) for PWDs, and the only viable path out of poverty and other resource constraints, it appeals and benefits the broader community in many ways. The obligations to ensure the sharing of scientific advances precede the CRPD; recently the voices for positive discrimination in applying these advances for the particularly marginalized, have grown.[12]

1.9 A Brief but Necessary Diversion: The Nexus of Poverty and Disabilities

There is a strong correlation between poverty and disabilities. Resource constraints exacerbate disabilities: PWDs are twice as likely as non-disabled persons to be impoverished, and poverty increases the likelihood of an impairment. Obviously, poverty in itself is usually a situation of resource constraints: it is aggravated in a country generally low in resources. The prevention of poverty and disability intersect but at the same time have to be thoughtfully distinguished.

"Prevention of impairment" carries very different meanings in diverse contexts and is easily prone to be perceived as a threat to PWDs. Prevention, understood as measures which reduce impairments that are solely related to poverty—malnutrition, poor or inexistent maternal care that increases the likelihood of injuries during delivery—are an extension of the human rights based approach, particularly to poverty reduction. "Attempts to reduce the numbers of disabled children being born are acceptable,

---

[12] A case in point: Special Rapporteur on the Right to Health, A/65/255, Para. 60.

if they are promoted in ways that do not threaten existing children and adults" (Shakespeare, 2015, p. 120).

Non-discrimination as a general human right sets a clear bar, strongly reinforced by the CRPD, to ensure that certain measures aimed at "prevention of impairment" are seen as unacceptable.[13] The adoption of the CRPD marks the decidedly overdue recognition distinguishing between poverty reduction and the right to life of PWDs: nesting road safety, adequate nutrition, skilled birth attendance, disaster prevention, disarmament, and social security, to name a few, in pertinent human rights treaties, including the CESCR; and firmly enshrining the right to life as well as the importance of poverty reduction (Article 28 [2][b] CRPD), social security (Article 28 CRPD) and the right to health, including prevention of secondary impairments (Article 25 CRPD).

Correspondingly, the World Health Organization and World Bank's seminal *World Report on Disability 2011* frames prevention as a development issue: "Prevention of health conditions associated with disability is a development issue. Attention to environmental factors—including nutrition, preventable diseases, safe water and sanitation, safety on roads and in workplaces—can greatly reduce the incidence of health conditions leading to disability" (WHO, 2011, p. 53).[14] Importantly, the report emphasizes the prevention of "disabling barriers" as the core prevention-theme of the CRPD (WHO, 2011, p. 59).

1.10 Conclusion: High Potential—The Promise(s) of Article 32, and Accessible Technology

The complexities of the world and the corresponding need for oversimplifications that fuel exclusion are particularly acute for PWDs, especially in contexts where resources are limited. There is a seemingly endless abundance of misconceptions piled up both for PWDs and for resource-constrained contexts, specifically "developing countries." The effects are perturbing—to say the least: the discrimination and exclusion of PWDs is amplified by resource-constraints, i.e., poverty, and the very essence of human rights, i.e., the right to life, is relentlessly in jeopardy for many. Poverty is a defining factor for most PWDs, particularly in "developing countries."

---

[13] Its outliers are eugenics; see, e.g., Article 3 European Union Fundamental Rights Charter on a prohibition. See, though, for a conflation of issues, the Venezuelan proposal on international cooperation in the CRPD negotiations (Lord, 2018). For a historical perspective at the UN level, compare the 1982 World Programme of Action A/RES/37/52 as well as the 1993 Standard Rules on the Equalization of Opportunities for PWDs, A/RES/48/96, Para. 22, among others.

[14] The report distinguishes levels of prevention (pp. 54–56) and primary prevention issues (pp. 57–58).

The CRPD brought about a stand-alone provision on the intersection between impairment-based discrimination and the role of providing support in resource-constrained contexts in addressing this very exclusion. Assistive technology is key in providing support for PWDs, also and particularly in resource-constrained contexts: the CRPD provisions intended to enable PWDs are highlighted to underscore their importance for inclusion.

In forging a path forward, the "increased nuance" (Schulze, 2014, p. 275) of the CRPD needs to be utilized. The concept and principle of accessibility is much broader than the needs of PWDs and extends far beyond the realm of physical access for persons with mobility impairments. The role and importance of technology are underscored and more refined in the most recent core human rights treaty of the UN. Accordingly, the rewards of these provisions need to be applied broadly and through processes that are based on the participatory approach enshrined in the CRPD, ensuring that users have a meaningful say from the start to ensure that standards are met and support needs are taken care of for the many.

Furthermore, the application of human rights principles has to be broadened in designing the use of technology. Particularly the four pillars: availability, accessibility, acceptability, and quality—also known as AAAQ—need to be applied to ensure accessible technology. Seen from a technological view, the human rights enshrined in the CRPD are high potentials.

References

Chapman, A. (1996). A "violations approach" for monitoring the covenant on economic, social and cultural rights. *Human Rights Quarterly*, *18*, 23–66.

Committee, Convention on the Rights of the Child. (2006). General Comment 9, Children with disabilities, CRC/C/GC/9.

Committee, Covenant on Economic, Social and Cultural Rights. (1990). General Comment 3, The nature of States parties' obligations.

Committee, Covenant on Economic, Social and Cultural Rights. (1994) General Comment 5, Persons with disabilities.

Committee, Covenant on Economic, Social and Cultural Rights. (2000) General Comment 14, The right to the highest attainable standard of physical and mental health, E/C.12/2000/4.

Committee, Covenant on Economic, Social and Cultural Rights. (2016). General Comment 22, The right to sexual and reproductive health.

Committee, Covenant on Economic, Social and Cultural Rights. (2017). General Comment 24, State obligations under the International Covenant on Economic, Social and Cultural Rights in the context of business activities.

Crewe, E., & Harrison, E. (1998). *Whose development? An ethnography of aid*. Zed.

Crisp, N. (2010). *Turning the world upside down*. CRC Press.

Deegan, P. (1996, September 16). *Dignity of risk and the conspiracy of hope* [Conference session]. The Sixth Annual Mental Health Services Conference of Australia and New Zealand, Brisbane, Australia.

Escobar, A. (1995). *Encountering development: The making and unmaking of the third world*. Princeton University Press.

European Union Parliament. (2006, January 19) Resolution on disability & development, P6_TA-PROV(2006)0033. Retrieved from http://www.europarl.eu.int/omk/sipade3?PUBREF=-//EP//TEXT+TA+P6-TA-2006–0033+0+DOC+XML+V0//EN&L=EN&LEVEL=0&NAV=S&LSTDOC=Y&LSTDOC=N.

Hickel, H. (2017). *The divide*. Windmill.

International Committee of the Red Cross (ICRC). (2017, October). Advisory Service, International humanitarian law and Persons with disabilities.

Kett, M. (2007). *Disability and disasters: towards an inclusive approach*. International Federation of the Red Cross, World Disaster Report.

Lord, J. E., & Stein, M. A. (2018). Article 32: International cooperation. In I. Bantekas, M. A. Stein, & D. Anastasiou (Eds.), *The UN Convention on the Rights of Persons with disabilities: A commentary* (pp. 955–977). Oxford University Press.

Marks, S. (2018). Development ethics as reflected in the right to development. In J. Drydyk and L. Keleher (Eds.), *Handbook of development ethics*. Routledge.

Office of the United Nations High Commissioner for Human Rights (OHCHR). (2006). Frequently asked questions on a human rights-based approach to development cooperation. Chapter II: Human rights and development. Retrieved from http://www.ohchr.org/Documents/Publications/FAQen.pdf.

Robinson, M. (2005). What rights can add to good development practice. In P. Alston and M. Robinson (Eds.), *Human rights and development: Towards mutual reinforcement* (pp. 25–44). Oxford University Press.

McInerney-Lankford, S. (2009). Human rights and development: A comment on challenges and opportunities from a legal perspective. *Journal of Human Rights Practice, 1*(1), 51–82.

Quinn, G., & Degener, T. (2002). *Human rights and disability: The current use and future potential of United Nations human rights instruments in the context of disability*. OHCHR.

Sabatello, M. (2013). A short history of the international disability rights movement. In M. Sabatello and M. Schulze, *Human rights and disability advocacy* (pp. 13–24). University of Pennsylvania Press.

Schulze, M. (2013). Monitoring the convention's implementation. In M. Sabatello and M. Schulze, *Human rights and disability advocacy* (pp. 209–221). University of Pennsylvania Press.

Schulze, M. (2014). The human rights of PWDs. In A. Mihr and M. Gibney, *Sage handbook of human rights* (pp. 267–283). Sage.

Shakespeare, T. (2015). *Disability rights and wrongs* (2nd ed.). Routledge.

Trani, J. F. (Ed.). (2012). *Development efforts in Afghanistan: Is there a will and a way? The case of disability and vulnerability.* Harmattan.

United Nations High Commission for Refugees, Executive Committee (UNHCR). (2010).

Conclusion on refugees with disabilities and other Persons with disabilities protected and assisted by UNHCR No. 110 (LXI) – 2010 Executive Committee 61st session. Contained in: United Nations General Assembly document A/AC.96/1095.

United Nations (UN), General Assembly. (1975, November 10). Resolution Declaration on the Use of Scientific and Technological Progress in the Interests of Peace and for the Benefit of Mankind, 3384(XXX).

United Nations (UN), General Assembly. (1986). Declaration on the Right to Development, A/RES/41/128.

United Nations (UN), General Assembly. (1993). Vienna Declaration and Programme of Action.

United Nations (UN), General Assembly. (2000). Millennium Declaration A/RES/55/2.

United Nations (UN), General Assembly. (2001). Resolution Comprehensive and Integral International Convention to Promote and Protect the Rights and Dignity of Persons with disabilities, A/RES/56/168.

United Nations (UN). (2008). Implementation of the Outcome of the World Summit for Social Development and of the Twenty-Fourth Special Session of the United Nations General Assembly, A/63/133.

United Nations (UN), General Assembly. (2009a). Resolution Realizing the Millennium Development Goals for Persons with disabilities, A/RES/64/131.

United Nations (UN), Security Council. (2009b) Resolution Protection of Civilians in Armed Conflict, S/RES/1894, 11 November 2009.

United Nations (UN), Human Rights Council. (2010). Thematic Study by the Office of the United Nations High Commissioner for Human Rights on the Role of International Cooperation in Support of National Efforts for the Realization of the Rights of Persons with disabilities, A/HRC/16/38.

United Nations (UN), General Assembly. (2011). Resolution International Cooperation on Humanitarian Assistance in the Field of Natural Disasters, From Relief to Development, A/RES/66/227.

United Nations (UN), Human Rights Council. (2012a). Report by the Special Rapporteur in the Field of Cultural Rights, A/HRC/20/26, 31.

United Nations (UN), General Assembly. (2012b). Resolution, Quadrennial Comprehensive Policy Review, A/RES/67/226.

United Nations (UN), General Assembly. (2013). Resolution Outcome Document of the High-Level Meeting of the General Assembly on the Realization of the Millennium Development Goals and Other Internationally Agreed Development

Goals for Persons with Disabilities: The way forward, a disability-inclusive development agenda towards 2015 and beyond, A/RES/68/3.

United Nations (UN), General Assembly. (2015a). Resolution, Sustainable Development Goals, A/RES/70/1.

United Nations (UN), General Assembly. (2015b). Standard Minimum Rules for the Treatment of Prisoners (the Nelson Mandela Rules), A/RES/70/175.

United Nations (UN), Human Rights Council. (2015c). Thematic Study on the Rights of Persons with disabilities Under Article 11 of the Convention on the Rights of Persons with disabilities, on Situations of Risk and Humanitarian Emergencies, A/HRC/30/31.

United Nations (UN), Department of Economic and Social Affairs, Disability and Development. (2018). Report, Realizing the Sustainable Development Goals by and for Persons with disabilities.

United Nations (UN), Security Council. (2019). Persons with disabilities as Civilians in Armed Conflict, S/RES/2475.

Wolfensohn, J. (2005). Some reflections on human rights and development. In P. Alston and M. Robinson (Eds.), *Human rights and development: Towards mutual reinforcement*. Oxford University Press.

Women's Committee for Refugee Women and Children (WCRWC). (2008). Disabilities Among Refugees and Conflict-Affected Populations.

World Health Organization (WHO). (2001). International Classification on Functionality, Disability and Health.

World Health Organization (WHO). (2010). Mental Health and Development: Targeting People with Mental Health Conditions as a Vulnerable Group.

World Bank. (2013). *Inclusion matters: The foundation for shared prosperity*. New Frontiers of Social Policy. World Bank.

World Bank and World Health Organization (WHO). (2011). *World report on disability*.

2

Addressing the Drivers of Digital Technology for Disability-Inclusive Development

Charlotte McClain-Nhlapo and Deepti Samant Raja
World Bank

2.1 Introduction

Digital technologies are transforming poverty reduction and development programs through providing new avenues to access markets and services, absorb and create information, and experience environments and communities. They can lower social and economic transaction costs to propel inclusion, increase productivity and efficiency, and drive innovation (World Bank, 2016). E-government, mobile payments, online and digital work, educational and financial technologies, and mobile health applications are dominant examples of economic development investments in low and middle-income countries.[1] The digital data revolution and its implications for data-driven services through artificial intelligence and smart technologies are changing how persons with disabilities (PWDs) interact with their surroundings.

This raises both opportunities and threats in ensuring the inclusion of PWDs, and is forefront in discussions on digital development. Over a billion people have a disability (World Health Organization & World Bank, 2011). The ability to use voice, text, touch, and gestures to receive, create, and communicate content fundamentally alters the ways in which public and private sector actors engage and serve PWDs (Raja, 2016). Incorporating accessibility into digital investments will propel disability-inclusive development with social, economic, and humanitarian benefits, addressing the needs of PWDs (Samant et al., 2012). Consequently, not

---

[1] This chapter follows the terminology used by the World Bank Group (2019a) for country income classifications. Low-income economies are defined as those with a Gross National Income per capita of $995 or less in 2017; middle-income economies combine lower and upper middle-income categories with a Gross National Income per capita of between $996 and $12,055; high-income economies are those with a Gross National Income per capita of $12,056 or more.

Accessible Technology and the Developing World. Michael Ashley Stein and Jonathan Lazar, Oxford University Press.
© Oxford University Press 2021. DOI: 10.1093/oso/9780198846413.003.0003

doing so will exacerbate existing inequities and could create new divides between persons with and without disabilities.

There are many singular examples of technology positively impacting the lives of PWDs in low- and middle-income countries. However, examples of scaled inclusive technology with accessibility baked into design and delivery are limited, or, when present, may not be associated with universal accessibility. There continues to be a wide gap between the potential of technology and its large-scale application in the daily lives of most PWDs.

The World Bank's *World Development Report 2016: Digital Dividends* concluded that "digital dividends," i.e., actual development benefits, have lagged the spread and adoption of digital technologies. Strengthening the "analog complements" of digital investments, i.e., "skills, institutions, and regulations" (World Bank, 2016, p. 5), is crucial for sustained benefits of technology-driven development for PWDs.

This chapter focuses on how digital ecosystems in low- and middle-income countries (LMICs) can evolve to scale digital opportunities for the social inclusion, economic self-sufficiency, and resilience of PWDs. It describes how information and communications technologies (ICTs) can benefit the lives of PWDs in developing countries. It also highlights the continuing inequities and gaps in access, affordability, and usage. Finally, it offers specific recommendations to strengthen capacity and institutions within the larger ecosystem to ensure that PWDs reap the digital dividends of ICT-enabled development.

2.2 The Promise and Potential of Technology in Development

Inclusive technology catalyzes access to markets, spaces, and services (World Bank, 2013). Development practitioners are increasingly using ICT solutions for disability inclusion across several sectors. The sub-sections below provide examples of accessible technology solutions advancing education and labor market participation and offering PWDs tools to access health and disaster management services, exercise their rights, and obtain public services.

2.2.1 Education

Children with disabilities in LMICs have lower rates of school participation, literacy, and learning outcomes as compared to their peers without disabilities (Male & Wodon, 2017). Accessible technologies can address delivery of appropriate learning materials, communication challenges, and difficulties in procuring assistive devices. The e-Disabled Project in Tunisia funded 24 technology centers in schools equipped with touchscreen devices, magnification software, text-to-speech, and sign language translation software (World Bank, 2015). Programs such as India's Sarva Shiksha Abhiyan (Government of India, 2018) and Open the Window's Active Inclusion project in Macedonia and Serbia provide accessible technologies and computer-assisted communication tools for learning (Fembek et al., 2016).

Digital books and content are attractive to students with diverse print disabilities, who can use their preferred devices and modes to access them. Users can download and access content through low-cost digital audio players, mobile phones, and personal computing devices, or convert into Braille (Fruchterman, 2015). Under the All Children Reading program, Bookshare by Benetech added human-narrated audio in Marathi, an Indian regional language, to facilitate the use of mother-tongue language educational content on low-cost DAISY audio players (Benetech, 2017). Bookshare subsequently added over 390,000 titles targeting users in India, including in the languages of Gujarati, Hindi, Tamil, and Telugu (Benetech, 2018).

Educational content for children using local sign languages is another innovation. A team of researchers from Center for Languages and Communication at the École Nationale Supérieure des Mines de Rabat in Morocco and the Institute for Disabilities Research and Training, Inc. in the United States is rolling out software that translates Arabic text to Moroccan Sign Language using graphics and video (Vinopol, 2015). The United Nations International Children's Emergency Fund's (UNICEF's) Digital Accessible Textbook initiative is testing tools to access textbooks in user-preferred formats, including narration, sign language, and audio description. eKitabu (2018), a Kenya-based company, has developed a toolkit and step-by-step guide to support users in creating accessible digital publications with "image descriptions, accessible navigation, dyslexic fonts, and optional sign language videos."

2.2.2   Jobs and Income Generation

The proliferation of ICTs, if accessible, in the world of work opens opportunities to break labor market inequities for PWDs. This includes jobs that use ICTs, online or microwork freelance platforms, and jobs within the technology industry. Disabled peoples' organizations, entrepreneurs, and innovators in developing countries are developing and marketing contextually relevant accessible technology solutions (GSMA, 2018).

Business process outsourcing centers, including call centers, have been used as a model of ICT-based employment for PWDs in developing countries (Kocak

& Kavi, 2011; Mundy, 2018). Impact sourcing social enterprises such as Digital Divide Data (2018) train PWDs and other marginalized groups to perform digital jobs in the outsourcing industry. Engelsiz Kariyer (https://www.engelsizkariyer .com/), a job seeking and matching platform in Turkey for PWDs, has been used by more than 10,000 applicants with disabilities (Fembek et al., 2017). Leonard Cheshire's (2015) JobAbility online jobs portal serves job seekers with disabilities in India, Sri Lanka, Bangladesh, Pakistan, the Philippines, and South Africa.

2.2.3 Disaster Management

PWDs experience higher mortality rates and higher vulnerability in disaster situations due to existing systemic inequities and lack of access to support services (Global Facility for Disaster Reduction and Recovery [GFDRR], 2017a). ICTs can support critical information exchange and service delivery for disaster mitigation, response, and recovery, helping to save lives and assets. Ensuring accessibility of these services will help save lives and connect PWDs with necessary preparedness, relief, and recovery services. Examples of ICT use for inclusive disaster management include SMS messages to contact emergency service centers (Vodacom South Africa, 2019) and training television broadcasters on good practices for accessibility in disaster situations (Global Alliance on Accessible Technologies and Environments, n.d.). CBM's (2017) Humanitarian Hands-On Tool mobile application provides step-by-step guidance to humanitarian workers on supporting the needs of PWDs in disaster situations and can work without an internet connection. The Higher Council of Affairs of PWDs in Jordan helped develop capabilities for the Jordan Command and Control Centre to receive video calls from persons who are deaf and trained staff at the center in sign language (Zero Project, 2016). India's National Disaster Management Authority's (2018) guidelines on disability-inclusive disaster risk management include recommendations on using accessible and inclusive ICTs and facilitating access to assistive technologies in disaster situations.

Geospatial mapping and imagery using data from satellites, aerial photography, and unmanned aerial vehicles such as drones are used to conduct spatial impact assessments before, during, and after disaster events (GFDRR, 2017b). Geographical information systems enable the layering of natural, infrastructural, and socioeconomic data collected from automated or manual sources (Fung, 2012). Responders can combine data on PWDs gathered through government systems or non-governmental organizations with spatial impact assessment data to gain a deeper understanding of the needs of PWDs for preparedness and response activities. Registries or databases with information about PWDs have provided time-critical evacuation and disaster response services in developing countries.

2.2.4   Financial Transactions

Accessible financial technologies (Fintech) allow PWDs independence, privacy, and autonomy in managing their own finances. PWDs can engage in commercial transactions and contribute to business development. Examples of accessible Fintech solutions in developing countries are limited but growing; most cater to the needs of persons with visual impairments. Safaricom's M-PESA mobile payments platform launched an interactive voice response (IVR) platform in 2017 to support users with visual impairments. Earlier, persons with visual impairments had to disclose their PIN to another individual to manage their accounts. The IVR facility allows them to call the service independently and access their account balances, bypassing the security and privacy risks of disclosing their PIN and financial information. Thailand's KBank's (2017) mobile application, K Plus Beacon, offers touch, voice, and vibration features. Talking ATMs are available in many countries, although coverage is far lower than needed and designs may be non-standard (Che, 2013; Raja et al., 2015; Singanamalla et al., 2019). Entrepreneurs in developing countries are using AI and image recognition to develop currency recognition mobile applications and tools (BarrierBreak, 2018; Yousry et al., 2018).

2.2.5   Health and Rehabilitation

PWDs in LMICs are at risk of worse health outcomes than persons without disabilities, and face information and infrastructure barriers to accessing healthcare services (Kuper & Heydt, 2019). E-health and mobile health services that enable access to medical information, practitioners, psychosocial counseling and related services such as pharmacy services can be extremely useful for persons living in remote areas or without convenient means of transport (Wootton & Bonnardot, 2015). Telerehabilitation services connect individuals with therapy services using video conferencing, motion detection, and even virtual reality and robotics (Narváez et al., 2017; Zahid et al., 2017).

Researchers are testing mobile applications for vision and auditory screenings for early identification and assessment of disability in LMIC settings. The United States Agency for International Development's Read project in Ethiopia tested two applications, *hearScreen* and *Peekvision*, in schools to identify students with difficulties in hearing (4%) and visual acuity (5.6%) and refer them for follow-up diagnoses and services (Strigel, 2017). Teachers had previously estimated that only 1.8% of their students had any disabilities that impacted learning.

The AT-Info-Map (2018) project, led by the Southern Africa Federation of the Disabled in partnership with the African Network for Evidence-to-Action in Disability (AfriNEAD), the University of Washington, and Dimagi, Inc., provides an example of using technology to increase information about available AT. AT-Info-Map uses digital platforms to capture, organize, and map assistive technologies in ten African countries; users can sort through the information using mobile and web-enabled interfaces.

2.2.6 E-Governance, Citizen Engagement, and Civic Participation

Increasingly, countries are using e-governance to deliver services and communicate with citizens. E-governance systems allow people to pay for utilities, settle fines, register new business or vehicles, and get information about stakeholder consultations, as well as other civic duties/engagements. The United Nations (UN) named India among the top half of 193 member states in the E-Government Development Index (EGDI) and marked four African countries—Ghana, Mauritius, South Africa, and Tunisia—with high EGDI for making select public services available online (United Nations Department of Economic and Social Affairs [UNDESA], 2018a). Thirty additional African countries, including Cameroon, Nigeria, Lesotho, Togo, and Rwanda, were rated as having made visible progress in e-governance. Effective e-governance reduces bureaucratic hurdles in accessing services, minimizes corruption, and allows for real-time feedback on service quality and consumer experience. In many developing countries, however, e-governance platforms have not reached the bottom 40% and are not reaching PWDs due to gaps in internet access and lack of accessible design.

Social accountability, civic participation, and voice are important to the development agenda, as engaging citizens improves accessibility, coverage, and quality of services. Development agencies such as the World Bank define citizen engagement as a two-way interaction between citizens and governments or private sectors (Manroth et al., 2014). There is scant evidence in LMICs of digital engagement and political participation for PWDs. This in part is due to the exclusion of PWDs from research on civic participation, but also, to their educational competencies and lower digital participation. Civic participation often spurs digital activism lauded for mobilizing and sharing images and information about human rights violations, protests, and lack of service delivery. However, in repressive regimes, there is a growing trend of shutting down digital spaces. The phenomenon of internet shutdown came to prominence in Egypt during the Arab Spring and has been reported in Cameroon, Ethiopia, Gambia, Uganda, Chad, Democratic Republic of Congo, Gabon, Mali, Zambia, and Zimbabwe (Mukeredzi, 2017). This uncertainty of internet access can chip away at the rights of PWDs.

2.2.7 Digital Data and Smart Technologies

There is growing enthusiasm to harness the digital data revolution and technologies such as AI to disrupt and alter international development methods and processes for greater efficiency and higher beneficiary impact (World Bank Group, 2019b). Digital data-enabled tools and applications can reshape how PWDs access their societies. However, it is worth re-emphasizing that the lack of attention to the accessibility dimensions of new and disruptive technologies will engulf new digital divides.

2.2.7.1 Big Data
Development practitioners see the ongoing data revolution as having an important role in improving the design and delivery of consumer-centered services. Big data combines vast data collected through traditional, active means (e.g., household surveys, government databases) with data collected passively through the multiple digital touchpoints that individuals encounter daily (Moorthy et al., 2015). Digital data is growing exponentially in the volume and variety, number of data providers, and speed of collection (Independent Expert Advisory Group Secretariat, 2014). Data on individuals' lives, purchasing preferences, grievances, emotions, and behaviors are collected through social media, online marketplaces, news sources, satellites, and sensors in our environments. Together, these data can facilitate descriptive, predictive, and prescriptive analysis in ways that were not possible earlier (Lopes & Bailur, 2018). Practitioners are using big data to improve urban planning, road safety, and traffic management (Monroe, 2017); track and improve the coordination of ambulances (Pollonais, 2018); and understand women's needs for and access to psychosocial care services (Data2x, 2017).

Notably, the UN Secretary General's Independent Expert Advisory Group on a Data Revolution for Sustainable Development underlined that big data can fill data-gaps on groups that remain underrepresented in development data such as PWDs or those living in hard-to-reach locations. However, as the Independent Expert Advisory Group cautioned, this requires deliberate attempts to invest in collecting disaggregated data on excluded groups—such as PWDs.

The paucity of traditional and digital data on the lives of PWDs in LMICs remains a significant challenge in inclusive development (UNDESA, 2018b). It is crucial to increase the collection of disability-disaggregated data through national and local population surveys. The disparities in ICT access and use by PWDs in resource-limited areas automatically limit their data on social media and other big data sources. This raises the risk that big data analytics used for development planning and policy decision making will overlook and exclude the needs and experiences of PWDs (Lerman, 2013).

Concerns regarding privacy, consent, anonymity, and confidentiality are important to consider when obtaining and analyzing user-generated digital data. In many countries, data on disability, age, race, and other characteristics are protected from collection and dissemination in many datasets (Veale & Binns, 2017). PWDs may choose not to disclose disability status where risks of discrimination are high (Trewin, 2018a). Analysts need to carefully consider the ethical, privacy, security, and societal risk dimensions of using digital data and adhere to strict ethical guidelines on using sensitive personal information (UN Development Group, 2017).

2.2.7.2 Artificial Intelligence

Artificial intelligence (AI) refers to computing methods that display facets of human intelligence, including recognizing sights and sounds, deciphering and discerning patterns, planning, reasoning, and problem solving. AI is powering products and solutions that surpass what assistive technologies have been able to achieve, particularly in the domains of computer vision, image recognition, and natural language processing. AI solutions such as Microsoft's (2019) Seeing AI (Singh, 2018) and Accenture's (2017) Drishti, which recognize people, expressions, objects, and text, and YouTube's (2017) automatic sound effect captioning system, which can recognize sounds such as applause and laughter, provide new and cost-effective ways of social engagement.

AI based on machine or deep learning analyzes and independently discerns patterns from large volumes of data of varying complexity and format, and uses what it has learned to make predictions or offer solutions to development problems. This raises opportunities to disrupt socioeconomic development for PWDs, and occasions questions to consider for inclusive and equitable outcomes (Stewart, 2018; Zurutuza, 2018).

AI needs accurate information on the lives, actions, and behaviors of PWDs in LMICs to make appropriate decisions that impact them (Li, 2018). Human coders may unconsciously transmit their prejudices and biases into their algorithms (European Commission, 2018; Vanian, 2018). Biases may also creep into AI due to inequities and gaps in the underlying data (European Commission, 2018).

AI applications may interpret PWDs as outliers (Trewin, 2018a) or end up mimicking the direct and indirect discrimination they face in society (Vanian, 2018). For example, AI programs that screen resumes of job applicants can interpret disclosed disability as a negative characteristic if applicants with disabilities represented in the databases were frequently screened out at early stages. AI recruitment programs have discriminated based on race and gender due to past hiring patterns without understanding the nuances of societal bias and discrimination (Dastin, 2018; Hao, 2018; Vanian, 2018). One can therefore extrapolate these biases to the exclusion of PWDs.

AI programs thus require deliberate actions by human coders to identify data-gaps on PWDs and manually correct or augment data for representativeness (Trewin, 2018a,b). Processes to manually "debias" datasets and algorithms by adding or encoding protected characteristics such as disability status are still emerging and in nascent phases (Veale and Binns, 2017). Trewin (2018a) proposes several steps to build fairness into AI—assessing whether PWDs may be excluded from the datasets; taking steps to mitigate bias throughout the AI development cycle; understanding how disability may impact other data variables and how the data is interpreted; handling outliers; and involving PWDs in design and testing.

2.2.7.3 Smart Cities

Urban planners and city leaders are using technology to improve evidence-driven resource allocation. City governments are embedding technology into infrastructure such as transport systems and soft services to enhance efficiency through automation, gather user data, and improve systems monitoring.

Smart cities can either improve service delivery and urban participation for PWDs or unintentionally exacerbate the exclusion they face through inaccessible design. For example, programs monitoring traffic patterns and tracking citizen mobility need to be cognizant of whether the data is inclusive of PWDs who may not use public transit due to inaccessibility. Technology interfaces such as kiosks or web-enabled applications need to be accessible to persons with different disabilities (Woyke, 2019). This is particularly crucial when the technology interface is the only way to access a service such as a ticketing counter (Woyke, 2019). Programs such as the Smart Cities for All (SC4A, https://smartcities4all.org) initiative are developing tools and solutions to support governments and urban planners in building disability-inclusive smart cities.

2.3 Scaling Accessible Technologies for Inclusive Development

The preceding sections illustrate how accessible technologies can drive and accelerate socioeconomic and civic participation of PWDs. It is important to evaluate these practices and invest in their scale and sustainability to make accessibility a standard feature in development programming.

Globally, development agencies are channeling financing toward digital infrastructure, digital economies, and digital skills to bridge divides such as those experienced by PWDs. This is an opportune moment for conscious and deliberate work to ensure that these efforts are accessible from the beginning and address societal and infrastructural inequities that hamper benefits for PWDs. Bridging digital divides requires complementary investments in core infrastructure, skills, institutions, and regulations (World Bank, 2016). The next section will look at why systemic gaps exist for PWDs in LMICs and their experiences in using digital services.

2.3.1 Digital Footprint of PWDs in LMICs

On average, 80% of the global population owns a mobile phone, with 101.6 mobile-cellular telephone subscriptions per 100 people (World Bank & International Telecommunication Union [ITU], 2018). Strikingly, more people have access to mobile phones than clean water or electricity (World Bank, 2016). However, internet access and use have not followed the rapid pace of mobile phone penetration. Internet-dependent development is still in its nascent phases in LMICs, as a large portion of their population remains offline and disconnected from the digital economy (World Bank, 2015). Figures 2.1 and 2.2 compare household ownership of computers, internet access, and use by geographic regions (Figure 2.1) and income categories (Figure 2.2). Globally, 45.9% of individuals use the internet, 46.6% of households have a computer, and 51.5% of households have internet access at home (World Bank & ITU, 2018). These numbers are particularly low in South Asia (only 14.7% of households have a

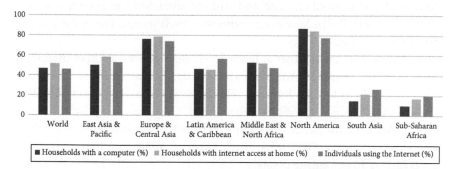

Figure 2.1 Regional Comparison of Household Ownership and Access to Computer and Internet (Source: World Bank and ITU, 2018)

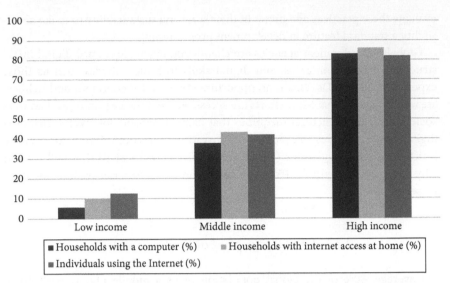

Figure 2.2 Computer and Internet Access by Income Groups (Source: World Bank and ITU, 2018)

computer and 21.4% of households having internet access at home) and Sub-Saharan Africa (9.9% of households with computers and 17% with internet access at home). Only 26.5% of individuals in South Asia and 20% of individuals in Sub-Saharan Africa use the internet. Less than a fifth of the global population has a fixed-broadband subscription.

Households with PWDs have lower access to ICT across the development or income spectrum (Figures 2.3–2.6). The figures, using data from Integrated Public Use Microdata Series and Demographic and Health Surveys datasets, show differences in ownership of internet facilities (Figure 2.3), mobile phones (Figure 2.4), computers (Figure 2.5), and landline phones (Figure 2.6) between households, with members with and without disabilities in several countries. Households with PWDs have lower ownership levels irrespective of the country's development status.

PWDs also use the internet at lower levels than persons without disabilities. Data from 14 countries (Antigua and Barbuda, Barbados, Belize, Cambodia, Costa Rica, Ecuador, El Salvador, Grenada, Guyana, Honduras, Maldives, Trinidad and Tobago, Uganda, and the United Kingdom) show that on average 19% of PWDs use the internet, as compared to 36% of persons without disabilities across the 14 countries (UNDESA, 2018). However, usage gaps in some countries are as high as 30 percentage points.

PWDs between the ages of 5 and 39 from Latin America and the Caribbean report higher levels of internet use than access to the internet at home (UNDESA, 2018b). This flips for ages 40 and above, with differences of up to 11 percentage

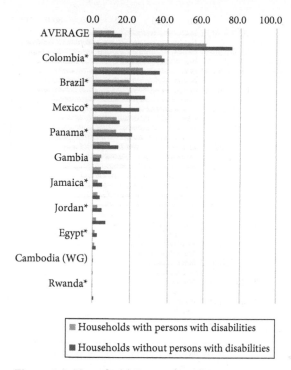

Figure 2.3 Household Ownership of Internet
Facility/Connection

points between having access to the internet at home and using it for persons of ages 75 and above.

Women with disabilities report lower access to ICTs and lower usage than men with disabilities. Figure 2.7 shows internet usage for men and women with and without disabilities using Demographic and Health Surveys data from Cambodia, the Maldives, and Uganda.

Internet affordability is a significant barrier, particularly in Sub-Saharan Africa and South Asia. Households must spend on average more than 5.5% of their monthly income to buy just 1GB of mobile broadband data (Alliance for Affordable Internet, 2018). Fixed broadband remains more expensive than mobile broadband (World Bank & ITU, 2018). Purchasing an average priced internet-enabled smartphone in India can cost 16% of the income for individuals on the lower end of the income spectrum (GSMA, 2017). PWDs are more likely to live in poverty and experience higher rates of multi-dimensional poverty than persons without disabilities, have lower employment rates and lower average wages, and bear additional out-of-pocket costs for disability-related expenses (Mitra et al., 2013; Mitra, 2018; UN, 2018; World Health Organization & World Bank, 2011). This increases the internet affordability gap for PWDs living in lower income groups in LMICs.

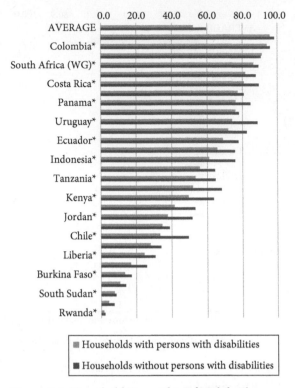

Figure 2.4 Household Ownership of Mobile Phones

2.3.2 Ecosystem Gaps and Barriers to Digital Accessibility

Technology use in education offers important lessons for development practitioners. Teachers' competencies and comfort in using digital and accessible technologies are strong determinants of how they will use and support technology use in the classroom, and whether technology will improve learning outcomes (Mavrou, 2011; Trucano, 2016; Wong & Cohen, 2012). For instance, during a visit to a public primary school in Lesotho that included some students with visual disabilities, one of the authors observed several boxes of *Jot a Dot* manual Braille writers stacked up in the principal's office. Students with visual disabilities on average face a lag of about two to three weeks to receive class notes in Braille. The low cost and light manual Braille writers could have been useful to the students, given the lack of accessible technology solutions. However, as often happens with assistive technology in international development, the manual devices were donated to the school without training the teachers and resource personnel. The writers remained in storage as a result.

Another illustrative example is the use of technology in providing unique identification numbers. Digital identification systems use biometrics (e.g., fingerprints,

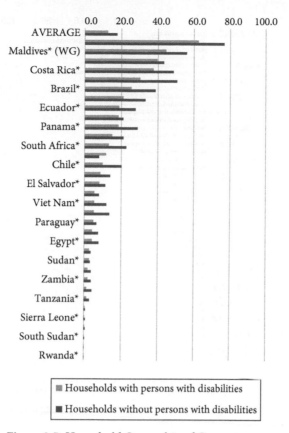

Figure 2.5 Household Ownership of Computer

iris scans, facial recognition) for identification. Most biometrics-based identification (ID) systems have exception management protocols when users face functional limitations in providing biometric data such as lack of fingerprints or lack
of iris pigmentation. During World Bank consultations by the authors in West
Africa on inclusive access to digital ID systems, participants with disabilities
shared several experiences of discrimination and exclusion. These included a face
scanning algorithm which returned an error message upon photographing a child
with atypical facial features; human operators did not use manual overrides,
which resulted in the child's ID not being generated; a person with albinism had
to return multiple times to register for an ID because the iris scan could not be
completed due to retinal pigmentation.

The most frequent barriers in digital ID programs stem from manual processes
and operators' lack of knowledge and training to serve individuals with different
types of disabilities, despite the existence of exception protocols on paper
(Hamid, 2017; Lord, 2019). Delays in receiving IDs are particularly concerning
when IDs are mandatory for receiving disability and other government benefits
(e.g., food rations, pensions) and services.

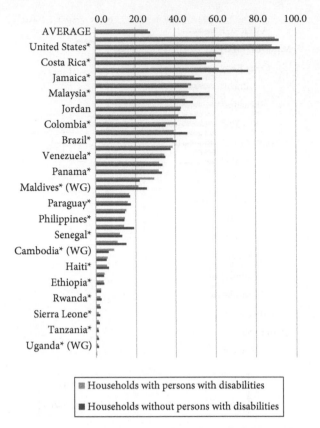

Figure 2.6 Household Ownership of Landline Phones

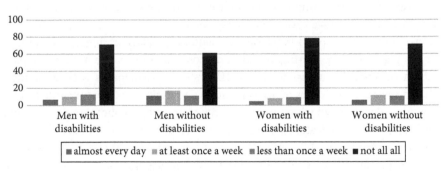

Figure 2.7 Internet Usage Differences Between Women And Men With Disabilities in Cambodia, Maldives And Uganda (Source: DHS data)

2.4 Strengthening Digital Ecosystems for Accessibility

Accessible technology can leapfrog development and support new avenues of serving PWDs in the developing world, but needs complementary investment in skills, institutions, and regulations (World Bank, 2016).

2.4.1 Skills

Skills include basic digital skills and digital literacy for PWDs and competencies to work in the ICT sector. They also include skills required by stakeholders such as technology developers, governments, teachers, and service providers to design, use, and support inclusive digital technologies.

PWDs need opportunities and tools to develop core digital literacy (such as using keyboards and touchscreens, simple word processing, using email) and task-oriented skills (e.g., digital publishing, analytical software) for jobs that depend on ICTs. And they need advanced competencies for employment in the ICT industry (ITU, 2018; Solutions for Youth Employment, 2018).

Digital skills should be a core competency of quality education and equitable learning to suitably prepare youth and adults with disabilities for the workforce demands of the future (Montoya, 2018).

Government-financed ICT skills development programs are important for the sustainable inclusion of PWDs. Egypt's Ministry of Communications and Information Technology, for instance, has trained over 600 youths with disabilities on basic and advanced digital skills and soft skills required for workforce transition (ITU, 2018).

Organizations such as inABLE in Kenya, v-shesh in India, and Catholic Relief Services in Vietnam, provide digital skills training for pathways into ICT-dependent jobs and jobs in the ICT industry. Programs support training on using assistive technology; for example, the POETA centers in Peru. Diversity initiatives by the digital technology industry open opportunities for PWDs to enter the technology industry.

Other actors within the digital ecosystem need skills in developing and applying accessible technologies. Development practitioners across sectors need to be cognizant of accessibility considerations in programs dependent on the use of technology. This will facilitate teams that potentially include members with and without disabilities with knowledge of inclusive designs and service delivery. Accessibility competency trainings and certifications for technology developers and content creators, such as ITU's (2019) Self Paced Online Training on ICT Accessibility, and trainings by the International Association of Accessibility Professionals (2021) can cultivate technical expertise in LMICs.

Pre-service and in-service curricula for engineers, designers, and product developers should include accessibility. Similarly, individuals with disabilities should have equal opportunities to participate in mainstream digital skills development programs.

2.4.2 Institutions

Capable and accountable institutions are essential to reaping the benefits of technology in development (World Bank, 2016). Technology in development programming can accelerate effective service delivery for all citizens, provide safe and reliable channels for beneficiary engagement, and improve government accountability. While technologies can bring efficiency to service provision, the underlying services themselves must exist and be stable and accessible themselves.

For example, smart transit services will at best offer stopgap solutions if public transport systems are not accessible or do not offer end-to-end connectivity for PWDs. Fintech will be of limited use if banking institutions have guidelines against the independent use of services by PWDs or do not train staff on serving PWDs.

Grievance redress mechanisms (GRMs) offer another case study for the importance of strong institutions. These aim to be accessible at the local level and offer multiple options for filing complaints—through a hotline, visiting branch offices, submitting online complaints, and/or reporting to community committees—however, many GRMs are not accessible or disability sensitive. For GRMs to be fully inclusive and accessible, they need to be more responsive to the needs of PWDs, have protocols or guidance on what accessible features are needed, and understand the degree of inequality and exclusion experienced by PWDs and how the right to access ICT invokes complimentary access to civil and political rights such as access to information, the right to vote, and related political rights. Democratization of digital technology and the meaningful inclusion of PWDs also depends on understanding the existence of elite capture by urban-based disability populations.

Hence, working toward universal and equitable access to digital technologies for PWDs needs continued actions toward improving the policy environment, range, and type of services for PWDs and elderly residents, and mechanisms to engage with public and private systems and services. Studies show that the success of e-governance projects is linked to the strength of a country's institutions (World Bank, 2018). Failing this, there will continue to be an implementation gap between the potential and application of technology for inclusion. The *2016 ICT Accessibility Progress Report* published by the Global Initiative for Inclusive ICTs (Gould and Montenegro, 2017) showed that one-third of 89 countries surveyed from the developing world had made legal and programmatic commitments toward ICT accessibility. Out of these 89 countries, only 18% were deemed by expert responders to have the capacity to implement ICT accessibility dispositions of the Convention on the Rights of Persons with Disabilities (CRPD). A little over one-third (35%) had implemented programs and policies to support inclusive technologies.

2.4.3 Regulations

As emerging technologies rapidly drive development and new business and delivery models, it is imperative that governments put in place and enforce regulations/guidance regimes. Laws and policies on the accessibility of electronic and digital devices, services, and content can shape the technology ecosystem for disability inclusion. Constitutional rights to inclusion, disability rights legislation, and direct references to inclusive technology in thematic laws such as those on education, labor market participation, or health data privacy provide the broader regulatory framework. Public procurement policies are an important tool through which governments can lead a shift toward inclusive technologies in development projects. By requiring accessibility in ICT procured by public agencies or government-funded contractors, governments can implement one method of top-down change. Evidence from the impacts of accessible public procurement in developed countries shows that it serves as an incentive for technology companies to gain a competitive edge in the bidding process and leads to innovative and inclusive products (G3ict, 2015). Similarly, donors, development actors, and producers must also develop policies that guide their work in this space.

Countries should adopt strong accessibility standards (such as Section 508 in the United States and the EN 301 549 standard in the European Union) for technology products. This is particularly important to ensure that technologies used for public service delivery and socioeconomic development meet minimum thresholds for access and inclusion of PWDs. Discussions on accessibility standards in the developing world have considered adopting one of the existing standards or creating a global standard to minimize the time and costs of creating country-specific standardization processes (G3ict, 2015).

The preeminent issue from a disability inclusion perspective is how to ensure that PWDs are included and innovations are impactful to them and that they do not exacerbate their exclusion. The main challenge for regulators is maintaining a balance between the rapid pace of innovation, protecting the beneficiary and the unintended consequences of disruptive technologies, as described earlier in Section 2.3 in the case of facial recognition and iris scans for IDs. Or consider 3D-printed products that are developing more affordable and customized prosthetics to the wearer: how should product liability laws and regulations be applied? Who is responsible if the artificial arm fails or causes further damage to the person with a disability?

Another important regulatory conundrum is who owns all the data that is being collected? This is a particularly sensitive issue for the privacy of PWDs.

What is clear, is that regulations cannot take the traditional route of being drafted over extended periods of time. The "pacing problem" created by disruptive technologies requires just-in-time policies (guidance/soft law) that are adaptive/ modifiable, rapidly enforceable, and widely communicated to the beneficiaries— both consumers and producers. As we extend the benefits of the digital age, the regulatory and policy objectives must focus on the consumer.

2.5 Conclusion

Technology holds great promise for development. Similarly, accessible technology has the power to include PWDs by giving the power to inspire, to organize, to communicate, to learn, and to influence change. Accessible technology is breaking barriers to participation in education, employment, financial management, access to critical services, and social and civic participation in LMICs (Raja, 2016). Its power and impacts are not without constraints.

The potential of digital development is not the sole panacea for changing social norms and realizing the rights of PWDs. For PWDs to reap the maximum and optimal dividends of digital dividends, practitioners have to pay attention to investments beyond technology and understand the constraints to consumption of digital platforms for PWDs.

This includes reducing the affordability and access gaps in digital infrastructure, devices, and services. It requires building the capacity of accessible technology users, providers, and facilitators such as teachers, banking officials, city planners, election officials, and emergency responders. It is important to design products in collaboration with the user to avoid flaws.

The rapid advancements in technology often outpace legislative and regulatory structures; regulations need updates on an ongoing basis to catch up with current trends and innovations. There needs to be a stronger shift toward building accessibility into products, content, and services right from the design stage. The popularity of devices such as the iPhone, which has in-built accessibility features, makes the business case for scale and return on investment (Donovan, 2018).

Regulating for and monitoring accessibility in the production, transmission, and rendering of ICT technology is fundamental; with obligations on both the supply and demand sides. If PWDs are to benefit from disruptive technologies, there needs to be a deliberate policy move to remove barriers that hinder the adoption of accessible technologies, build in competitive incentives to incorporate universal design, and offer clear channels for remediation.

It is important to recognize that injustice against PWDs in the application of new technologies may exist and that there needs to be a better understanding of how disability-biased algorithms can result in discriminatory allocation of resources and/or access to services. Technological solutions to detect and mitigate disability bias created by data-driven approaches in AI are still emerging and need enhancement. Accessible ICTs are good for business, and companies with higher percentages of PWDs have reported higher rates of return for their shareholders (Accenture et al., 2018). Businesses should procure more accessible ICT and deliberately work to ensure that PWDs are trained to work using ICTs.

The divides in technology use are rooted in underlying inequalities and inequities in societies. It is necessary to address inequality at every level and identify pathways for PWDs to benefit from and be part of the digital revolution. This is both possible and scalable; however, more deliberate attention needs to go into removing regulatory, attitudinal, or structural barriers that exclude PWDs. This needs to be proactive, and an integral part of the onset discussion about bridging the digital divide. It must never lose sight of the fact that innovation itself, whether digital or not, does not guarantee change, but can be a facilitator to change the fundamentals of behavior, leaving no one behind.

References

Accenture, American Association of People with Disabilities, & Disability:IN. (2018). *Getting to equal 2018: The disability inclusion advantage*. Retrieved from https://www.accenture.com/t20181029T185446Z__w__/us-en/_acnmedia/PDF-89/Accenture-Disability-Inclusion-Research-Report.pdf.

Accenture. (2017). *Accenture develops artificial intelligence-powered solution to help improve how visually impaired people live and work*. Updated July 28. Retrieved February 12, 2021, from https://newsroom.accenture.com/news/accenture-develops-artificial-intelligence-powered-solution-to-help-improve-how-visually-impaired-people-live-and-work.htm.

Alliance for Affordable Internet. (2018). *2018 affordability report*. Retrieved January 19, 2019, from http://1e8q3q16vyc81g8l3h3md6q5f5e-wpengine.netdna-ssl.com/wp-content/uploads/2018/10/A4AI-2018-Affordability-Report.pdf.

AT-Info-Map. (2018). Retrieved February 12, 2021, from https://assistivetechmap.org/index.php.

Barrier Break. (2018). *6 by 6*. Retrieved February 12, 2021, from https://www.barrierbreak.com/6by6/.

Benetech, Inc. (2017). *All children reading: A grand challenge for development, round 2 (Final project report)*. Retrieved February 12, 2021, from https://benetech.org/wp-content/uploads/2017/09/Benetech-ACR-Final-Report-082117.pdf.

Benetech. (2018, June 27). *Benetech expands Bookshare's international reach in India*. Globe Newswire. Retrieved from: https://globenewswire.com/news-release/2018/06/27/1530508/0/en/Benetech-Expands-Bookshare-s-International-Reach-in-India.html.

CBM. (2017). *Humanitarian hands-on tool*. Retrieved February 12, 2021, from https://hhot.cbm.org/.

Che, C. (2013). *Finally, an ATM that can talk to the blind*. South China Morning Post. Retrieved February 12, 2021, from https://www.scmp.com/news/hong-kong/article/1293152/finally-atm-can-talk-blind.

Dastin, J. (2018). *Amazon scraps secret AI recruiting tool that showed bias against women*. Reuters. Retrieved February 12, 2021, from https://www.reuters.com/article/us-amazon-com-jobs-automation-insight/amazon-scraps-secret-ai-recruiting-tool-that-showed-bias-against-women-idUSKCN1MK08G.

Data2X. (2017). *Big data and the well-being of women and girls: Applications on the social scientific frontier*. Retrieved February 12, 2021, from https://data2x.org/resource-center/big-data-and-the-wellbeing-of-women-and-girls/.

Digital Divide Data. (2018). *Success stories*. Retrieved February 12, 2021, from https://www.digitaldividedata.com/impact/success-stories/.

Donovan, R. 2018. *Unleash different: Achieving business success through disability*. ECW Press.

eKitabu LLC. (2018). *Accessible EPUB toolkit*. Retrieved February 12, 2021, from https://www.ekitabu.com/index.php/impact/toolkit/.

European Commission, High-Level Expert Group on Artificial Intelligence. (2018). *Draft ethics guidelines for trustworthy AI*. Retrieved January 5, 2019, from https://ec.europa.eu/futurium/en/system/files/ged/ai_hleg_draft_ethics_guidelines_18_december.pdf.

Fembek, M., Heindorf, I., Kainz, W., & Saupe, A. (2016). *Zero Project report 2016*. Focus: Education and information & communication technologies. Retrieved February, 12, 2021, from https://zeroproject.org/wp-content/uploads/2016/02/ZeroProjectReport_2016_barrierfree.pdf.

Fembek, M., Heindorf, I., Kainz, W., and Saupe, A. (2017). *Zero project report 2017*. Employment: Work and vocational education & training. Retrieved March, 16, 2021 from https://s3.amazonaws.com/zeroproject-uploads/zeroproject/wp-content/uploads/2017/02/Zero+Project+Report+2017+on+Employment_Vocational+Education+and+Training_accessible+format_English_13MB.pdf.

Fruchterman, J. (2015). *The state of the world's children 2015: Reimagine the future*. Accessible e-books for equal opportunity. UNICEF. Retrieved February 12, 2021, from https://sowc2015.unicef.org/stories/accessible-e-books-providing-equal-opportunity-for-all-children/.

Fung, V. (2012, April 24). *Using GIS for disaster risk reduction*. United Nations Office for Disaster Risk Reduction (UNISDR). http://www.unisdr.org/archive/26424.

Global Facility for Disaster Reduction and Recovery (GFDRR). (2017a). *Disability inclusion in disaster risk management*. GFDRR.

Global Facility for Disaster Reduction and Recovery (GFDRR). (2017b). *Spatial impact assessment*. GFDRR Innovation Lab. Retrieved January 20, 2019, from https://www.gfdrr.org/sites/default/files/4_Innovation_lab_Factsheet_SIA_rev. pdf.

Global Initiative for Inclusive Information and Communication Technologies (G3ict). (2015). *CRPD implementation: Promoting global digital inclusion through ICT procurement policies & accessibility standards*. Retrieved January 15, 2019, from https:// g3ict.org/publication/tags/public-procurement.

Google Play. (2018). *Hand talk translator* [Apparatus]. Retrieved February 12, 2021, from https://play.google.com/store/apps/details?id=br.com.handtalk&hl=en_US.

Gould, M., & Montenegro, V. 2017. *2016 CRPD ICT accessibility progress report*. G3ict. Retrieved February 12, 2021, from https://g3ict.org/publication/2016-crpd-ict -accessibility-progress-report.

GSMA. (2017). *Accelerating affordable smartphone ownership in emerging markets*. Retrieved January 19, 2019, from https://www.gsma.com/mobilefordevelopment/ wp-content/uploads/2017/07/accelerating-affordable-smartphone-ownership- emerging-markets-2017.pdf.

GSMA. (2018). *Leveraging the potential of mobile for persons with disabilities. Scoping study*. Retrieved January 19, 2019, from https://www.gsma.com/mobilefordevelop- ment/wp-content/uploads/2018/12/Disability_Report_V03_SinglePages.pdf.

Hamid, Z. (2017, September 9). *Getting Aadhaar is harrowing for persons with disability*. The Hindu. https://www.thehindu.com/news/cities/chennai/getting-aadhaar- is-harrowing-for-persons-with-disability/article19647191.ece.

Hao, K. (2018, November 29). Can you make an AI that isn't ableist? *MIT Technology Review*.https://medium.com/mit-technology-review/can-you-make-an-ai-that-isnt- ableist-7218aa11f378.

Independent Expert Advisory Group Secretariat. (2014). *A world that counts. Mobilising the data revolution for sustainable development*. Retrieved January 15, 2019, from http://www.undatarevolution.org/wp-content/uploads/2014/11/A- World-That-Counts.pdf.

International Association of Accessibility Professionals. (2021). *Educational training database*. Retrieved February 12, 2021, from https://www.accessibilityassociation. org/trainingdatabase.

International Telecommunication Union (ITU). (2018). *Digital skills toolkit*. Retrieved January 11, 2019, from https://www.itu.int/en/ITU-D/Digital-Inclusion/Documents/ ITU%20Digital%20Skills%20Toolkit.pdf.

International Telecommunication Union (ITU). (2019). *Self paced online training on ICT accessibility*. Retrieved October 25, 2019, https://www.itu.int/en/ITU-D/

Digital-Inclusion/Persons-with-Disabilities/Pages/Self-Paced-Online-Training-on-ICT-Accessibility.aspx.

KBank. (2017, November 15). *KBank introduces 'K PLUS Beacon'*. https://www.kasikornbank.com/en/sustainable-development/news/Pages/K-PLUS-Beacon.aspx.

Kocak, O., & Kavi, E. (2011). The employment opportunities of disabled people at call centers: Some cases from Turkey. *European Integration Studies, 5*, 205–212.

Kuper, H., & Heydt, P. (2019). *The missing billion. Access to health services for 1 billion people with disabilities*. LSHTM.

Lerman, J. (2013). Big data and its exclusions. *Stanford Law Review Online, 66*, 55. https://www.stanfordlawreview.org/online/privacy-and-big-data-big-data-and-its-exclusions/

Lopes, C. A., & Bailur, S. (2018). *Gender equality and big data*. UN Women. Retrieved January 19, 2019, from http://www.unwomen.org/-/media/headquarters/attachments/sections/library/publications/2018/gender-equality-and-big-data-en.pdf?la=en&vs=3955.

Lord, J. (2019). *Guidance note on disability and identification*. World Bank Group.

Male, C., & Wodon, Q. (2017). *Disability gaps in educational attainment and literacy*. Global Partnership for Education and The World Bank. Retrieved February 12, 2021, from https://documents.worldbank.org/en/publication/documents-reports/documentdetail/396291511988894028/disability-gaps-in-educational-attainment-and-literacy.

Manroth, A., Hernandez, Z., Masu, H., Zakhour, J., Rebolledo, M., Mahmood, S., Seyedian, A., Hamad, Q., & Peixoto, T. 2014. *Strategic framework for mainstreaming citizen engagement in World Bank Group operations: Engaging with citizens for improved results (English)*. World Bank Group.

Mavrou, K. (2011). Assistive technology as an emerging policy and practice: Processes, challenges and future directions. *Technology and Disability, 23*(1), 41–52.

Microsoft Inc. (2019). *Seeing AI*. Retrieved January 25, 2019, https://www.microsoft.com/en-us/seeing-ai.

Mitra, S. (2018). *Disability, health and human development*. Palgrave MacMillan.

Mitra, S., Posarac, A., & Brandon, V. (2013). Disability and poverty in developing countries: A multidimensional study. *World Development, 41*, 1–18.

Monroe, T. (2017). *Big data and thriving cities: Innovations in analytics to build sustainable, resilient, equitable and livable urban spaces*. World Bank. https://openknowledge.worldbank.org/handle/10986/26299.

Montoya, S. (2018, August 8). *Meet the sdg 4 data: Skills for a digital world*. Global partnership for education. https://www.globalpartnership.org/blog/meet-sdg-4-data-skills-digital-world.

Moorthy, J., Lahiri, R., Biswas, N., Sanyal, D., Ranjan, J., Nanath, K., & Ghosh, P. (2015). Big data: Prospects and challenges. *VIKALPA: The Journal for Decision Makers, 40*(1), 74–96.

Mukeredzi, T. (2017). *Uproar over internet shutdowns.* Africa Renewal. August–November. https://www.un.org/africarenewal/magazine/august-november-2017/uproar-over-internet-shutdowns.

Mundy, S. (2018, January 23). *Why start-ups are hiring India's disabled workers.* Financial Times. https://www.ft.com/content/ee83b842-f3ba-11e7-88f7-5465a6ce1a00.

Narváez, F., Marín-Castrillón, D., Cuenca, Ma. C., & Latta, Ma. A. (2017). Development and implementation of technologies for physical telerehabilitation in Latin America: A systematic review of literature, programs and projects. *TecnoLógicas, 20*(40), 155–176.

National Disaster Management Authority, Government of India. (2019). *Guidelines on disability inclusive disaster risk reduction.* Retrieved February 12, 2021, from https://ndma.gov.in/sites/default/files/PDF/Guidelines/DIDRR.pdf.

Pollonais, J. (2018, August 13). *Ambulance tracking tool helps improve coordination of emergency service vehicles in Uganda.* United Nations Global Pulse. https://www.unglobalpulse.org/news/ambulance-tracking-tool-helps-improve-coordination-emergency-service-vehicles-uganda.

Raja, D. (2016). *Bridging the disability divide through digital technologies.* World Bank Group. Retrieved December 5, 2018, from http://pubdocs.worldbank.org/en/123481461249337484/WDR16-BP-Bridging-the-Disability-Divide-through-Digital-Technology-RAJA.pdf.

Raja, D., Narasimhan, N., D'Intino, P., Maheshwari, V., & Montenegro, V. (2015). *Inclusive financial services for seniors and persons with disabilities: Global trends in accessibility requirements.* G3ict. Retrieved January 13, 2019, from https://g3ict.org/publication/inclusive-financial-services-for-seniors-and-persons-with-disabilities-global-trends-in-accessibility-requirements.

Samant, D., Matter, R., & Harniss, M. (2013). Realizing the potential of accessible ICTs in developing countries. *Disability and Rehabilitation: Assistive Technology, 8*(1), 11–20.

Singanamalla, S., Potluri, V., Scott, C., & Medhi-Thies, I. (2019). PocketATM: Understanding and improving ATM accessibility in India. *ICTD '19.* Retrieved January 20, 2019, from https://www.microsoft.com/en-us/research/uploads/prod/2018/12/ictdx19.pdf.

Singh, N. (2018, June 12). *Microsoft Seeing AI's latest update will help the blind identify Indian currency.* Entrepreneur India. https://www.entrepreneur.com/article/314900.

Solutions for Youth Employment (S4YE). (2018). *Digital jobs for youth: Young women in the digital economy.* World Bank Group.

Strigel, C. (2017, February 13). *Invisible in the classroom: Addressing reading hurdles for low vision and hard of hearing students.* SHARED. https://shared.rti.org/content/invisible-classroom-addressing-reading-hurdles-low-vision-and-hard-hearing-students-0.

Trewin, S. (2018a). *AI fairness for people with disabilities: Point of view.* Retrieved January 12, 2019, from https://arxiv.org/abs/1811.10670v1.

Trewin, S. (2018b, December 3). *How to tackle AI bias for people with disabilities.* VentureBeat. https://venturebeat.com/2018/12/03/how-to-tackle-ai-bias-for-people -with-disabilities/.

Trucano, M. (2016). *SABER-ICT Framework Paper for Policy Analysis: Documenting national educational technology policies around the world and their evolution over time.* World Bank Education, Technology & Innovation: SABER-ICT Technical Paper Series (#01). World Bank Group.

United Nations Department of Economic and Social Affairs (UNDESA). (2018a). *United Nations e-government survey 2018.* Retrieved January 10, 2019, from https:// publicadministration.un.org/egovkb/Portals/egovkb/Documents/un/2018- Survey/E-Government%20Survey%202018_FINAL%20for%20web.pdf.

United Nations Department of Economic and Social Affairs (UNDESA). (2018b). *Realization of the sustainable development goals by, for and with persons with disabilities. UN flagship report on disability and development 2018.* Retrieved January 10, 2019, from https://www.un.org/development/desa/disabilities/publication-disability-sdgs.html.

United Nations Development Group. (2017). *Data privacy, ethics and protection guidance note on big data for achievement of the 2030 agenda.* Retrieved February 12, 2021, from https://unsdg.un.org/resources/data-privacy-ethics-and-protection-guidance-note-big-data-achievement-2030-agenda.

Vanian, J. (2018, July 1). *Unmasking AI's bias problem.* Fortune. https://fortune.com/ longform/ai-bias-problem/.

Veale, M., & Binns, R. (2017). Fairer machine learning in the real world: Mitigating discrimination without collecting sensitive data. *Big Data & Society,* 4(2), 1–17.

Vinopol, C. (2015, December 2). *Technology opens literacy opportunities for Moroccan children who are deaf/hard of hearing.* All Children Reading. https://allchildrenreading.org/technology-opens-literacy-opportunities-for-moroccan-children-who-are-deafhard-of-hearing/.

Vodacom South Africa. (2019). *Vodacom SMS emergency service for the deaf.* https:// www.vodacom.co.za/vodacom/services/emergency-service-for-the-deaf?cid= MG_dfpp_9_nmp_7695.

Wong, M. E., & Cohen, L. (2012). *Assistive technology use amongst students with visual impairments and their teachers: Barriers and challenges in special education. Research Brief [12-005].* Retrieved January 25, 2019, from https://repository.nie.edu.sg/ handle/10497/6173.

Wootton, R., & Bonnardot, L. (2015). Telemedicine in low-resource settings. *Frontiers in Public Health,* 3(3). https://doi.org/10.3389/fpubh.2015.00003.

World Bank. (2013). *Inclusion matters: The foundation for shared prosperity.* World Bank.

World Bank. (2015). *Social inclusion through ICT for Tunisian disabled.* Retrieved March 23, 2015, from http://web.worldbank.org/archive/website01055/WEB/0__ CO-93.HTM.

World Bank. (2016). *World development report 2016: Digital dividends*. World Bank. 10.1596/978-1-4648-0671-1.

World Bank and International Telecommunication Union. (2018). *The little data book on information and communication technology 2018*. World Bank Group.

World Bank Group. (2019a). *World Bank country and lending groups*. https://datahelp-desk.worldbank.org/knowledgebase/articles/906519-world-bank-country-and-lending-groups.

World Bank Group. (2019b). *Disruptive technologies for development*. Retrieved January 12, 2019, from https://worldbankgroup.sharepoint.com/sites/wbsites/disruptivetechnologies/Pages/Home_New.aspx.

World Health Organization & World Bank. (2011). *World report on disability*. World Health Organization.

Woyke, E. (2019, January 9). *Smart cities could be lousy to live in if you have a disability*. MIT Technology Review. https://www.technologyreview.com/s/612712/smart-cities-coule-be-lousy-if-you-have-a-disability/.

Yousry, A., Taha, M., & Selim, M. (2018). Currency recognition system for blind people using ORB algorithm. *International Arab Journal of e-Technology*, 5(1), 34–40.

YouTube. (2017, March 23). *Visualizing sound effects*. https://youtube-eng.googleblog.com/2017/03/visualizing-sound-effects.html.

Zahid, Z., Atique, S., Saghir, M. H., Ali, I., Shahid, A., & Malik, R. A. (2017). A commentary on telerehabilitation services in Pakistan: Current trends and future possibilities. *International Journal of Telerehabilitation*, 9(1), 71–76.

Zero Project. (2016). *Video emergency line for the hearing-impaired*. Innovative Practices 2016 on Education and ICT. https://zeroproject.org/practice/jordan-higher-council-of-affairs-of-persons-with-disabilities-hcd/.

Zurutuza, N. (2018). Information poverty and algorithmic equity: Bringing advances in AI to the most vulnerable. *Artificial Intelligence for Global Good. ITU News Magazine*. Retrieved February 12, 2021, from https://www.itu.int/en/itunews/Documents/2018/2018–01/2018_ITUNews01-en.pdf.

3

Global Trends for Accessible Technologies in the Developing World

An Analysis of the Results of the Digital Rights Evaluation Index

H. E. Luis Gallegos, Martin Gould, and Axel Leblois

Global Initiative for Inclusive Information and Communication Technologies (G3ict)

3.1 Introduction

The Global Initiative for Inclusive Information and Communication Technologies (G3ict) was founded in December 2006 at the initiative of the United Nations (UN) Department of Economic and Social Affairs (DESA) as a global multistakeholder advocacy initiative to promote the information and communications technology (ICT) accessibility and assistive technology dispositions of the Convention on the Rights of Persons with Disabilities (CRPD). The Initiative's driving principle is that unrestricted access to digital devices, contents, information, and services is a necessary condition for persons with disabilities (PWDs) to enjoy their full rights. This includes unrestricted access to everything digital, such as television, e-books, computers, websites, or mobile devices and services.

The Digital Accessibility Rights Evaluation (DARE) Index (https://g3ict.org/digital-accessibility-rights-evaluation-index/) measures the degree to which States Parties to the CPRD are successful at implementing digital accessibility by collecting and analyzing three categories of variables in each country: Country Commitments (legal, regulatory, policies, and programs), Country Capacity to Implement (organization, processes, resources), and Actual Digital Accessibility Outcomes for PWDs in ten areas of digital products and services (see Section 3.4.2).

The DARE Index three-step analysis is consistent with human rights monitoring principles and allows for useful gap analysis and linkages between variables. Variables were derived from the *Model ICT Accessibility Policy Report* (Msimang et al., 2014) developed by G3ict and the International Telecommunication Union as well as the Digital Accessibility Decennial Call for Action issued at the UN on

Accessible Technology and the Developing World. Michael Ashley Stein and Jonathan Lazar, Oxford University Press.
© Oxford University Press 2021. DOI: 10.1093/oso/9780198846413.003.0004

December 3, 2016 by the International Disability Alliance, Disabled Peoples' International (DPI), and G3ict.

Results show positive advances of countries' legislations supporting the accessibility of ICTs but also underline significant gaps in countries' capacities to implement, which result in significant shortcomings in making digital products and services accessible to PWDs. Overall, the DARE Index shows considerable progress in the development of national legislations, reflecting the core dispositions of the CRPD: 88% of the countries surveyed have a constitutional article, law, or regulation defining the rights of PWDs, and 68% of the countries have a definition of "Reasonable Accommodation" included in a law or regulation regarding the rights of PWDs. However, capacity to implement is lagging with the most significant opportunities for progress in the following two areas: first, ensuring the consultation and participation of PWDs in developing and monitoring ICT accessibility policies and programs; and second, capacity building of local stakeholders in matters of digital accessibility. More than 78% of the States Parties surveyed reported no or minimal levels of implementation of policies or programs promoting accessibility in critical areas, such as web accessibility, television and audio-visual accessibility, inclusive ICTs for all in education, and procurement of accessible goods and services, among others.

Local advocates and policymakers will be able to analyze differences between their own country score cards (G3ict, 2020b) and worldwide averages, as well as peer group averages by level of income and region, and identify areas of opportunity to promote digital accessibility in their countries. The country profile score cards include a feedback mechanism for suggestions by viewers. Those suggestions are then processed by G3ict's analysts, allowing for a systematic grassroots-based update and validation process. Success stories from most advanced countries suggest that closing gaps requires more than governments' advocacy and resources. Indeed, it requires a long-term partnership between the public sector, industry, disabled peoples' organizations (DPOs), and non-governmental organizations. The participation and continuous involvement of PWDs in policymaking, development, and monitoring processes is vital to building a fully accessible information society that ensures the right to communicate and the use of knowledge for all.

3.2 Background

The DARE Index (G3ict, 2018) was created during the 2017 to 2018 period by G3ict to provide a global benchmark for governments, advocates, and

Table 3.1 Leading Human Rights, Economic, Political, and Social
Conditions Indexes

Human Rights Index (HRI). The HRI is a tool whose aim is to quantify the level of
protection of human rights in European countries. HRI includes all European countries
(apart from Belarus, Kosovo, and several micro states: the Vatican, Monaco, San Marino,
Andorra, and Liechtenstein), Turkey, countries from the Caucasus, coupled with
Tajikistan and Kyrgyz Republic. There are precisely 45 countries within the scope of the
HRI. The HRI measures focus on: personal safety, access to education, tolerance and
inclusion, and personal rights (Gajić & Gamser, 2018).

Human Rights Measurement Initiative (HRMI). The HRMI is a free access database of
metrics, summarizing human rights performance in countries around the world. The
HRMI produces metrics that cover the 12 key rights embodied in international law,
particularly the collection of international treaties known as the International Bill of
Human Rights. These involve five economic and social human rights and seven civil and
political human rights (https://humanrightsmeasurement.org/).

Index of Economic Freedom. The Heritage Foundation measures economic freedom
based on ten quantitative and qualitative factors, grouped into four broad categories, or
pillars, of economic freedom: Rule of Law (property rights, freedom from corruption);
Limited Government (fiscal freedom, government spending); Regulatory Efficiency
(business freedom, labor freedom, monetary freedom); and Open Markets (trade
freedom, investment freedom, financial freedom) (Heritage Foundation, 2020).

Corruption Perception Index. Transparency International's index ranks countries and
territories based on how corrupt their public sector is perceived to be. It is a composite
index drawing on corruption-related data collected by a variety of reputable institutions.
The index reflects the views of observers from around the world, including experts living
and working in the countries and territories evaluated (Transparency International, 2018).

Human Development Index (HDI). The HDI is a summary measure of average
achievement in key dimensions of human development, including long and healthy life,
being knowledgeable, and having a decent standard of living (UNDP, n.d.).

Democracy Index. The Democracy Index is an index compiled by the Economist
Intelligence Unit (EIU) that measures the state of democracy in 167 countries, of which
166 are sovereign states and 165 are United Nations member states. The index is based on
60 indicators grouped into five different categories measuring pluralism, civil liberties,
and political culture. In addition to a numeric score and ranking, the index categorizes
countries as one of four regime types: full democracies, flawed democracies, hybrid
regimes, and authoritarian regimes (EIU, 2019).

Freedom Index by the CATO Institute. This index gives countries a total press freedom
score from 0 (best) to 100 (worst) based on a set of 23 methodology questions divided
into three subcategories. It also designates countries according to category: "Free," "Partly
Free," or "Not Free" (Vásquez & Porcnik, 2019).

The Web Index (WWW). The Web Index is designed and produced by the World Wide
Web Foundation. It is the world's first measure of the World Wide Web's contribution to
social, economic, and political progress in countries across the world (World Wide Web
Foundation, 2014).

Digital Access Index (DAI). ITU's Digital Access Index measures the overall ability of
individuals in a country to access and use new ICTs. The DAI is built around four
fundamental vectors that impact a country's ability to access ICTs: infrastructure,
affordability, knowledge, and quality and actual usage of ICTs. The DAI has been
calculated for 181 economies where European countries were among the highest ranked.
The DAI allows countries to compare to peers and their relative strengths and
weaknesses. It provides a transparent and globally measurable way of tracking progress
toward improving access to ICTs (ITU, 2012).

---

private-sector organizations to assess their progress and identify opportunities for promoting and implementing digital accessibility. By compiling data across different CRPD digital accessibility requirements and making it freely available, the DARE Index helps to deepen and broaden our understanding of how ratifying countries can maximize the impact of ICTs and assistive technologies (ATs) on peoples' lives. Taking the format of an annual country ranking, it will eventually allow for comparisons of trends over time and the benchmarking of performance within and across countries, continuously improving our understanding of the value of implementing CRPD's articles about ICT accessibility for humanity.

In fact, there are a range of global indices that measure different conditions within and/or across countries (e.g., human rights, economic, political, and social conditions). Some of these indices are associated with the UN, like the Human Development Index (UN Development Programme [UNDP], n.d.); some indices are associated with independent/non-UN organizations, like the Human Freedom Index by the Cato Institute (Vásquez & Porcnik, 2019). These global social indicator systems allow us to compare, contrast, and rate countries by specific country conditions or attributes. They provide us with a sense of how countries in the global community rank against one another.

It must be noted, however, that none of the indices mentioned in Table 3.1 measures progress by ratifying countries in implementing the ICT accessibility dispositions of the CRPD—hence the need for the DARE Index.

3.3 A Tool for Advocates

By now, 13 years after the adoption of the text of the CRPD by the UN General Assembly and its subsequent ratification by 181 countries, the accessibility of ICTs has gained significant momentum around the world: About two-thirds (65%) of States Parties to the CRPD define it in their laws and regulations. As a result, in its 11th year of actual CRPD operation, policies and programs for making TV, websites, mobile phones and services, e-books, electronic kiosks, and everything digital accessible to PWDs are on the agenda of many governments.

Actual progress in implementing those, however, is still very limited. Data collected by the G3ict Research Team in collaboration with organizations of PWDs and accessibility experts in 137 countries show that, while good intentions are progressively translated into policies, laws, and regulations, systemic issues limit their implementation. For instance, while 61% of countries have TV accessibility policies or regulations in place, 39% have not started implementing them or remain at a minimum level of implementation. A similar picture appears for web accessibility: 58% of countries have web accessibility policies or regulations in place but 42% have no or minimum levels of implementation. Across the ten ICT accessibility sectors measured for outcomes by the DARE Index (G3ict, 2020c), only 21% of countries reach a partial or substantial level of implementation to date.

What are the systemic issues limiting implementation? How do individual countries perform compared to global benchmarks and peer countries' results? Are there specific action steps that advocates and governments can take to accelerate progress? The DARE Index offers unique insights and answers based on real-world data. For each of the 137 countries surveyed, it provides scores for *commitments* (laws, policies, and regulations identified by the International Telecommunication Union and G3ict as critical foundations for ICT accessibility in their *Model ICT Accessibility Policy Report* [Msimang et al., 2014]), *capacity and process to implement* (identified by G3ict over seven years of research as critical success factors), and *actual outcomes* for each of the ten categories of ICTs identified by the Decennial Call for Action jointly issued by the International Disability Alliance, DPI, and G3ict on the occasion of the celebration of the tenth anniversary of the CRPD at the UN (Msimang et al., 2014). This three-step analysis (*commitments, capacity to implement,* and *outcomes*), is consistent with human rights monitoring principles and allows for useful gap analysis and linkages between variables.

The DARE Index, developed by G3ict in cooperation with DPI, serves as a resource for advocates and policymakers to benchmark progress in making ICTs accessible in compliance with the CRPD.

As there are no consistent data sources available around the world on digital accessibility, G3ict collects and verifies standardized implementation data in close cooperation with DPI and PWDs worldwide, considering their best position to assess and report on digital accessibility matters. Research panel members are identified in each country and are responsible for providing data on their own country's ICT accessibility profile. When DPI does not have local representatives available, G3ict calls on other advocacy organizations, researchers, disability community leaders, or accessibility experts. G3ict and DPI work to ensure all country responses are evidence-based, relying on supporting documentation and references to archived information materials. Most panel members are experienced advocates who volunteer their time and effort and are unlikely to volunteer erroneous data. However, due to language, terminology, and other unforeseen data input factors, a systematic pre-publication validation process takes place by analyzing data variances and anomalies such as discontinuity of historical data or unusual relations to the Human Development Index and income per capita. When such anomalies are detected, G3ict analysts double-check through desk research and discuss with national panels how to interpret the data received. Ultimately, all country evaluation data is posted in the "Country Dashboard" section of the G3ict website, with an option for visitors to provide feedback for updates or anomalies.

3.4 Methodology

In 2008, G3ict benchmarked implementation of the dispositions of the CRPD on ICT accessibility and AT by States Parties. G3ict partnered with DPI to research and produce the CRPD ICT accessibility progress reports. Between 2010 and 2016, an extraordinary group of volunteer country panelists contributed to the CRPD progress reports. The publication, with its four editions, established itself as one of the few sources allowing for global monitoring of the implementation of the digital accessibility provisions of the CRPD (Gould & Montenegro, 2017; Gould et al., 2014; Gould et al., 2012; Gould & Studer, 2011).

It appeared, however, that advocates were looking for a simpler benchmarking tool that would allow for quick comparisons between publicly available individual countries' datasets. During the 2017–2018 period, G3ict took the next step in benchmarking with the launch of the DARE Index, which ranks countries' actual implementation of key dispositions of the CRPD on ICT accessibility (G3ict, 2018). More specifically, the DARE Index survey is designed to evaluate, track, and rank the progress made by CRPD signatory countries in the implementation of key digital accessibility treaty provisions (e.g., CRPD Article 9).

As noted above, the DARE Index builds on data collection and analysis experience gained through the past editions of the G3ict CRPD progress reports. The first edition covered 121 countries (31 developing countries, 90 developed countries). The 121 countries surveyed had a combined population of 6.7 billion, which is 89.4% of the world population.

The 2020 second edition of the DARE is derived from one set of questionnaires completed by 160 local correspondents in 137 countries during the 2019–2020 period. As was the case with the first edition, data collection for the second edition of the DARE Index was completed in cooperation with the DPI and various DPOs and experts in countries around the world where DPI correspondents were not available. The second edition covers 137 countries (25 developing countries, 112 developed countries). The 137 countries surveyed had a combined population of 7 billion, which is 90% of the world population. G3ict analysts use the regional definitions of the Office of the UN High Commissioner for Human Rights, and World Bank and International Monetary Fund country classifications for income per capita.

An overall comparison analysis showing the trend between the years 2018 and 2020 is not possible, since the 2020 DARE Index edition includes three new variables and covers a higher number of countries. However, a direct comparison of the exact same set of variables among the 105 countries that participated in both

Table 3.4 Global Averages for Outcomes

Average outcomes across ten categories of ICTs	Global results
No policy	51%
Policy but no implementation	3%
Minimum level of implementation	24%
Partial level of implementation	19%
Substantial level of implementation	2%
Full level of implementation	1%

2018 and 2020 shows steady progress in all three dimensions of the DARE Index. For instance, among those 105 countries' commitment variables, 59% now incorporate in their legislation or regulations a definition of accessibility that includes ICTs, compared to 49% in 2018. Similarly, 39% now include PWDs as beneficiaries of their universal service obligation, vs. 34% in 2018. Those trends illustrate the positive impact of digital accessibility advocates and international organizations such as G3ict and the International Telecommunication Union (ITU) that have heavily promoted those specific dispositions among states seeking to align their legislations with CRPD obligations.

A comparison analysis of those same 105 countries also shows a slightly better performance in 2020 vs. 2018 for their capacity to implement, while their actual implementation and outcome, as measured by a digital accessibility policy in the process of being implemented, experienced significant improvements across all areas. Five out of the ten accessibility areas measured experienced a marked increase in two years: web accessibility (8% increase), e-government and smart cities (8% increase), television (10% increase), public procurement (11% increase), and availability and internet usage (14% increase).

Those trends are encouraging and seem to indicate that countries do implement ICT accessibility policies once they are adopted, even though their levels of implementation are still very limited, as shown in Table 3.4, "Global Averages for Outcomes," and Table 3.5, "Global Level of Implementation and Outcomes."

3.4.1 Human Rights Monitoring Framework

Furthermore, the DARE Index's conceptual framework is consistent with the Office of the UN High Commissioner for Human Rights' member states' treaty reporting guidelines (OHCHR, 2012): *Human Rights Indicators—A Guide to Measurement and Implementation reporting*. This OHCHR conceptual framework uses a configuration of structural, process, and outcome indicators, aiming to measure acceptance, intent, or commitment to human rights standards; the

Table 3.5 Global Level of Implementation and Outcomes

Global % level of ICT accessibility implementation and outcomes (Ranked by percentages of "No policy")	No policy %	No implementation %	Minimum level %	Partial level %	Substantial level %	Full level %
TV	39%	7%	26%	24%	3%	1%
Inclusive ICTS in education	46%	0%	31%	21%	1%	1%
Web	42%	4%	20%	30%	4%	1%
E-books	49%	4%	26%	19%	3%	0%
Enabling ICTs for employment	53%	2%	26%	16%	3%	0%
E-government and smart cities	50%	4%	23%	18%	4%	0%
Promotion of internet usage by PWDs	51%	1%	27%	18%	2%	1%
ATs and ICTs for independent living	61%	3%	23%	12%	1%	0%
Mobile	62%	4%	18%	13%	1%	2%
Public procurement	54%	4%	24%	16%	2%	0%
Global average all areas of ICTs	51%	3%	24%	19%	2%	1%

efforts required to make that commitment a reality; and the results of those efforts in terms of the increased enjoyment of human rights over time. Each category, through its information sets, brings to the fore an assessment of the steps taken by the States Parties to meet their obligations, whether in terms of respecting, protecting, or fulfilling a human right.

Once a state has ratified a human rights treaty, there is a need to assess its commitment to implementing the standards it has accepted.

Structural indicators (commitments) help in such an assessment. They reflect the ratification and adoption of legal instruments and the existence, as well as the creation, of basic institutional mechanisms deemed necessary for the promotion and protection of human rights. Some common structural indicators are: international human rights treaties, relevant to the right to adequate housing, ratified by the state; timeframe and coverage of national policy on vocational and technical education; and date of entry into force, and coverage, of formal procedure governing the inspection of police cells, detention centers, and prisons by independent inspection entities.

Process indicators (efforts) measure duty bearers' ongoing efforts to transform their human rights commitments into the desired results. Unlike with structural indicators, this involves indicators that continuously assess the policies and specific measures taken by the duty bearer to implement its commitments on the ground. Some common process indicators are: coverage of targeted population groups under public programs; indicators based on budget allocations; and incentive and awareness measures extended by the duty bearer to address specific human rights issues.

Outcome indicators (results) capture individual and collective attainments that reflect the state of enjoyment of human rights in a given context. An outcome indicator consolidates over time the impact of various underlying processes (that can be captured by one or more process indicators). Some common examples are: proportion of labor force participating in social security scheme(s); reported cases of miscarriage of justice and proportion of victims who received compensation within a reasonable time; and educational attainments (e.g., youth and adult literacy rates) by targeted population group.

In keeping with the OHCHR conceptual framework on human rights reporting, the DARE Index survey is divided into three dimensions or legs, as follows:

- **Leg #1 Country Commitments (Structure):** Monitors essential laws, regulations, or policy programs in place in the country. This section measures the formal status of the country's government, legal, and policy regime in relation to those ICT commitments—the political commitments made with respect to national laws, policies, programs, and plans of action that are relevant to the ICT provisions. Commitment variables are consistent with, and derived from, the *Model ICT Accessibility Policy Report* (Msimang et al., 2014) jointly published by the ITU and G3ict.
- **Leg #2 Audit Tool or Country Capacity to Implement (Process):** Perceived as representing the basic capacity of a country to implement the ICT provisions of the CRPD. It refers to the required government agencies, organizational and institutional resources involved, the digital/technology resources available, and the standard bodies the country follows. Those variables are derived from multiple years of analysis of the success factors supporting the effective implementation of ICT accessibility policies and programs.
- **Leg #3 Audit Tool or Country Actual Implementation (Outcomes):** Checks the systemic and/or individual impact(s) of a country's fulfillment of the ICT provisions of the CRPD. Shows the level of actual implementation and outcomes of core policies and programs in ten essential ICT accessibility areas by evaluating the status of development, application level, availability, accessibility, and population coverage of ICTs and AT in the country. Those ten variables are derived from the Decennial Call for Action for ICT accessibility issued by the IDA, DPI, and G3ict in December 2016 at the UN.

3.4.2 List of Variables and Index Calculation

The DARE Index is calculated with 20 variables: five for leg #1 (Commitments), five for leg #2 (Capacity to implement), and ten for leg #3 (Outcomes). Each variable is valued at five points, so for all 20 variables the maximum number of points is $20 \times 5 = 100$.

Country Commitments—Legislation and Regulation in Place:

(1) CRPD ratification completed
(2) General law protecting the rights of PWDs
(3) Reasonable accommodation defined
(4) Definition of accessibility includes ICTs
(5) Universal service obligation includes PWDs

Country Capacity to Implement—Country Has:

(1) Government agency for PWDs
(2) Government agency for ICTs
(3) Process to involve PWDs in policymaking on ICT accessibility
(4) Participation in standard development organizations
(5) ICT accessibility courses available at major universities in the country

Country ICT Accessibility Outcomes for PWDs:

(1) Web
(2) TV
(3) Mobile
(4) E-books
(5) Promotion of internet usage by PWDs
(6) Inclusive ICTs in education
(7) Enabling ICTs for employment
(8) E-government and smart cities
(9) ATs and ICTs for independent living
(10) Public procurement

The DARE Index results allow quick comparison among countries and linkages between those three groups. Complete results of the DARE Index can be viewed on the DARE Index page of the G3ict website (G3ict, 2020c). Detailed individual DARE Index scores for 137 countries are available on the G3ict Country Dashboard (G3ict, 2020b).

3.4.3   Scoring Method

For the first two legs, points are calculated as 0 (item not present in country) or 5 (item present in country). For the third leg, points are calculated on a scale of 0 to 5 as follows: 0 = no program or policy; 1 = policy or program but no implementation; 2 = minimum level of implementation (pilot project only); 3 = policy or program being rolled out but less than half implemented (examples provided to respondents); 4 = policy or program being rolled out and more than half implemented (examples provided to respondents); and 5 = full implementation.

How is the evaluation conducted at a country level? Local panel members are asked to fill in a questionnaire and provide back-up information and references in support of their answers to the greatest possible extent. For country commitments, links to laws, regulations, or program webpages are often used; and for country capacity to implement, links to webpages, reports, or articles are likewise cited. To evaluate the degree to which digital products and services are made accessible to PWDs, questions include a normalized scale to guide respondents in their answers to be as close to reality as possible. For instance, "Minimum level of implementation" would be described as "pilot projects only." "Partial level of implementation" would be defined as "a national program is being deployed but does not yet reach most beneficiaries," and so on and so forth.

How are response data validated? Most panel members are experienced advocates who volunteer time and effort, and are unlikely to offer erroneous data. However, due to language, terminology, and other unforeseen data input factors, a systematic pre-publication validation process occurs by analyzing data variances and anomalies such as discontinuity of historical data or unusual relations to Human Development Index and income per capita. When such anomalies are detected, G3ict analysts double-check through desk research and with national panels how to interpret the data received. All country evaluation data is posted in the "Country Profiles" section of the G3ict website with an option for visitors to provide feedback for updates or anomalies.

3.4.4   Country Classification

A classification approach is used for each country surveyed for the DARE Index, both to provide context for the ranking and information displayed under each country profile and to facilitate cross-tabulations by peer countries' groupings. Countries have been organized by region, income level, and Human Development Index ranking (see World Bank Country Classification, 2020; UNDESA/ Development Policy & Analysis Division Country Classification, 2019), defined as follows:

- **Regions:** Geographic classification includes eight main regions: Africa, East Asia and Pacific, Europe, Central Asia, Latin America and Caribbean, Middle East and North Africa, Northern America, and South Asia.
- **Income per capita:** Countries have been categorized into four income level groups (low, lower-middle, upper-middle, and high) consistent with the World Bank classification of economies and its Atlas gross national income per capita estimates.
- **Human Development Index (HDI):** Ranking for each country is derived from the latest UNDP statistics. HDI country scores are based on a composite statistic of life expectancy, education, and income per capita indicators used to rank countries into four tiers of human development: low, medium, high, and very high.

Detailed lists of countries for each of the three classifications listed above are available on the G3ict website (G3ict, 2020a).

3.5 Global Results

Results show positive advances of countries' legislations supporting the accessibility of ICTs but also underline significant gaps in countries' capacities to implement. This results in significant shortcomings in making digital products and services accessible to PWDs.

3.5.1 Country Commitments Analysis

The following overall global results (see Table 3.2) show the considerable impact of the CRPD on the progress of countries' legislations, policies, and regulations. For instance, whereas a minority of countries had general legislation protecting the rights of PWDs prior to 2006, today 88% do, which is remarkable progress achieved in a few years. Similarly, legal definitions such as "Reasonable Accommodation," which merely existed in the legislation of five countries prior to the CRPD launch, is now present in almost two-thirds of the 137 countries surveyed for the DARE Index.

Commitments specific to digital accessibility (or accessibility of ICTs) are also progressing, with 61% of countries formally including ICTs in their definition of accessibility requirements, compared to a handful when the CRPD was launched back in 2006. Continuous efforts are applied by DPOs, G3ict, and UN-affiliated organizations to promote the adoption of such definitions by more countries, consistent with Article 9 of the CRPD.

Table 3.2 Country Commitments

Country commitments	Global average % of 137 countries with law/regulation/policy in place
Marrakesh Treaty ratification/accession	37%
CRPD ratification	93%
General law protecting the rights of persons with disabilities	88%
Reasonable accommodation defined	68%
Definition of accessibility includes ICTs	61%
Universal service obligation includes PWDs	41%
Global average in percentage	65%

Finally, only 41% of countries follow the recommendation of the ITU–G3ict *Model ICT Accessibility Policy Report* (Msimang et al., 2014) to include PWDs in their universal service obligation (USO) legislation. USOs are meant to equalize access to telecommunication services among all citizens, nitially addressing coverage gaps in rural areas. Coverage of rural areas, however, is better today, given the progress of wireless networks. Funding for universal services has been historically justified as in the public interest, because it solves important issues of inclusiveness, equity, and social justice. Ongoing justification for government intervention in global telecommunications industries are premised on the idea of a digital divide between those with access to global information networks and those without. This understanding of the public interest has resulted in several attempts by governments in different parts of the world to promote the development of broadband telecommunications networks and to solve the problem of the digital divide. For instance, in 2009, the U.S. Federal Communications Commission (FCC), helped create a national broadband plan "ensuring that every American has 'access to broadband capability'" (FCC, 2010, p. 3).

More recently, the European Commission launched "The Europe 2020 Strategy," which included an initiative referred to as "The Digital Agenda for Europe." Among the objectives of the digital agenda are the creation of a single European digital market, promotion of interoperability and standards, attraction of investment in networks, facilitation of "innovation efforts," combating the "lack of digital literacy" and the "digital divide," and encouragement of the development of next generation networks.

Both the United States and the EU government USO interventions address the digital divides, aiming for affordability and ubiquity regarding broadband access by all citizens. For example, "universal service funds" set up by USO legislations can re-allocate funds toward programs in support of accessibility, serving as very powerful funding mechanisms to close the accessibility gap for millions of PWDs. In most countries, universal service funds are financed by taxes on calls levied by telecom service providers, creating a predictable recurring revenue stream rather than annual budget items—a more reliable mode of funding for critical ongoing services such as relay services for the deaf or supporting mobile services for various types of disabilities.

3.5.2 Country Capacity to Implement Analysis

The 2020 DARE Index reflects the same areas for improvements under capacity to implement as the 2018 DARE Index. Capacity building is a more complex challenge due to the cross-sector collaboration and coordination needed among the public, academia, and private sectors to address the limited international cooperation, resources, and language barriers affecting access to technical information. Among others, these indicators of progress are as follows: Lack of a process to involve PWDs in policy making on ICT accessibility; lack of country references to international ICT accessibility standards; and lack of availability in ICT accessibility courses at country level. These reflect the main issues slowing down the implementation of digital accessibility in many countries (see Table 3.3), and will be outlined here in more detail.

(1) Lack of a process to involve PWDs in policy making on ICT accessibility

While many countries have government agencies overseeing PWDs (88%) and ICTs (39%), slightly more than a quarter (26%) involve PWDs in the policy making and monitoring for digital accessibility. This situation is inconsistent with CRPD Article 4.3:

> In the development and implementation of legislation and policies to implement the present Convention, and in other decision-making processes concerning issues relating to PWDs, States Parties shall closely consult with and actively involve PWDs, including children with disabilities, through their representative organizations.

This situation is also inconsistent with the ITU–G3ict *Model ICT Accessibility Policy Report* (Msimang et al., 2014), which details processes that can be implemented to involve PWDs in the development and monitoring of ICT accessibility policies and programs.

Table 3.3 Country Capacity to Implement

Capacity to implement	Global average % of 137 Countries with key implementation resources or processes
Government agency for PWDs	88%
Government agency for accessible ICT	39%
Process to involve PWDs in policy making on ICT accessibility	26%
Participation in standard development organizations	44%
ICT accessibility courses available at major universities in the country	38%
Country capacity to implement global progress average in percentage	46%

Furthermore, statistical analysis and anecdotal evidence show that unless organizations of PWDs are involved in the development of policies and programs, those are likely to fail to address their requirements adequately. And without the involvement of PWDs in monitoring progress, those programs and policies are likely to fail and be discontinued. As part of its strategy, G3ict promotes the participation of organizations of PWDs in developing policies and programs and organizes capacity-building programs to facilitate such participation.

(2) Lack of country participation in standards developing organizations

The involvement of countries in standards developing organizations is often correlated with country size and level of economic development. Accessibility standards are a critical foundation for any program or policy in support of ICT accessibility. Furthermore, given the global nature of the ICT marketplace, it is in the best interests of countries to align themselves with well-recognized international standards such as WCAG or EN 301 549. It should be a priority for all stakeholders promoting digital accessibility to promote those standards and ensure that they are referenced by national accessibility policies and programs.

Some ministerial departments or national communications authorities for regulations of ICT devices and services with accessibility features support international standards and good practices. In general, legislators pass laws, while government regulators issue regulations to carry out or implement laws. It is in the best interests of regulatory authorities to support international standards and good practices, since this enables the country to be competitive in the global economy. For example, in the United States, the FCC was established by the Communications Act of 1934 and charged with regulating interstate and international communications by radio, television, wire, satellite, and cable. It was not until 1990, when legislators enacted the Americans with Disabilities Act, that persons with hearing and speech disabilities were provided with the ability to use telecommunications services. As the regulatory authority, the FCC was authorized to establish and manage the telecommunications relay services (TRS) program. Today there are nine types of TRS calls that can be made by a person with a disability, depending on the needs of the user and the equipment available (FCC, 2020).

In other instances, private-sector initiatives are supportive of international standards and good practices. As a developer and marketer of ICT products and services, the private sector is directly impacted by government requirements to acquire accessible products and services. It is expected that there will be increased efforts toward harmonization of international ICT standards in the accessibility arena. This is due to the global growth of ICT and consumer electronic markets and the acknowledgment that innovation is the foundation of the global economy. According to the Japan/US/EU Trilateral IT Electronics Associations (JEITA, 2007), compliance with international standards helps to "promote technology diffusion, production efficiency, product compatibility, interoperability, enhanced competition, consumer choice, and lower costs." Collaborative ventures between the private sector and governments will continue, such as internet infrastructure development in Africa and other regions of the world. Tele-centers and community multimedia centers will be upgraded and built with ICT accessibility in mind. The possibilities for private-sector initiatives and collaborations with non-governmental organizations are unlimited.

(3) Lack of availability in ICT accessibility courses at country level

The fifth DARE Index capacity to implement variable checks if ICT accessibility courses are available at major universities in the country. Only 38% of countries have these available. For the other 62%, the level of know-how in matters of ICT accessibility in the country is likely very limited. This means, among various factors, that students in two-thirds of countries continue to graduate in computer sciences or any other related discipline without having ever heard of ICT accessibility. Furthermore, employers who need to implement ICT accessibility need to close the gap by ensuring that their personnel are trained in ICT accessibility. This also affects schools of education preparing future teachers, which must create or use routinely digital education material without having acquired the basic knowledge of accessibility issues they should have to accommodate students with disabilities.

Several organizations are trying to close this gap at an academic level. In the United States, Teach Access, for example, develops a college-level accessibility curriculum, and the Alternative Media Accessibility Center's Massive Open Online Course (MOOC) is fully accessible. Meanwhile, in Europe, MOOCs have also been launched for university-level students and cooperate in matters of curriculum.

For organizations and professionals seeking training and certification in ICT accessibility, G3ict promotes the programs of the International Association of Accessibility Professionals and its certifications: Certified Professional in Accessibility Core Competencies, Web Accessibility Specialist, and, new in 2019, Public Procurement Specialist in ICT Accessibility. These certification frameworks are now embraced by government agencies, academic institutions, and corporations to ensure that they have the appropriate ICT accessibility skillset available within their organization. Countries seeking to promote digital accessibility can explore those various sources to jumpstart their efforts in developing their ICT accessibility know-how.

3.5.3   Country Outcomes Analysis

Data presented in this chapter show the global levels of implementation by ICT accessibility area. Across the board, 51% of countries do not have any policy in place for implementation of ICT accessibility. However, interestingly, only 3% report that they have a policy but that it is not implemented. This is, in fact, an encouraging sign which seems to indicate that countries do try to implement ICT accessibility policies once they are adopted.

Levels of implementation are globally very low, with only 46% of countries in the process of implementing policies at various stages. The clear majority of those are either at a minimum level of implementation, such as pilot projects (24%), or at a partial level of implementation (19%), where a deployment program is proceeding but still with limited impact.

These numbers show momentum and are encouraging. However, the reality is that across all sectors, substantial or full levels of implementation only average 3%. This reflects the considerable gap that exists globally between the promises of Article 9 of the CRPD and the actual availability of accessible digital products and services.

The most advanced areas include TV, inclusive ICTs or education, web and e-books, all with more than 40% of countries in the process of implementing policies. ATs and ICTs for independent living remain behind mostly because of the lack of delivery and support ecosystems in most countries. Indeed, more than 59% of the countries reported no policy or program undertaken to promote accessibility in the following five ICT areas: e-government and smart cities; mobile telephony; promotion of the internet among people with disabilities; ATs and ICTs for independent living; and procurement of public goods and services. In each of those ten areas (see Section 3.4.2), specific programs and resources to promote policies and good practices are offered by international organizations such as the ITU, the UN Educational Scientific and Cultural Organization (UNESCO), the World Health Organization, and the International Labour Organization.

3.6 Lessons Learned: Developing Countries

DARE Index country scores, rated on the scale of 25 points for commitments, 25 points for capacity to implement, and 50 points for actual implementation and outcomes, provide useful insight into assessing the relative progress accomplished by developing countries. Global results by region and levels of economic development are closely interrelated.

3.6.1 Relationship Between DARE Index Scores and Level of Economic Development

As Figure 3.1 shows, countries' overall scores are related to their levels of economic development, as measured by levels of income per capita. However, while average scores for commitments and capacity to implement do not depict important variations by levels of economic development, actual implementation and outcome scores do vary considerably.

Three main conclusions can be derived from these numbers in relation to developing nations:

(1) Thanks to the legislative, regulatory, and policy development momentum generated by the CRPD, most countries, including developing countries, are making progress in establishing the legal and regulatory foundations for ICT accessibility and, to a lesser extent, in establishing the capacity to implement.

(2) There is no "one size fits all" analytical profile regarding developing countries. Outcomes and overall scores vary considerably between middle-income and low-income countries. Quite a few low-income countries are doing better than middle-income countries or even high-income countries in specific areas.

(3) Scores for outcomes are also influenced by the level of penetration of various ICTs in countries. While TV and mobile are omnipresent in all countries, other areas of ICT accessibility may not be as relevant, given the lesser penetration of ICT infrastructure and services.

3.6.2 Regional Analysis of DARE Index Scores

Progress by region covers a variety of situations related to penetration of digital technologies and services, level of economic development, and the period when efforts to implement digital accessibility started.

To assess the global progress accomplished, it is worth noting the significant level of commitments made by countries around the world. However, for a clear majority of countries, capacity to implement is lagging.

The net result is a weak level of outcomes across the board, with Northern America and Europe outperforming the rest of the world both in terms of actual outcomes for PWDs and overall DARE Index scores.

As shown by the following bar chart (Figure 3.2), overall scores and their components vary considerably by region, due in large part to the average level of economic development of individual countries in each region. Northern America and Europe are clearly ahead in all areas of commitments, capacity, and outcomes. In regions such as the Middle East and North Africa, country scores can vary considerably, with very high-income countries with small populations such as Qatar and Oman among the top performers globally.

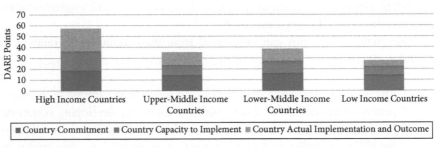

Figure 3.1 Average DARE Index Country Scores by Levels of Economic Development

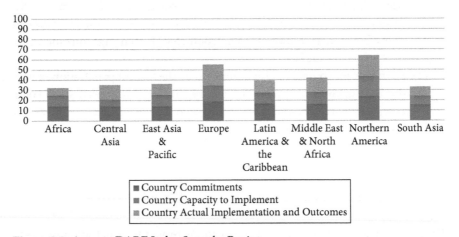

Figure 3.2 Average DARE Index Score by Region

3.6.3 Top Performers Among Developing Countries

A careful analysis of individual countries in each region shows emerging economies and developing countries with good performance and impressive global rankings, such as Brazil in Latin America, Jamaica in the Caribbean, or South Africa in Africa. Country commitments to disability rights, participation of PWDs in ICT accessibility policy making, and local technology ecosystems are important success factors.

How do these three countries compare to the rest of the world? As shown in Table 3.6, these three countries have a high level of commitment to ICT accessibility—in fact, higher than most countries around the world, including high-income countries. This translates into strong foundations for ICT accessibility policies and programs and better translation of accessibility imperatives in all domains of technology policy.

Similarly, in Table 3.7, Brazil and South Africa are seen to have key elements of capacity to implement, which explains their relative leadership versus most other countries.

Brazil ranks first in Latin America, Jamaica ranks first in the Caribbean, and South Africa ranks first in Africa with respect to their level of implementation and outcomes (see Table 3.8). Remarkably, those three countries rank higher than many high-income countries, especially Brazil and South Africa, which have strong commitments aligned with good capacity to implement. While not high performers globally in any specific category, those countries show steady progress across the board, with policies and programs in the process of implementation in most categories of ICTs.

Among the least developed countries, Malawi and Ethiopia reach overall scores of 45 and 30.5, respectively. Malawi ranks 57 out of 137 countries, with 20/25

Table 3.6 Comparison of Brazil, Jamaica, and South Africa Commitments With Global Averages

Country commitments	Brazil	Jamaica	South Africa	Global average
Marrakesh Treaty ratification/accession	2.5	0	0	0.93
CRPD ratification (UN Treaty)	2.5	2.5	2.5	2.3
General law protecting the rights of persons with disabilities	5	5	5	4.4
Definition of accessibility including ICTs in laws or regulations	5	5	5	3.1
Definition of reasonable accommodation	5	5	5	3.4
Universal service obligation includes PWDs	5	5	5	2.1
Total Score out of 25	25	22.5	22.5	16.1

Table 3.7 Comparison of Brazil, Jamaica, and South Africa Capacity to Implement With Global Averages

Country capacity to implement	Brazil	Jamaica	South Africa	Global average
Government agency for PWDs	5	5	5	4.4
Government agency for accessible ICTs	0	5	5	1.9
Process to involve DPOs in ICT accessibility policy making	5	0	5	1.2
Country refers to international ICT accessibility standards	5	0	5	2.2
ICT accessibility courses available at universities	5	5	5	1.9
Total score out of 25	20	15	25	11.6

Table 3.8 Comparison of Brazil, Jamaica, and South Africa Levels of Implementation and Outcomes With Global Averages

Country level of implementation and outcomes	Brazil	Jamaica	South Africa	Global average
Website accessibility	3	3	1	1.5
TV and multimedia	3	2	3	1.5
Mobile telephony	2	0	3	0.94
E-books and digital contents	3	0	3	1.2
Promoting the internet usage among persons with disabilities	3	3	3	1.2
Inclusive ICTs for all in education	3	3	3	1.3
Enabling ICTs for all in employment	3	0	3	1.1
E-government and smart cities for all	3	3	3	1.2
Enabling assistive technologies and ICTs for independent living	2	2	3	0.9
Procurement of accessible public goods and services	2	3	3	1.1
Total score out of 50	27	19	28	12.1

scores for commitments and 10/25 for capacity, sectorial policies and programs in place and initial implementation for mobile, ICTs in education, e-books, ATs, and TV. Overall, Malawi ranks better than many middle-income and even high-income countries.

The above-mentioned Country Profiles and examples show that positive steps can be accomplished at various levels of economic development, provided strong commitments are made and key implementation processes are in place.

3.7 Conclusion

Each of the critical gap areas in results regarding the CRPD's digital accessibility provisions present opportunities for improvement by ratifying countries, particularly in their capacity for implementation and the involvement of PWDs. Bridging those vital gaps requires establishing, guiding, and deploying long-term and focused multi-stakeholder partnerships between the public sector, industry, nonprofit sector, DPOs, and non-governmental organizations.

Tracking and knowing how much progress is accomplished by States Parties in ICT accessibility is an essential step forward for all stakeholders to address gaps and opportunities in their own countries. Local advocates and policymakers may want to analyze differences between their own country's score cards and worldwide averages and peer group averages by level of income and region, and identify areas of opportunity to promote digital accessibility in their countries. The Country Profiles score cards posted on the G3ict website Country Dashboard (2020b) include a feedback mechanism for suggestions by viewers. Those suggestions are then processed by G3ict's analysts, allowing for a systematic, grassroots-based update and validation process.

People across the globe will continue to look to leaders to rise to the challenge of achieving a digital accessibility agenda that is both universal and adaptable to the conditions of each country. Their voices have underscored the need for sustainability, livability, and more effective governance and capable institutions; for new and innovative partnerships, responsible businesses, and local authorities; and for an enabling ICT strategy, accountability mechanisms, and vigorous partnerships.

References

Economist Intelligence Unit (EIU). (2019). *Democracy index 2019* [Report]. https://www.eiu.com/topic/democracy-index.

Federal Communications Commission (FCC). (2020, April 8). *Telecommunications relay service—TRS*. https://www.fcc.gov/consumers/guides/telecommunications-relay-service-trs.

Gajić, M., & Gamser, D. (2018). *Freedom barometer human rights index* [PDF]. Friedrich Naumann Foundation. https://esee.fnst.org/sites/default/files/uploads/2018/12/12/humanrightsindex2018final.pdf.

Gould, M., Leblois, A., Bianchi, F. C., Montenegro, V., & Studer, E. (2012, January). *2012 CRPD ICT accessibility progress report* [PDF]. G3ict Publications. https://g3ict.org/publication/crpd-2012-ict-accessibility-progress-report.

Gould, M., Leblois, A., Bianchi, F. C., & Montenegro, V. (2014, March). *Convention on the rights of persons with disabilities 2013 ICT accessibility progress report* [PDF]. G3ict Publications. https://g3ict.org/publication/convention-on-the-rights-of

-persons-with-disabilities-g3icts-2013-ict-accessibility-progress-report-survey-conducted-in-cooperation-with-dpi.

Gould, M., & Montenegro, V. (2017, October). *2016 CRPD ICT Accessibility progress report* [PDF]. G3ict Publications. https://g3ict.org/publication/2016-crpd-ict-accessibility-progress-report.

Gould, M., & Studer, E. (2011, March). *2010 CRPD progress report on ICT accessibility—text version* [PDF]. G3ict Publications. https://g3ict.org/publication/text-version-crpd-progress-report-on-ict-accessibility-2010.

Global Initiative for Inclusive ICTs (G3ict). (2018, November). *DARE Index 2017–2018: Global progress by CRPD state parties.* https://g3ict.org/publication/global-progress-by-crpd-states-parties.

Global Initiative for Inclusive ICTs (G3ict). (2020a). *Country classification.* https://g3ict.org/digital-accessibility-rights-evaluation-index/country_classification.

Global Initiative for Inclusive ICTs (G3ict). (2020b). *Country dashboard.* https://g3ict.org/country-profile.

Global Initiative for Inclusive ICTs (G3ict). (2020c). *Digital accessibility rights evaluation index (DARE index).* https://g3ict.org/digital-accessibility-rights-evaluation-index/.

Heritage Foundation. (2020). *About the index* [Overview of the 2020 Index of Economic Freedom book]. https://www.heritage.org/index/about.

International Telecommunication Union (ITU). (2012). *Digital Access Index (DAI).* https://www.itu.int/ITU-D/ict/dai/.

Japan Electronics and Information Technology Industries Association (JEITA). (2007, April 27). *Third Japan/US/EU Trilateral IT–Electronics associations meeting is held in Brussels* [Press release]. https://www.jeita.or.jp/english/press/2007/0427/index.htm.

Msimang, M., Leblois, A., & Schorr, S. (2014, November). *Model ICT accessibility policy report* [PDF]. G3ict Publications. https://g3ict.org/publication/model-ict-accessibility-policy-report.

Office of the United Nations High Commissioner for Human Rights (OHCHR). (2012). *Human Rights Indicators: A guide to measurement and implementation* [Summary PDF]. https://www.ohchr.org/Documents/Issues/HRIndicators/Summary_en.pdf

Transparency International. (2018). *Corruption perceptions index.* https://www.transparency.org/en/cpi/2018.

United Nations Department of Economic and Social Affairs (DESA). (2014). *World economic situation and prospects (WESP) report* [Statistical annex]. https://www.un.org/en/development/desa/policy/wesp/wesp_current/2014wesp_country_classification.pdf.

United Nations Development Programme (UNDP). (n.d.). *Human development index (HDI)* [Report]. http://hdr.undp.org/en/content/human-development-index-hdi.

Vásquez, I., & Porcnik, T. (2019). *Human freedom index* [Report]. CATO Institute. https://www.cato.org/human-freedom-index-new.

World Wide Web Foundation. (2014). *Web index* [Homepage]. https://thewebindex.org/.

World Bank. (2020). *Countries* [Interactive database]. https://projects.worldbank.org/en/country?lang=en&page.

4

Digital Accessibility Innovation in Latin America and the Caribbean

Francesca Cesa Bianchi
Vice President of Institutional Relations and Advocacy
Global Initiative for Inclusive Information and Communication Technologies (G3ict)

4.1 Introduction

The progress of accessible technology in Latin America and the Caribbean is driven by a combination of the region's strong historic involvement with the Convention on the Rights of Persons with Disabilities (CRPD), legislative and regulatory commitments among countries in support of the rights of persons with disabilities (PWDs) slightly ahead of global averages, and a high usage of information technologies, notably the internet and mobile devices. An additional positive factor is the national languages shared with other regions of the world, allowing users and local developers of assistive solutions to benefit from advanced accessibility features, such as text-to-speech and speech recognition, in the four major languages of the region (Spanish, Portuguese, French, and English). In comparison to other countries where no such accessibility features are available, Latin American and Caribbean app developers are in a better position to innovate, and users to benefit from accessible digital interfaces.

From a policy standpoint, however, Latin America and Caribbean (LAC) countries are lagging in their capacity to implement digital accessibility and, as a result, are also lagging in availability of services for PWDs compared to global averages. In addition, the socioeconomic fabric of the region is characterized by strong income inequalities and high proportions of populations living in poverty. While few countries in the region qualify as least developed countries, the World Bank Gini Index of inequalities across the region is one of the highest in the world, suggesting that a large population of PWDs lives in situations of poverty or extreme poverty equivalent to those in the environments of least developed countries.

Latin American and Caribbean countries' commitments to promoting disability rights and innovation for PWDs has a long track record: In 1999, the Inter-American Convention on the Elimination of All Forms of Discrimination

Accessible Technology and the Developing World. Michael Ashley Stein and Jonathan Lazar, Oxford University Press.
© Oxford University Press 2021. DOI: 10.1093/oso/9780198846413.003.0005

Against PWDs was adopted within the Organization of American States as a regional human rights instrument, calling on States to facilitate the full integration of PWDs into society through legislation, social initiatives, and educational programs. In 2001, Mexico proposed the establishment of an Ad Hoc Committee to the CRPD, chaired by the then-Permanent Representative of the Republic of Ecuador, which ultimately led to the adoption of the text of the Convention in December 2006. In 2007, Jamaica was the first ratifying country of the CRPD. Latin America and the Caribbean nations can, therefore, be credited for providing the initial impetus and framework that has had a profound impact on the promotion of technology for PWDs during the past decade among the 181 countries that have ratified the CRPD, as of June 2020.

While cultures and languages inherited from colonial times suggest strong commonalities among nations, Latin American countries, and their close neighbors in the Caribbean, are very diverse and span all four categories of economic development defined by the International Monetary Fund/World Bank based on GDP per capita—from least developed to developed economies.

With a few exceptions, however, the region has one of the highest Gini coefficients as calculated by the World Bank to measure inequality of income, comparable to those of several African nations and Saudi Arabia, the latter scoring 45.9 as of 2013 (IndexMundi, 2019). Table 4.1 shows the Gini Index calculated by the World Bank. A level of zero would reflect complete income equality among citizens of a country, while high scores reflect very high levels of income inequality.

Twenty-three LAC countries have Gini Index scores greater than 40, making the region second only to Africa for income inequalities. Among the top 20 Gini Indexes in the world, 11 countries are from the region: Suriname, Brazil, Belize, St. Lucia, Honduras, Panama, Colombia, Paraguay, Mexico, Guatemala, and Costa Rica. Those data points suggest that large segments of the population of LAC countries, including PWDs who typically have less access to economic opportunities, are living in conditions of poverty or extreme poverty, with little, if any, disposable income available to acquire assistive solutions.

Counterbalancing the effect of inequality and poverty is the positive impact of fast-growing internet and mobile ecosystems, with considerable economies of scale and declining costs. According to a recent report by the Global System for Mobile Communications Association (GSMA Intelligence, 2017), there were 451 million unique mobile subscribers in the region in 2016, projected to reach 511 million in 2020 for a market penetration rate of 76%. Among those, broadband connections are projected to rise from 62% to 79%—and smartphone adoption

Table 4.1 Latin America and the Caribbean
Countries Gini Inequality Index

Country	Gini Index Value
Suriname	57.6
Brazil	53.9
Belize	53.3
Honduras	52.1
St. Lucia	51.2
Colombia	50.4
Panama	49.2
Guatemala	48.3
Costa Rica	48.0
Venezuela, RB	46.9
Nicaragua	46.2
Paraguay	46.2
Jamaica	45.5
Ecuador	45.4
Mexico	45.4
Guyana	44.6
Chile	44.4
Dominican Republic	43.7
Peru	42.8
Bolivia	42.2
Argentina	41.4
Haiti	41.1
Trinidad & Tobago	40.3
Uruguay	39.7
El Salvador	38.6

Source: The World Bank, Gini Index, https://data.worldbank.
org/indicator/SI.POV.GINI

from 55% to 71%—between 2016 and 2020: higher rates than worldwide average rates for the same period, and ahead of most middle-income countries.

Those statistics translate into a mobile internet penetration growing from 51% to 63% of the population, a sea change in how mobile-based assistive and accessible digital applications and services become available to PWDs, leveraging the embedded accessibility features available in Spanish, Portuguese, French, and English in global mobile operating systems. Making assistive apps and services and internet resources available and accessible to PWDs represents a major opportunity for the LAC region to leapfrog past generations of assistive and accessible technologies. This opportunity is clearly supported by the fact that younger generations of PWDs in LAC countries tend to use the internet more than young persons without a disability.

How can governments and civil society help leverage such opportunity? From a digital accessibility rights standpoint, data derived from the Global Initiative for Inclusive Information and Communication Technologies (G3ict)' Digital Accessibility Rights Evaluation Index 2020, a survey conducted in 137 countries worldwide, covering 27 countries in LAC countries and representing 96% of the LAC population, shows that overall the region is ahead for countries' legislative, regulatory, and program commitments, but slightly lagging in capacity to implement. For instance, 42% of countries in the region (vs. 44% globally) refer to international digital accessibility standards, and only 11% involve PWDs in digital accessibility policymaking and monitoring (vs. 26% globally). As a result, outcomes for PWDs, as measured by levels of implementation of accessibility in 10 key areas of digital technologies, show a 47% completion rate, versus an average performance of 49% among the 137 countries monitored globally by the Digital Accessibility Rights Evaluation Index across all levels of development and covering 91% of the world population.

Section 4.2 analyzes those gaps, benchmarking LAC countries' legislative and regulatory dispositions in support of information and communication technology (ICT) accessibility policies and programs as well as their capacity to implement.

The five country case studies—presented in this chapter as examples of successful innovation which are potentially replicable across the region—are meant to support the thesis that progress can be achieved through political will and a more systematic approach to policymaking, higher levels of awareness and education about digital accessibility, cross-fertilization of innovations among countries, and multi-stakeholder cooperation that includes PWDs.

4.2 Background on Digital Accessibility Implementation in Latin America and the Caribbean

With the previously mentioned metrics regarding economic development as a background, it is useful to look at the detailed results for the LAC region in the Digital Accessibility Rights Evaluation (DARE) Index.

The DARE Index measures three categories of variables in each country: Country Commitments (legal, regulatory, policies, and programs); Country Capacity to Implement (organization, processes, resources); and Actual Digital Accessibility Outcomes for PWDs in 10 areas of products and services. This three-step analysis—Commitments, Capacity to Implement, and Outcomes—is consistent with human rights monitoring principles and allows for useful gap analysis and linkages between variables. Variables were derived from the International Telecommunication Union–G3ict *Model ICT Accessibility Policy Report* and the joint Digital Accessibility Decennial Call for Action issued at the

United Nations (UN) on December 3, 2016 by the International (CPI) Disability Alliance, Disabled Peoples' International, and G3ict.

4.2.1 Regional Latin America and the Caribbean Digital Accessibility Rights Benchmarks

Overall, according to Figure 4.1 (which shows the average DARE Index country scores by region of the world), LAC countries' legislative and regulatory commitments to digital accessibility are comparable to those of other regions with similar levels of economic development. However, countries' capacities to implement those commitments and actual outcomes, as measured by levels of implementation of accessible digital technologies for PWDs, remain weak, placing the region in fourth position out of eight regions after Northern America, Europe, and the Middle East and North Africa.

4.2.2 Latin America and the Caribbean Countries' Legislative and Regulatory Commitments to Digital Accessibility

More specifically, levels of commitments reflect legislative and regulatory dispositions or programs supporting the adoption of accessible digital technologies, as shown in Table 4.2, which illustrates the average country commitments to ICT accessibility in LAC. Regional results are generally aligned with global results. The table lists the average performance of countries in the region compared to global averages

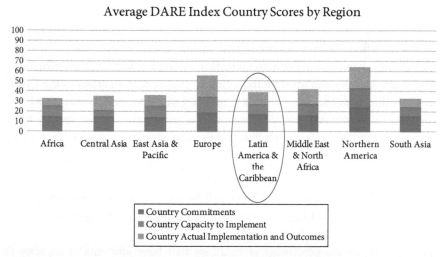

Figure 4.1 Average DARE Index Country Scores by Region of the World *(Source: G3ict DARE Index 2020)*

Table 4.2 Average Country Commitments to ICT Accessibility in LAC

Country commitments	Latin America & the Caribbean results	Global results
CRPD ratification	93%	93%
Marrakesh Treaty Ratification/Accession	67%	37%
General law protecting the rights of persons with disabilities	81%	88%
Law defines reasonable accommodation	73%	68%
Definition of accessibility for ICTs in legislation	63%	61%
Universal service obligation includes PWDs	44%	41%
Average country commitment	68%	65%

Source: G3ict DARE Index 2020

for five legislative and regulatory elements deemed essential for full implementation of the CRPD in matters of digital accessibility.

Examples of laws reflecting country commitments, as indicated by countries' responses to the DARE Index, include: the Colombian Statutory Law 1618 (2013), "Through which the provisions are established to guarantee the full exercise of the rights of PWDs"; the Peruvian regulation of Law No. 29973, which establishes the definition of "Reasonable Accommodation"; Article 17 of the Nicaraguan Law 763 on the Rights of PWDs, which includes a definition of accessibility that covers ICTs or electronic media; and Jamaica's act establishing its Universal Service Fund (2017), which includes PWDs as a primary target population.

Progress is obvious compared to pre-CRPD years, with 63% of LAC countries having a definition of accessibility that includes ICT as per Article 9 of the CRPD. This data point reflects an increasing awareness and focus on promoting digital accessibility and assistive technologies among States Parties to the CRPD in the region. Indeed, a legal definition makes it easier for policymakers and civil society to advocate and justify the adoption of programs to make specific technologies accessible, such as websites, mobile, television, e-books, and their many sectorial applications.

4.2.3 Latin America and the Caribbean Countries' Capacity to Implement Digital Accessibility Policies and Programs

Table 4.3 shows the percentage of countries that have elements of capacity to implement, reflecting their ability to achieve results. Latin American and Caribbean countries are lagging overall compared with world averages in their capacity to implement (44% vs. 46%), and their main weakness is the lack of processes to

Table 4.3 Average Country Capacity to Implement ICT Accessibility in LAC

Country with:	Latin America & the Caribbean results	Global results
Government agency for PWDs	85%	88%
Government agency for accessible ICT	38%	39%
Process to involve PWDs in policymaking on ICT accessibility	11%	26%
ICT accessibility international standards	42%	44%
ICT accessibility courses available at major universities	44%	38%
Average capacity to implement	44%	46%

Source: G3ict DARE Index 2020

involve PWDs in policymaking and monitoring. However, countries in the region have made progress in adopting international accessibility standards and in expanding the availability of educational resources in digital accessibility. Those data points are particularly important to foster innovation and the promotion of digital accessibility solutions for PWDs.

The following are good examples of capacity-building steps taken by countries in LAC, showing that progress is achievable: (1) the Dominican Republic government body specifically dedicated to PWDs (Consejo Nacional de Discapacidad, 2020), which initiated ICT accessibility awareness-raising activities in the country; (2) the Mexican government agency for ICT (Instituto Federal de Telecomunicaciones, 2020), very involved in promoting digital accessibility policies and programs; (3) Colombia's reference to international ICT Accessibility International Standards (on June 15, 2011, the Colombian Institute of Technical Standards and Certification, ICONTEC, ratified Technical Standard 5854 [Norma Técnica Colombiana, 2011]), which seeks to establish the accessibility requirements that webpages in Colombia must implement—this standard is equivalent to Web Content Accessibility Guidelines (WCAG) 2.0; (4) the availability of ICT Accessibility courses available at major universities in Panama at the Universidad Santa Maria La Antigua, Universidad Latina, and Universidad de Panamá.

4.2.4 Latin America and the Caribbean Countries' Actual Implementation of Digital Accessibility, as Evaluated by Local Advocates and Organizations of PWDs

For the purpose of evaluating regional outcomes for the present study, Table 4.4 displays the percentage of countries in the LAC region that are in the process of implementing programs or policies promoting digital accessibility, even at a very

Table 4.4 Averages of Countries in Process of Implementing ICT Accessibility Programs in LAC by Sector

Areas of ICT accessibility % countries in the process of implementation	Latin America & the Caribbean results	Global results
TV	56%	61%
E-books	56%	51%
Promotion of internet availability usage by PWDs	56%	49%
Web	59%	58%
Assistive technologies and ICTs for independent living	44%	39%
Inclusive ICTs in education	59%	54%
Enabling ICTs for employment	30%	47%
E-government and smart cities	44%	50%
Mobile	22%	38%
Public procurement	41%	46%
Average outcome	47%	49%

Source: G3ict DARE Index 2020

limited level. While levels of implementation vary widely, the fact that implementation has started for a category does indicate momentum backed by some level of local political consensus.

With a negative variance of two points versus world average outcomes, LAC countries are increasingly catching up. Initiatives in various sectors show that progress has been achieved by different countries. Among the most advanced countries are Brazil, Colombia, and Nicaragua for accessible television; Brazil, Peru, and Uruguay for web accessibility; Brazil for mobile telephony; and Brazil and Jamaica for promoting inclusive ICTs in education.

Of interest for this survey of innovation in ICT accessibility in LAC is the relative lack of focus by most policymakers in the region on promoting mobile solutions for PWDs. As mentioned in our introduction, unlike in other regions of the world, language commonalities allow for considerable economies of scale for mobile user interfaces. Core enabling accessibility features such as text-to-speech and speech recognition are available, for instance on the operating systems of mainstream smartphones and tablets in the four major languages of the region (Spanish, Portuguese, French, and English). Compared with other countries, where no such accessibility features are available, the LAC entrepreneurs who develop assistive solutions and their users clearly benefit from such situations. Non-mainstream tribal or local minority languages in LAC, however, will likely remain excluded.

Of equal interest for the propagation of innovations are the gaps observed in promoting accessible ICTs in education, for the workplace, and for e-government and smart cities. Several countries are showing considerable innovation and ingenuity in promoting these, as well as mobile solutions, as illustrated by the following examples (Brazil, Colombia, Costa Rica, Ecuador, and Guyana) of effective country-wide policies, entrepreneurial initiatives, and innovative low-cost solutions.

4.3 Country Case Studies

4.3.1 Brazil: Leveraging Mobile Communications and the Internet for Innovative Apps and Services

As noted earlier, the DARE Index (G3ict, 2018a) ranks Brazil as the leading country in Latin America from an ICT accessibility implementation standpoint. Besides a strong legal and regulatory framework, several key areas, such as accessible websites and accessible television, have been proactively dealt with by the government. For instance, the Brazilian Law for Inclusion of PWDs (Brazil Law 13.146, 2015),which covers many of the dispositions of the CRPD, has been in force since 2015. One of its articles (Article 63) makes accessibility mandatory for websites maintained by government agencies or by companies with headquarters or commercial representation in the country. Similarly, the promotion of assistive technologies in education and at the local level has been effective in many areas. Recently, Brazil enacted a law requiring sign language interpretation of all films released theatrically in the country.

As a result of the above legislative mandates, the awareness and interest in all sectors of society for accessible technology that serves PWDs are high. Besides government and advocacy organizations, industry is also very aware of Brazil's advances in accessibility. For instance, in the fall of 2018, Verifone launched Navigator, a Payment Card Industry-compliant full touchscreen feature for the blind and low vision that enables merchants to transform the non-tactile touchscreen into a universal keypad with audio confirmation (Verifone Editorial Board, 2018). It allows blind and low vision consumers to enter a PIN or other numerical data to easily make purchases. Verifone chose to launch Navigator first in Brazil and North America.

Innovation for PWDs in Brazil is also a function of its high degree of ICT and internet usage. In 2018, Brazil counted 138.4 million internet users, 85% of whom use it daily. With a population of 213 million, this represents a penetration of 65% of the overall population. Similarly, in 2018 there were approximately 207 million mobile subscribers in Brazil for a population of 211 million, half of those using smartphones (World Bank, 2018).

The combination of a comprehensive Brazilian disability legislative framework and a strong mobile and internet ecosystem have led to the launch of multiple apps and services for PWDs that are the hallmark of the creativity of Brazilian developers. Examples include:

- **The Federal Government Portal for PWDs:** This portal, launched as early as 2011, aims to ensure that all essential public information, including disability-related information, is accessible to PWDs (Ministry of Women, Family & Human Rights Brazil, 2020). It includes, for instance, retirement guidance and other benefits for PWDs. The portal offers accessibility features, as provided for in the Brazilian Inclusion Law, and is compliant with Web Content Accessibility Guidelines and the Brazilian e-Government Accessibility Model. It also includes a tool that translates the text into Brazilian Sign Language.
- **The National Portal for Assistive Technologies:** This website provides comprehensive resources and references on all aspects of assistive technologies and services, and serves as a platform to share information and knowledge about research, development, application, and dissemination of assistive technology in Brazil (Instituto de Tecnologia Social Brasil, 2016).
- **The Portal for Blind and Low-Vision Persons:** A public–private partnership with a non-profit organization, this portal (Portal da Deficiência Visual, n.d.) aims to provide resources and training in assistive technology, computer usage, and programming, and multiple courses and resources to prepare blind and low-vision persons to gain employment.
- **Signa for Deaf Persons:** This is a privately funded start-up (2018) that provides a comprehensive library of courses in Brazilian Sign Language. Similar to the portal for blind and low-vision persons, Signa covers several disciplines aimed at helping deaf and hard-of-hearing persons get education and gain employment.
- **Livox:** Benefiting from the large Brazilian-installed base of mobile devices and using artificial intelligence and machine learning, this award-winning Brazilian application (Google Play, 2020) aids persons with speech impairments and without functional writing to communicate and develop abilities to express their needs and wishes. The founder, Carlos Pereira, created this alternative communication platform to speak with his daughter, who has cerebral palsy and is unable to walk or talk. The app functions on Android and is designed with multiple interface options to let users of all abilities handle the device. The smart keyboard also automatically corrects mistakes and transcribes whatever the person says, and the program adapts itself to those who cannot read or write. Livox, which was initially launched in Brazil, is available in 25 other languages, and is notably on the U.S. Google Play store. Due to its many innovative educational and training functionalities, driven by artificial intelligence, Livox was voted "World's Best Social Inclusion Application" by the World Summit Award WSA; "Technological Innovation with the Greatest Impact in 2014" by the Inter-American Development Bank; and "Winner of the Silicon Valley Technology World Cup in the U.S., Category Education."

4.3.2 Colombia: Leadership in Mobile Video
Relay Services for the Deaf

Colombia is another exemplary case study of how the government, in cooperation with civil society, was able to leapfrog traditional telecommunications services for the deaf by leveraging the country's fast-expanding mobile and internet infrastructure (G3ict, 2018b).

In order to close the communication gap between deaf and hearing persons across Colombia, in 2001 the Federacion Nacional de Sordos de Colombia (National Deaf Federation of Colombia), a non-governmental organization founded in 1984 and based in Bogotá (Federacion Nacional de Sordos de Colombia, 2020b), launched a project called "Centro de Relevo" (Relay Service), which enables telephone communication between deaf and hearing people by means of Colombian Sign Language interpreters, through mediated bidirectional communication (Ministry of Information Technologies & Communications, Colombia, 2020a). The need originated because deaf persons in Colombia—an estimated two million (2013)—were unable to access telephone communications due to a scarcity of interpreters, with only one interpreter available for every 239 deaf persons, according to a factsheet by the Federacion Nacional de Sordos de Colombia (Zero Project, 2018).

Over the years, jointly with the Ministry of Information Technologies and Communications, the "Centro de Relevo" has progressed successfully, positioning itself as a leading initiative in Colombia and in the Americas. In 2009, with the expansion of broadband in Colombia, the service was implemented online, via a web platform. More recently, as the technology evolved, the "Centro de Relevo Live Connect" app for mobile devices was launched. The app is available in iOS and Android (Ministry of Information Technologies & Communications, Colombia, 2020b).

Despite the income inequality in Colombia as among the worst in the world, with half the population living below the poverty line and a Gini Index of 50.4 in 2018 (World Bank Development Research Group, 2018), the telecommunications system is modern in many respects, with a nationwide internet user penetration of more than 64% (Navarro, 2020) and mobile penetration of 129.5% (2018) for a total estimated 2018 population of 49.46 million (Statista Research Department, 2020).

In 2017, according to the Federacion Nacional de Sordos de Colombia (FENASCOL), the annual cost of the "Centro de Relevo" service was $827,500, with 95% of the funds provided by the government through the Colombian Universal Service Fund. A study is currently being carried out, seeking to achieve future self-sustainability (Zero Project, 2018).

The "Centro de Relevo" services are provided at no cost and currently include a call relay service, online interpretation services, and video messages via

WhatsApp and PQRS ("Peticiones, quejas, reclamos, sugerencias"—Petitions, complaints, claims, suggestions). The services allow deaf persons to find an interpreter—via a text message or the internet—who will provide them with free translation services over a fixed line or mobile phone within Colombia for a maximum of 15 minutes. The same interpreter can also provide a variety of other communication services, such as for medical appointments, work-related consultations, legal formalities, and other similar situations for a maximum of 30 minutes.

The project is a public–private partnership between FENASCOL and the Ministry of Information Technologies and Communications, Colombia (MinTIC) in which the functions of the operator and the project leader are filled by deaf persons themselves (MinTIC, 2020c). Using the technology, the project's 50 working interpreters can operate from home in all areas of the country. As it is based on an economy of scale model, the interpreters, who often charge expensive rates, can provide the service many times a day. The services are available every day of the week, including Sundays and holidays. The "Centro de Relevo" offers virtual training for its interpreters on topics ranging from communication and translation to sign language, Deaf culture, and customer service. As a result, users can always count on qualified interpreters (FENASCOL, 2020a).

In terms of impact, the "Centro de Relevo" website reports that the service counts 53,597 users; more than 3 million free calls have been placed; and more than 78,000 online interpreting services have been provided, as of June 2020 (MinTIC, 2020a). Following its initial launch in Colombia's capital city, Bogotá, the service is now offered nationwide.

In 2013, the service was replicated in Paraguay, and at present FENASCOL provides consultancy services to Ecuador, Guatemala, and Mexico. The organization states that in order to operate this service successfully, it needs the backing of public finance in the country concerned, as well as a public policy that supports the implementation of such a project (Zero Project, 2018).

"Centro de Relevo" received international recognition: It was awarded the Innovative Practice 2018 on Accessibility (with the name "Video-sign language service throughout the country") by the Zero Project at the UN headquarters in Vienna. It was also named a finalist at the World Summit on the Information Society Prizes 2016 contest, among 88 champions (International Telecommunication Union, 2016).

4.3.3 Costa Rica: Universal Service Fund Policy for PWDs

According to the DARE Index (2020), only 44% of LAC countries have a universal service obligation that includes PWDs. Yet universal service funds are extremely suitable sources of funding for programs in support of PWDs which can be leveraged by governments to fulfill their obligation toward making all ICT-based applications and services accessible and promoting assistive technologies. universal service funds, which are funded by contributions levied on phone bills, provide a regular and recurring source of income for accessibility programs and services that is not reliant on annual government budgets.

A good practice of universal service fund (USF) management is found in Costa Rica (G3ict, 2018c). The need for bringing unprofitable telecommunication to both rural communities and vulnerable groups, as a shared responsibility of all operators in the telecommunication market, originated in Costa Rica in the last decade because of market liberalization, following ratification of the free trade agreement with the United States in 2007 and the end of Costa Rica's state monopoly in the telecommunications sector in 2008 (FONATEL, 2020).

The 2008 General Law of Telecommunications (*Ley General de Telecomunicaciones*, No. 8642), in its Article 32, contains detailed provisions for universal service, which includes PWDs (Sistema Costarricense, 2008). It is arguable that the ratification of the CRPD by the government of Costa Rica, also in 2008, has contributed to the alignment of the country's telecommunications law to support PWDs.

With a well-defined legislative framework in place, the stage was set in 2009 for the creation of the National Telecommunications Fund as the designated universal service provider, with the mandate to promote solidarity and development according to the General Law of Telecommunications and public policies. The National Telecommunications Fund is an independently operated fund, but state-affiliated as part of the Telecommunications Superintendence, the Costa Rican telecommunications regulator.

The USF in Costa Rica is financed by mandated contributions from telecommunications operators (between 1.5% and 3.0% of their annual gross income), as well as spectrum allocation tenders. Breach funds imposed on operators and interest charges are also added to the "solidarity" fund.

From a socioeconomic infrastructure perspective, Costa Rica is a highly concentrated country, with most of its population of 5.05 million (July 2019 est.) centered in the city of San José, its capital. While in 2017 the World Bank (2020) re-designated the country as an upper middle-income nation, poverty and income inequality persist. Costa Rica's poverty rate is lower than in most Latin American countries, stalling at around 20% for almost two decades (Central Intelligence Agency, 2020). A quarter of the total population continues to live in poverty, especially in rural areas. Yet Costa Rica's broadband market is the most advanced in Central America, with the highest broadband penetration for this sub-region. As of 2019, Costa Rica counted 4.3 million internet users, 86% of the total population. Internet penetration, however, is lower in rural regions when compared to urban areas. Also, among the poorest households, only 19.3% own a computer, according to the Encuesta Nacional de Hogares (2017; National Household Survey).

In this context, the National Telecommunications Fund (FONATEL) has been deploying a variety of programs to reduce the digital divide and universalize access to internet connectivity, in the context of the 2015–2021 National Telecommunications Development Plan and the 2018–2022 Digital Transformation Strategy toward the Costa Rica of Bicentennial 4.0, which are spearheaded by the Ministry of Science, Technology and Telecommunications (2019). The plans seek to reduce digital gaps with respect to the access, use, and adoption of technologies to improve the quality of life of vulnerable populations, as well to accelerate productivity, competitiveness, and socioeconomic development by capitalizing on the fourth industrial revolution. Including PWDs in the USO of Costa Rica is an innovative mechanism that allows for funding of assistive and accessible services. Examples include the "Hogares Connectedos" (Connected Homes Program) and the three-way partnership between FONATEL, the Consejo Nacional de Personas con Discapacidad (National Council of PWDs), and the Centro Nacional de Recursos para la Educación Inclusiva (National Resource Center for Inclusive Education).

The Connected Homes Program is one of FONATEL's flagship programs and a wider private–public initiative in Costa Rica (which includes the Vice Presidency, the Rector, and Regulator for Telecommunication), implemented by telecommunications companies and supported by non-governmental organizations and other institutions. This program provides low-income households with significantly subsidized, fixed internet connections and laptop computers. The objective is to subsidize up to 80% of computer and broadband costs for almost 140,000 low-income families, around 15% of Costa Rican homes. FONATEL provides the financial support for the subsidy, which lasts three years, and covers the cost of a basic computer and internet service. The telecommunications service providers supply both the internet service and the computer resources and software licenses; engage in program promotion; and provide the requisite e-government applications and digital literacy training. The program, which officially started disbursing subsidies in June 2016, will invest $128 million over the course of five years, until 2021. As of 2018, 46,322 households in extreme poverty nationwide have been provided with subsidized internet and computers (Presidencia de la República, 2018).

The partnership between FONATEL, the Consejo Nacional de Personas con Discapacidad (the government body specifically dedicated to PWDs), and the Centro Nacional de Recursos para la Educación Inclusiva (a national resource center for inclusive education in Costa Rica) is an additional program funded by FONATEL in favor of PWDs. The partnership includes a supply of assistive technology products for about 318 students and staff with visual, hearing, physical, and cognitive disabilities and with Autism Spectrum Disorders (ASD; Vazquez Chaves, 2017).

4.3.4 Ecuador: Emergency GPS Localization Services for PWDs

As pointed out in Stein and Lazar's introduction to this book, Ecuador could be considered an "example of the underexplored phenomena of developing States" which, because of the lack of resources, develop creative and cost-effective uses of ICT to include and enable PWDs, thus becoming a "source of significant innovation" (Stein & Lazar, 2021).

Because of its geographical position, lying along the Pacific Ring of Fire, Ecuador experiences frequent volcanic activity and several earthquakes of various magnitudes each year, with the added potential danger of tsunamis in coastal areas. Moreover, Ecuador is a country prone to other natural hazards, including drought, torrential rains, floods, and landslides due to the El Niño phenomenon, as well as forest fires, all of which pose serious threats to the population.

According to the United Nations Department of Economic and Social Affairs (UNDESA) (n.d.), "while different populations may face similar risks of exposure to the negative effects of environmental and man-made disasters, their actual vulnerability is dependent on their socioeconomic conditions, civic and social empowerment, and access to mitigation and relief resources." Experience reveals that disasters affect PWDs much more than the general populations. PWDs are more likely to be left behind or abandoned during evacuations; to experience inaccessible facilities such as shelters and transportation services; and to be discriminated against when resources are scarce.

In past years, Ecuador has played a leadership role in the protection of PWDs (G3ict, 2018d), whose estimated number is 471,615 (Secretaría Técnica, 2018). The first national institution serving PWDs, the Consejo Nacional de Discapacidad (National Council on Disabilities), was created in 1992. Over the last decades, the country has established a comprehensive platform of laws and regulations to guarantee the exercise of the rights of Ecuadorians with disabilities: the Constitution of the Republic (2008); the Organic Law on Disabilities (2012) and its regulations; the CRPD (ratified by Ecuador in 2008); and the Inter-American Convention on the Elimination of All Forms of Discrimination Against PWDs (OAS-1999). The personal commitment by former Vice President (2007–2013) and President of the Republic (2017–2021), Lenín Voltaire Moreno Garcés, who lives with a disability, has also been instrumental. President Moreno was appointed as Special Envoy of the UN Secretary-General on Disability and Accessibility (2013–2016).

Facilitated by both technological advances and the legal framework that supports the inclusion of PWDs in society, Ecuador has pursued a public policy of "Inclusive Disaster Risk Management (DRM)." According to the Servicio Nacional de Gestión de Riesgos y Emergencias (2019b; National Service for Risk and Emergency Management), this recently gained the country a place among the top-performing nations worldwide, upon presenting its experience at the Global Platform for Disaster Risk Reduction conference, organized by the UN Office for Disaster Risk Reduction in Geneva in May 2019 (UNDRR, 2019). The UN Office for Disaster Risk Reduction is the focal point of the UN system for disaster risk reduction and the custodian of the Sendai Framework, supporting countries and societies in its implementation, monitoring, and review of progress.

One of the most innovative technological solutions implemented in Ecuador to locate PWDs, in the context of the "Inclusive DRM" policy, is georeferencing (Tripathi, 2013), which technically defines the location of an object in a three-dimensional physical space. Through the satellite location system, during emergencies, first responders are informed of needs and can act in real time to reach out to all those registered georeferentially.

The practice of "georeferencing" PWDs in the country was deployed in July 2009 as part of the "Manuela Espejo Solidarity Mission," a scientific and medical research study aimed at determining the causes of disabilities and, consequently, outlining and implementing adequate State policies and programs, including technical support, medical care, housing, rehabilitation, and nutrition. Initiated in Cuba in 2001, this psychopedagogical, social, and clinical genetic study of disabilities was similarly carried out in Venezuela (2007), Bolivia (2009), Nicaragua and San Vicente and the Grenadines (2010). For the first time in the history of Ecuador, several medical brigades, composed of doctors, geneticists, psychologists, and health specialists, travelled to the most remote areas of the country to identify and collect data, provide medical care to PWDs who are living in poverty and extreme poverty, and ultimately geoposition them.

In 2011, the government conducted, through the Operational Emergency Committee "Manuela Espejo," a high-tech tsunami evacuation drill during an emergency warning alert. Thanks to the satellite location system, which registered 294,000 PWDs nationwide in the year prior, it was possible, in a very short period of time, to georefer the 1,737 persons with severe physical and intellectual disabilities inhabiting the country's 86 coastal parishes. About 80 mobile vehicles were available to the brigades, as well as a team of 59 doctors, support, and military personnel, guided from a satellite control room based at the Vice Presidential Office in Quito, which worked to evacuate 603 PWDs in critical condition and safely re-locate them into shelters (Vicepresidencia, 2011).

More recently, starting in 2016, an Early Tsunami Warning System was implemented in the coastal areas, including the Galapagos Islands (SNGRE, 2019a). Furthermore, in December 2018, the Servicio Nacional de Gestión de Riesgos y Emergencias announced that a new geodemographic digital platform was put in place. The platform, a georeferential information system, will serve to explore, visualize, and produce reports (with disaggregated data by age, sex, and geospatial information) in order to plan targeted emergency response actions (El Universo, 2018). At the local level, additional technological improvements were announced in March 2019, such as the installation of a new Cartographic Information Center of the Military Geographical Institute (IGM) to integrate and georeference flooded areas, and those highly vulnerable to flooding, in the coastal parish of Manabí (SNGRE, 2019c).

The use of advanced technology is complemented by a series of institutional, educational, and informational activities to strengthen communities which are carried out in the context of the Strategic Risk Management Plan 2019–2030, the National Response Plan to Disasters (2018), contingency plans, and the recently launched new social intervention program, Plan Toda una Vida, overseen by Ecuador's First Lady, which includes seven missions—among them the rebranded "Misión Las Manuelas," conceived as a public policy aimed at PWDs (Secretaría Técnica, n.d.).

Interagency relations, decentralization, inclusion of PWDs on an interinstitutional coordination basis (in respect of institutional competences), and promotion of joint action are at the core of the Servicio Nacional de Gestión de Riesgos y Emergencias (SNGRE) "Inclusive DRM" strategy. Among relevant activities and tools are, at the local level, the creation of Community Committees, trained in prevention and self-protection measures, with the participation of PWDs (SNGRE, 2018a) and their care givers (SNGRE, 2018b); the release of a "Guide for the Protection of PWDs in Emergencies and Disasters"; and the launch of the Ecuador Listo y Solidario (Ecuador Ready and Caring) emergency response campaign (Gobierno de la Republica del Ecuador, n.d.) via web and social networks (Twitter hashtags: #Riesgos Ecuador; #Ecuador Prevenido; #SNGRECuidaDeTi).

4.3.5 Guyana: Tackling Web Accessibility from the Ground Up

Guyana is the only English-speaking country in South America and shares cultural and historical bonds with the Anglophone Caribbean. It has a small population of 740,685 inhabitants (July 2018 est.), with about one-third living below the poverty line (G3ict, 2018e). A recent survey, conducted by the National Commission on Disability of Guyana, has revealed that there are more than 11,000 PWDs in the country (Guyana Times, 2019). The survey is part of the government's effort to ascertain the needs of this demographic and validate its spending. In the 2019 budget, in fact, funds were made available to promote the use of ICTs for a range of e-government, e-health, and tele-education services, as well as to build ICT hubs in remote areas affecting several communities (Lancaster, 2020).

Backed by political consensus, thanks to the commitment expressed by the Ministry of Public Telecommunications, the University of Guyana, and the National Commission on Disability, Guyana was recently selected by the International Telecommunication Union (ITU) to participate in its "Internet for @ll" national program, a five-day training course developed by the ITU Development Sector in four languages (Arabic, English, French, and Spanish) as part of the member states' drive to create capabilities in web accessibility. The program consists of five steps, including a high-level meeting of political buy-in, aimed at raising awareness among government representatives, national stakeholders, and decision makers about ICT accessibility. The next step is to "train the trainers," as well as technical training sessions for government information technology and communications staff during which participants learn to build capacity in designing and developing accessible websites according to international standards (i.e., Web Content Accessibility Guidelines and principles), validated by users with disabilities.

The training in Guyana was held in May 2019. As part of the technical training for web developers, participants had to prove, within a month of work, the accessibility remediation of their respective organizations' websites. As a result, according to the ITU, more than 10 official websites were deemed more accessible at the end of the program. Furthermore, the participation of the University of Guyana in the program and the partnership between the university and the government were crucial to ensure the continuity of the trainings in the country and the development of ICT professionals that understand the standards of web accessibility. At the end of the program, in fact, Ministry of Public Telecommunications signed an agreement with both the University of Guyana and the National Commission on Disability to not only create capabilities for government officials and website developers, but also allow PWDs to get digital skills.

Ultimately, enabling accessible digital environments aims to position Guyana as a role model in web accessibility. The implementation of the program is expected to have an impact in the Caribbean countries. Considering the close relationship of these nations, as well as the partnership that exists between Caribbean universities, ITU forecasts that the experiences gained and the capabilities created in Guyana will be exported to its fellow countries, contributing to achieving a more inclusive digital environment for PWDs and those with age-related disabilities. With the support of ITU, the Ministry of Public Telecommunications (MOPT) in Guyana intends to review the pertinent legislation and make the necessary changes, while inspiring other countries to implement similar policies (ITU Telecommunication Development Sector, n.d.). In the long run, the training program was a good way of building awareness and competencies in the country in view of establishing a web accessibility policy in the future.

4.4 Conclusion

As demonstrated by the above case studies, progress is being made in LAC, leveraging widespread accessible features in the regions' main languages and their fast-expanding mobile and internet infrastructure. In Brazil, both the government and entrepreneurs take advantage of the strong mobile and internet infrastructure and levels of adoption to develop innovative apps and services for PWDs, including apps that are now exported globally. Colombia's experience shows how the leadership of organizations of PWDs made it possible to launch and operate one of the most innovative mobile video relay services for the deaf in the world. In Costa Rica, universal service legislation opened the door for the USF to support programs promoting the availability of services for PWDs. In Ecuador, innovation in disaster and emergency management is driven by the government, which developed a unique Emergency GPS localization service for PWDs. In Guyana, despite limited means, the government is tackling web accessibility from the ground up, showing that web accessibility for e-government services can be achieved across all levels of economic development.

Those examples support the notion that progress can be achieved across the region by leveraging a fast-expanding mobile and internet ecosystem despite social and economic inequalities. This is important, as assistive features of mainstream technology products and services are available in the main languages of the region for core technology platforms such as mobile, web, TV, e-books, and computer. What are needed are stronger country commitments to establish and promote digital accessibility rights, a greater focus on ICT accessibility skill-building, multi-stakeholder cooperation, and most importantly, involvement of PWDs in formulating policies and programs addressing their needs.

Promoting digital accessibility in key areas of application, such as digital workplaces, cultural activities, and education institutions, is an essential step toward truly inclusive societies. While large gaps still exist, good practices can be found in most areas across the region. As a result, cross-fertilization of innovations among countries on how to promote programs and multi-stakeholder cooperation that includes PWDs can go a long way toward achieving progress.

International organizations such as the International Telecommunication Union and the UN Educational, Scientific and Cultural Organization; regional commissions such as the UN Economic Commission for LAC; civil society organizations such as G3ict, the Digital Accessible Information System Consortium, the World Wide Web Consortium, the International Association of Accessibility Professionals; and organizations of PWDs such as Red Latinoamerica de Organizaciones no Gubernamentales de Personas con Discapacidad y sus Familias (the Latin American Network of Non-Governmental Organizations of People With Disabilities and Their Families) or the Unión Latinoamericana de Ciegos (Latin American Union of the Blind), to name a few, are important knowledge-sharing hubs for the region and should be supported in their efforts. As the author witnessed in many circumstances across the region, peer-to-peer sharing from one country to another, demonstrating that progress can be achieved in similar social and economic conditions, is a powerful motivator for all stakeholders.

References

Brazil Law 13.146 [Statute]. (2015). Retrieved February 18, 2021, from Planalto—Governo Federal [Federal Government of Brazil website]: http://www.planalto.gov.br/ccivil_03/_Ato2015-2018/2015/Lei/L13146.htm

Central Intelligence Agency. (2020, June 10). *World factbook—Costa Rica.* Retrieved Retrieved February 18, 2021, from https://www.cia.gov/library/publications/the-world-factbook/geos/cs.html

Colombian Statutory Law 1618, Diaro Oficial No. 48.717 [Statute]. (2013, February 27). Retrieved February 18, 2021, from Secretaría General del Senado [General Secretary of the Senate]: http://www.secretariasenado.gov.co/senado/basedoc/ley_1618_2013.html

Consejo Nacional de Discapacidad ([CONADIS]—National Council on Disability, Dominican Republic). CONADIS [Homepage]. (2020). Retrieved February 18, 2021, from http://conadis.gob.do

Costa Rica Telefonos. (2020). *FONATEL (Fondo de Telecomunicaciones de Costa Rica [National Telecommunications Fund])* [Homepage]. Retrieved February 18, 2021, from https://costaricatelefonos.com/fonatel-fondo-nacional-de-telecomunicaciones.htm

Federación Nacional de Sordos de Colombia ([FENASCOL]—National Deaf Federation of Colombia). (2020a). *Cursos LSC (LSC courses).* Retrieved Retrieved February 18, 2021, from https://www.fenascol.org.co/cursos-lsc/

Federación Nacional de Sordos de Colombia ([FENASCOL]—National Deaf Federation of Colombia). (2020b). [Homepage]. Retrieved February 18, 2021, from http://www.fenascol.org

Global Initiative for Inclusive ICTs (G3ict). (2018a). *Brazil, country dashboard.* Retrieved February 18, 2021, from https://g3ict.org/country-profile/brazil

Global Initiative for Inclusive ICTs (G3ict). (2018b). *Colombia, country dashboard.* Retrieved February 18, 2021, from https://g3ict.org/country-profile/colombia

Global Initiative for Inclusive ICTs (G3ict). (2018c). *Costa Rica, country dashboard*. Retrieved February 18, 2021, from https://g3ict.org/country-profile/costa-rica

Global Initiative for Inclusive ICTs (G3ict). (2018d). *Ecuador, country dashboard*. Retrieved February 18, 2021, from https://g3ict.org/country-profile/ecuador

Global Initiative for Inclusive ICTs (G3ict). (2018e). *Guyana, country dashboard*. Retrieved February 18, 2021, from https://g3ict.org/country-profile/guyana

Global Initiative for Inclusive ICTs (G3ict). (2020). *Digital Accessibility Rights Evaluation Index (DARE Index)*. Retrieved February 18, 2021, from https://g3ict.org/digital-accessibility-rights-evaluation-index/

Gobierno de la Republica del Ecuador [Government of the Republic of Ecuador]. (n.d.). *Ecuador Listo y Solidario* [Emergency response campaign]. (n.d.). Retrieved February 18, 2021, from http://www.ecuadorlistoysolidario.gob.ec/

Google Play. (2020). *Livox*. Retrieved February 18, 2021, from https://play.google.com/store/apps/details?id=br.com.livox&hl=en

GSMA Intelligence. (2017). *Economía móvil 2017 América Latina y Caribe* [The mobile economy, Latin America and the Caribbean; PDF]. Retrieved February 18, 2021, from https://www.gsma.com/latinamerica/resources/mobile-economy-latin-america-caribbean-2017/

Guyana Times. (2019, June 28). *Over 11,000 persons living with disabilities in Guyana*. Retrieved from: https://guyanatimesgy.com/over-11,000-persons-living-with-disabilities-in-guyana/

IndexMundi. (2019, December 17). *Saudi Arabia distribution of family income—Gini Index*. Retrieved February 18, 2021, from https://www.indexmundi.com/saudi_arabia/distribution_of_family_income_gini_index.html

Instituto Federal de Telecomunicaciones ([IFT]—Federal Telecommunications Institute, Mexico). (2020). *IFT* [Homepage]. Retrieved February 18, 2021, from http://www.ift.org.mx

Instituto de Tecnologia Social Brasil ([ITS]—Institute of Social Technology, Brazil). (2016, 29 November). *Porto nacional de tecnologia assistiva* [National portal for assistive technologies, Brazil; Homepage]. Retrieved February 18, 2021, from https://assistivaitsbrasil.wordpress.com/

International Telecommunication Union (ITU). (2016). *Champion Projects: WSIS project prizes 2016*. Retrieved February 18, 2021, from http://groups.itu.int/stocktaking/WSISPrizes/WSISPrizes2016.aspx#champion-projects

ITU Telecommunication Development Sector (ITU-D). (n.d.). *The ITU-D national programme in web accessibility: "Internet for @ll."* Retrieved February 18, 2021, from https://www.itu.int/en/ITU-D/Digital-Inclusion/Persons-with-Disabilities/Pages/Internet-for-%40ll.aspx

Lancaster, H. (2020, April 22). *Guyana—telecoms, mobile and broadband—statistics and analyses* [Report summary]. Retrieved from BuddeComm: https://www.budde.com.au/Research/Guyana-Telecoms-Mobile-and-Broadband-Statistics-and-Analyses?r=51

Ministerio de Ciencia, Tecnología y Telecommunicaciones ([MICITT]—Ministry of Science, Technology and Telecommunications). (2019). *Estrategia de Transformación Digital hacia la Costa Rica del Bicentenario 4.0* [Bicentennial digital transformation strategy; PowerPoint report]. Retrieved February 18, 2021, from https://www.micit.go.cr/documentos/micitt_estrategia_transformacion_digitaldel_bicentenario.pdf

Ministry of Information Technologies and Communications, Colombia (MinTIC). (2020a). *Centro De Relevo* [Homepage]. Retrieved February 18, 2021, from http://www.centroderelevo.gov.co

Ministry of Information Technologies and Communications, Colombia (MinTIC). (2020b). *Mobile app* [Press release]. Retrieved February 18, 2021, from Centro De Relevo: https://centroderelevo.gov.co/694/w3-propertyvalue-47571.html

Ministry of Information Technologies and Communications, Colombia (MinTIC). (2020c). *Novedades: FENASCOL operará el Centro de Relevo durante 2020* [Press release: FENASCOL will operate Centro de Relevo during 2020]. Retrieved February 18, 2021, from https://centroderelevo.gov.co/632/w3-article-126057.html

Ministry of Women, Family and Human Rights Brazil. (2020). *Pessoa com deficiência* [Federal portal for persons with disabilities, Brazil]. Retrieved February 18, 2021, from https://www.gov.br/mdh/pt-br/navegue-por-temas/pessoa-com-deficiencia

National Institute of Statistics and Census of Costa Rica (INEC). (2017, July). *Encuesta Nacional de Hogares* ([Enaho]—National Household Survey) [Report]. Retrieved February 18, 2021, from https://www.inec.cr/multimedia/enaho-2017-encuesta-nacional-de-hogares-2017

Navarro, J. G. (2020, March 2). *Colombia: Internet penetration 2000–2018*. Retrieved February 18, 2021, from Statista: https://www.statista.com/statistics/209109/number-of-internet-users-per-100-inhabitants-in-colombia-since-2000/

Norma Técnica Colombiana ([NTC]—Colombian Technical Standard) 5854, Accessibilidad Web. (2011, June 15). Retrieved from Instituto Colombiano de Normas Técnicas y Certificación ([ICONTEC]—Colombian Institute of Technical Standards and Certification): https://www.icontec.org

Presidencia de la República. (2018, March 9). *En Puntarenas: 300 familias en pobreza extrema reciben beneficio Avancemos y Hogares Conectados* [Press release: In Puntarenas: 300 families in extreme poverty benefit from the 'Avancemos' (Move Forward) and 'Hogares Connectados' (Connected Homes) programs.]. Retrieved February 18, 2021, from https://www.presidencia.go.cr/comunicados/2018/03/en-puntarenas-300-familias-en-pobreza-extrema-reciben-beneficio-avancemos-y-hogares-conectados/

Portal da deficiência visual (Portal for blind and low vision persons, Brazil). (n.d.). [Homepage]. Retrieved February 18, 2021, from https://www.deficienciavisual.com.br/

El Universo. (2018, December 18). Se actualiza plan de contingencia por posible Desarollo del El Niño [Newspaper article: Contingency plan for possible child development is updated]. Retrieved February 18, 2021, from https://www.eluniverso. com/noticias/2018/12/18/nota/7103550/se-actualiza-plan-contingencia -posible-desarrollo-nino

Secretaría Técnica Plan Toda una Vida (Lifelong Plan Technical Secretariat). (n.d.). *Misión Las Manuelas*. Retrieved February 18, 2021, from Gobierno de la Republica del Ecuador: https://www.todaunavida.gob.ec/lasmanuelaslamisioncontinua/

Secretaría Técnica Plan Toda una Vida. (2018). *Plan Toda Una Vida* [Homepage]. Retrieved from Gobierno de la Republica del Ecuador: https://www.todaunavida. gob.ec/plan-toda-una-vida/

Servicio Nacional de Gestión de Riesgos y Emergencias ([SNGRE]—National Risk and Emergency Management Service). (2018a, October). *Riesgos incluye a personas con discapacidad* [SNGRE news: Risks include people with disabilities]. Retrieved February 18, 2021, from https://www.gestionderiesgos.gob.ec/riesgos-incluye-a-personas-con-discapacidad/

Servicio Nacional de Gestión de Riesgos y Emergencias (SNGRE). (2018b, December). *SNGRE capacitó a cuidadores de personas con discapacidad sobre gestión inclusiva del riesgo* [SNGRE news: SNGRE train caregivers of people with disabilities ono inclusive risk management]. Retrieved February 18, 2021, from https://www. gestionderiesgos.gob.ec/sngre-capacito-a-cuidadores-de-personas-con-discapacidad-sobre-gestion-inclusiva-del-riesgo/

Servicio Nacional de Gestión de Riesgos y Emergencias (SNGRE). (2019a, May). *Declaración Oficial Ecuador* [Global Platform for Disaster Risk Reduction Conference Proceedings; PDF]. Retrieved February 18, 2021, from United Nations Office for Disaster Risk Reduction: https://www.unisdr.org/files/globalplatform/ ecuador%5b1%5d.pdf

Servicio Nacional de Gestión de Riesgos y Emergencias (SNGRE). (2019b, June). *Ecuador en el top mundial de la gestion inclusiva* [SNGRE news: Ecuador leading the world in inclusive management]. Retrieved February 18, 2021, from https:// www.gestionderiesgos.gob.ec/sexta-edicion-de-punto-de-encuentro/

Servicio Nacional de Gestión de Riesgos y Emergencias (SNGRE). (2019c, March 14). *En el COE Manabí se instaló Centro de Información Cartográfica* [SNGRE news: Cartographic information center was installed at COE Manabí]. Retrieved February 18, 2021, from https://www.gestionderiesgos.gob.ec/en-el-coe-manabi-se-instalo-centro-de-informacion-cartografica/

Signa Edu. (2018). Signa for Deaf Persons, Brazil [Homepage]. Retrieved from: http:// www.signaedu.com/index.html

Sistema Costarricense de Información Jurídica, Artículo 32: Objetivos del acceso universal, servicio universal y solidaridad [Costa Rica General Telecommunications Law of 2008—Article 32]. (2008, June 4). Retrieved February 18, 2021, from Sistema Costarricense de Información Juridica, Attorney General of the Republic:http://www.pgrweb.go.cr/scij/Busqueda/Normativa/Normas/nrm_articulo. aspx?nValor1=1&nValor2=63,431&nValor3=91,176&nValor5=33

Statista Research Department. (2020, March 3). Colombia: Mobile services penetration 2012–2018. Retrieved February 18, 2021, from https://www.statista.com/statistics/622690/mobile-phone-penetration-in-colombia/

Stein, M. A., & Lazar, J. (2020). Introduction, in M. A. Stein & J. Lazar (Eds.), *Accessible technologies and the developing world.* Oxford University Press (pp. XXX).

Tripathi, N. K. (2013, September). Geospatial database for the GeoDRM system [PDF]. Retrieved from United Nations Economic and Social Commission for Asia and the Pacific (UNESCAP) GEO-REF Information Sharing Platform for Disaster Management: https://www.unescap.org/sites/default/files/UN_cook_sep2013_3_Tripathi.pdf

United Nations Department of Economic and Social Affairs (UNDESA). (n.d.). Disability-inclusive disaster risk reduction and emergency situations. Retrieved February 18, 2021, from https://www.un.org/development/desa/disabilities/issues/disability-inclusive-disaster-risk-reduction-and-emergency-situations.html

United Nations Office for Disaster Risk Reduction (UNDRR). (2019, May). *Global Platform for Disaster Risk Reduction* [Conference homepage—13–17 May 2019]. Retrieved February 18, 2021, from https://www.unisdr.org/conference/2019/globalplatform/home

Universal Service Fund Jamaica. (2017). [Homepage]. Retrieved February 18, 2021, from www.usf.gov.jm

Vazquez Chaves, A. P. (2017, November 21). *El esfuerzo nacional en Costa Rica para el fomento de la accesibilidad: tendencias; reformas legales y regulatorias realizadas en materia de telecomunicaciones y la accesibilidad en la región.* [The national effort in Costa Rica to promote accessibility: trends; legal and regulatory reforms in telecommunications and accessibility in the region; Paper presentation]. ITU Accessible Americas IV Conference, San José, Costa Rica. Retrieved February 18, 2021, from Centro Nacional de Recursos para la Educación Inclusiva—CENARAC: Presentación_CENAREC.pdf (itu.int)

Verifone Editorial Board. (2018, October 22). *Introducing Verifone Navigator* [Press release]. Retrieved February 18, 2021, from https://www.verifone.com/en/us/press-release/introducing-verifone-navigator

Vicepresidencia de la República del Ecuador. (2011, March 12). *Brigades "Manuela Espejo" functionaron según lo previsto en la tarea de evacuar a las personas con discapacidad de la zona constanera* [News: "Manuela Espejo" brigades functioned as planned in the task of evacuating people with disabilities from the zona constanera]. Retrieved February 18, 2021, from https://www.vicepresidencia.gob.ec/brigadas-manuela-espejo-funcionaron-segun-lo-previsto-en-la-tarea-de-evacuar-a-las-personas-con-discpacidad-de-la-zona-costanera/

World Bank. (2018). *Mobile cellular subscriptions* [World Telecommunication/ICT Development report and database]. (2018). Retrieved February 18, 2021, from https://data.worldbank.org/indicator/IT.CEL.SETS

World Bank. (2020, June 8). Costa Rica country overview. Retrieved February 18, 2021, from https://www.worldbank.org/en/country/costarica/overview

World Bank Development Research Group. (2018). GINI index (World Bank estimate). Retrieved February 18, 2021, from https://data.worldbank.org/indicator/SI.POV.GINI

Zero Project. (2018). *Innovative practice 2018 on accessibility: Video-sign language service throughout the country*. Retrieved February 18, 2021, from https://zeroproject.org/practice/pra181337col-factsheet/

5

Digital Accessibility in the Asia-Pacific Region

Nirmita Narasimhan
Global Initiative for Inclusive Information and Communication
Technologies (G3ict)

5.1 Introduction

The Asia-Pacific region covers a vast area and has many definitions. For the purpose of this chapter, the Asia-Pacific region is considered to include the countries of Central, South, and East Asia and Oceania, ranging from Armenia in Central Asia to Australia and New Zealand in the South Pacific, and is home to over 65% of the world's population with disabilities, or around 690 million people. It is a region of dramatic contrasts and complexities, with incredibly different languages, cultures, terrains, and economies, ranging from poor to middle- and higher-income economies as well as those in between. The economies in several countries, such as those which form part of the Association of South East Asian Nations are growing at a very rapid pace, while others are slower (McKinsey Global Institute, 2018). As a result, there are wide regional differences in infrastructure, resources, development, policy, and accessibility of technologies and content. While accessibility policy and implementation vary from country to country, there are several regional treaties and commitments which encapsulate the commitment of the region on this issue.

Across the region, there have been instances where successful initiatives have been backed by partnerships between the government and civil society. While the involvement of the private sector as a partner is desirable in order to have good results, this does not always happen. There are, however, a few good examples of this as well, as in the case of the Accessible Online Library in India described later in this chapter, which has a partnership between disabled peoples' organizations (DPOs), government, and industry. Universities can also be a great hub for accessibility. Unfortunately, these partnerships are not explored as much as they potentially could be, which is perhaps one of the reasons for fewer implementation initiatives.

Looking at technology accessibility in the region from a Global South perspective, there are many technologies which are available in the Global North

Accessible Technology and the Developing World. Michael Ashley Stein and Jonathan Lazar, Oxford University Press.
© Oxford University Press 2021. DOI: 10.1093/oso/9780198846413.003.0006

that are theoretically available to persons with disabilities (PWDs) in the Global South as well. However, it is important to remember that technologies which are tied to knowledge of the English language can only be used by a sub-set of the people in the region who are familiar with English. Until they are both affordable and available in local languages, they cannot be deployed on a large scale in different countries. Knowledge of these technologies are mostly concentrated in urban areas. Furthermore, it is to be seen whether these technologies are relevant in a Global South/country-specific setting. For instance, in countries where broadband internet access is still not ubiquitous, technologies which require connectivity will not be appropriate. Daily use apps and services would become more relevant if developed in a local context and in local language. Also, the proprietary nature of many technologies makes localization difficult.

In some cases, some of the technology developed in a country of the Global South can be transferred to the Global North or to other South countries. The Braille display Braille Me is an excellent example of technology which can be, and is being, transferred to other countries and can bring down the cost of Braille displays globally. Even in the Global North, the price of these Braille displays is high. Again, innovations have value in a context. There is lesser funding of innovative technologies in the Global South, and many innovations remain in labs because of lack of resources to see them to the global market. So while innovation may be limited, commercialization is even more constrained. More encouragement, prioritization, incentivization, and partnerships would lead to more innovation.

This chapter seeks to give a regional overview of digital accessibility implementation based on insights from a survey conducted across 34 countries in the region and provides brief descriptions of a few countries by way of case studies. It does not purport to be an exhaustive or in-depth research on the region but rather an indicative report on prevailing levels of digital accessibility and associated adoption and implementation challenges in these countries.

5.2 Regional Treaties and Commitments

As of September 2019, most countries in the Asia-Pacific (APAC) region had signed or ratified the United Nations (UN) Convention on the Rights of Persons with Disabilities (CRPD) (2017), with the exception of Timor-Leste. Approximately half the countries in the region are also signatories to the Marrakesh Treaty to Facilitate Access to Published Works for Persons Who Are Blind, Visually Impaired or Otherwise Print Disabled (World Intellectual Property Organization, 2013). Many countries, including Armenia, Laos, Malaysia, New Zealand, Pakistan, and Vietnam, are yet to sign the Marrakesh Treaty.

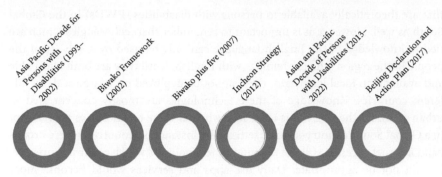

Figure 5.1 Timeline View of APAC Initiatives for PWDs

The United Nations Economic and Social Commission for Asia and the Pacific (UNESCAP) (https://www.unescap.org/) has spearheaded multiple regional initiatives around disability rights. Figure 5.1 shows a chronological view of the various regional accessibility initiatives.

5.2.1 Biwako Millennium Framework for Action

The Asian and Pacific Decade of Disabled Persons 1993–2002 was launched at the end of the UN Decade of Disabled Persons (1983–1992). A majority of the estimated 400 million PWDs in the region at the time were excluded from education, employment, and other economic and social opportunities, and constituted around 20% of the poorest people. At the launch of the decade, the countries of the UNESCAP adopted the proclamation on the Full Participation and Equality of People with Disabilities in the Asian and Pacific Region and the Agenda for Action for the Asian and Pacific Decade of Disabled Persons 1993–2002.

In 2002, the UNESCAP established the Biwako Millennium Framework on "Promoting an inclusive, barrier-free and rights-based society for PWDs in the Asian and Pacific region in the twenty-first century" (UNESCAP, 2011a).

Access to information and communications, including information, communications, and assistive technologies, was identified as one of its seven priority areas for action (Article 6).

The Biwako Framework was supplemented by the Biwako Plus Five (UNESCAP, 2011b): further efforts toward an inclusive, barrier-free, and rights-based society for PWDs in Asia and the Pacific. This provided additional actions in the seven priority areas, reconfigured the strategy areas, and included additional strategies around cooperation and support and monitoring and review.

5.2.2 Incheon Strategy to "Make the Right Real" for PWDs in Asia and the Pacific

Building upon the CRPD and the Biwako Millennium Framework, the UNESCAP countries launched the Incheon Strategy in 2012, at the start of the new Asian and Pacific Decade of PWDs for the period 2013 to 2022. The Incheon Strategy has 10 goals, 27 targets, and 62 indicators, enabling the region to track progress toward improving the quality of life, and the fulfillment of the rights, of the region's 690 million PWDs, most of whom live in poverty (UNESCAP, n.d.a). Information and Communication Technologies (ICT) accessibility is referenced in Goal 3:

- Goal 3: Enhance access to the physical environment, public transportation, knowledge, information, and communication
  - ◦ Target 3.C: Enhance the accessibility and usability of information and communications services
  - ◦ Target 3.D: Halve the proportion of PWDs who need but do not have appropriate assistive devices or products

The Incheon Strategy strengthens the Sustainable Development Goals and the 2030 Agenda, as both are based on respect for human rights and take a people-centered and gender-sensitive approach to development (UNESCAP, n.d.b). All 10 goals of the Incheon Strategy have synergies with goals 4, 8, 10, 11, and 17 of the Sustainable Development Goals, which explicitly mention PWDs, as well as with six additional Sustainable Development Goals (SDGs) that are implicitly linked to PWDs through their mandate to leave no one behind.

The Beijing Declaration and Action Plan to accelerate the implementation of the Incheon Strategy was adopted in 2017 and specifies a set of policy actions to be taken by governments, civil society stakeholders, and UNESCAP for each goal of the Incheon Strategy. Taken together, the Incheon Strategy and the Beijing Declaration provide strategic guidance to governments in support of their efforts toward the full and effective implementation of the Decade.

Thus, there is a strong regional commitment toward information and communication technology accessibility and making assistive technologies available. The challenge lies in translating this into action on the ground.

5.2.3 Digital Accessibility Rights Evaluation Index

To get an overview of digital accessibility in countries across the APAC region, we can look at the Digital Accessibility Rights Evaluation Index, a tool to globally benchmark and track the performance of countries in implementing the dispositions of the CRPD. This annual survey is a collaboration between the Global Initiative for Inclusive Information and Communication Technologies (G3ict) (n.d.) and Disabled Peoples' International (DPI) (n.d.), which reaches out to multiple stakeholders such as disability advocates, PWDs, researchers, and the community. More information about the G3ict can be found in Chapter 3: Global Trends for Accessible Technologies in the Developing World.

The Digital Accessibility Rights Evaluation (DARE) Index evaluates the performance of countries under three parameters—(1) Country Commitments (legal, regulatory, policies, and programs); (2) Country Capacity to Implement (organization, processes, resources); and (3) Actual Digital Accessibility Outcomes for PWDs in 10 areas of products and services. While the first two parameters carry five questions with five marks for each question, totaling 25 marks for these two sections, Actual Digital Accessibility Outcomes covers the accessibility of 10 key technologies, to also be rated on a scale of 1 to 5, totaling 50 marks for this section. Hence the total achievable score for each country is 100. In 2019, the DARE Index survey was filled out by more than 150 respondents in 140 countries.

The DARE Index survey included 34 countries from the APAC region:

- **Central Asia:** Armenia, Azerbaijan, Kazakhstan, Kyrgyz Republic, Russian Federation, Tajikistan, Turkmenistan, Uzbekistan
- **Asia & Pacific:** Australia, Cambodia, China, Cook Islands, Fiji, Indonesia, Japan, Laos PDR, Malaysia, Mongolia, Myanmar, New Zealand, Papua New Guinea, Philippines, Samoa, Singapore, Thailand, Tuvalu
- **South Asia:** Afghanistan, Bangladesh, Bhutan, India, Maldives, Nepal, Pakistan, Sri Lanka

Overall, countries in the APAC region have among the lowest DARE Index scores, as can be seen in Figure 5.2. South Asia has the lowest DARE Index scores, at around 33, followed by Central Asia, and then East Asia & Pacific, at around 36. In contrast, Europe has a DARE Index score of 55 and North America scores around 64. The best performing country in Central Asia is the Russian Federation (61), and in Asia & Pacific it is Australia (80), followed by China (49.5).

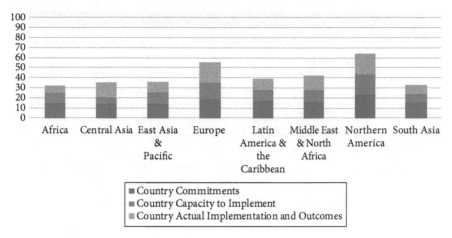

Figure 5.2 Average DARE Index Country Scores by Region

The APAC region has made good progress on country commitments, with 85% of countries having a constitutional article, law, or regulation defining the rights of PWDs and 62% of countries including a definition of "reasonable accommodation" in any law or regulation regarding the rights of PWDs. However, progress has been slower on the ICT front, with only 53% of countries having a definition of accessibility that includes ICTs or electronic media in the country laws or regulations, and only 24% with a definition of universal service obligation in telecommunication legislation that includes PWDs.

When it comes to Capacity to Implement, as well as Actual Implementation Outcomes, progress has been much slower, which is consistent with the situation in many parts of the world. Only 21% of countries in the APAC region have a systematic mechanism for involving DPOs working in the field of digital accessibility in the drafting, designing, implementation, and evaluation of laws and policies, and only 26% of countries have ICT accessibility courses in their universities. Lack of involvement of DPOs is a global phenomenon, with the global percentage for involvement of DPOs in policymaking being 26%, which is only slightly higher than the APAC average. This is due to a multitude of reasons, some of which are also country-specific. Key drivers for the low involvement of DPOs in policymaking include the fact that a majority of DPOs work on a shoestring budget with less human resources and are unable to devote their time/expertise to policymaking. The failure to recognize accessibility as a multi-disciplinary issue also results in the non-participation of DPOs in domains other than disability welfare. Additional obstacles are the requirement of constant follow-up and lengthy or inconvenient mechanisms for government engagement. There is seldom a legal/procedural requirement to solicit opinions of DPOs and to act on

Sector	Implementation progress
TV	68%
Web	53%
Inclusive ICTs in education	50%
E-books	50%
Enabling ICTs for employment	50%
e-Government and Smart Cities	53%
Assistive Technologies and ICTs for independent living	38%
Promotion of internet usage	38%
Mobile	35%
Public procurement	41%

Figure 5.3 APAC Digital Accessibility Implementation and Outcomes Progress by Sector

their inputs, to have a set timeframe and targets to meet, and to involve DPOs in policy formulation, audit, and implementation. Often, once the policy is formulated, it is assumed that it will run the course of implementation like other policies do, which unfortunately does not happen. Often, there are no mandatory measures to ensure implementation.

The slow progress in Country Capacity to Implement has a consequential impact on DARE Index Actual Implementation Outcomes, as can be seen in Figure 5.3. Web accessibility and usage of inclusive ICTs in education have seen the most traction, while mobile telephony, public procurement, promotion of internet among PWDs, and assistive technology have seen the least progress. Even in areas where progress has been made, the levels of implementation outcomes are low. For instance, of the 17 countries which have policies for the implementation of accessibility of websites, only one (the Russian Federation) has a substantial level of implementation covering both public and private sectors. Australia, China, Japan, Kazakhstan, Mongolia, Philippines, and Uzbekistan have a partial level of implementation, while the remaining nine countries with policies, such as India and New Zealand, claim to have a minimum level of implementation. Around 16 countries, including Indonesia, Malaysia, Singapore, and Thailand reported not yet having a policy in place for website accessibility.

5.3 Regional Challenges

By and large, many countries in this region face common challenges. While there is a strong national commitment and belief that the needs of PWDs must be addressed, it does not receive due priority in national governance. Among many national issues, digital accessibility ranks fairly low in the priority ladder and is always only addressed after years of advocacy by DPOs, and when there is a legal obligation to do so. The DARE Index gives ample evidence that accessibility measures have mostly been implemented by countries only when required by national legislation and policy. The low prioritization of accessibility leads to the other key challenges in the region—lack of progress in adopting policies or standards related to digital accessibility and fewer resources allocated or made available for accessible technologies, and hence lower technical capability to develop local assistive technologies.

5.3.1 Adoption of Policies and Standards Related to Digital Accessibility

Approximately 76% of countries in APAC do not have a government body specifically dedicated to accessible ICTs. This also appears to have slowed down their progress with respect to digital accessibility and standardization; most of these countries do not follow any international accessibility standards or have any policies in place for web accessibility. This includes countries such as Laos, Bangladesh, Mongolia, Myanmar, and Cambodia.

In countries which have government bodies dedicated to ICT, progress has been better but nevertheless slow. Around 59% of APAC countries do not have policies mandating ICT accessibility. This is an area where DPOs often have less expertise than in environmental accessibility. In such cases, where DPOs themselves do not have complete knowledge of the extent and urgency with which issues need to be addressed as well as possible measures that can be taken, they will find it difficult to know what to ask for. In the absence of active involvement and advocacy by this key stakeholder group, the issue remains completely neglected. There have been several regional efforts to reach out to disability advocates in different countries which have met with some degree of success.

A common challenge faced throughout is the difficulty in communication and the language barrier. Most of the existing resources for ICT accessibility are in the most-spoken languages worldwide, such as English, Spanish, and French. However, in countries like Laos and Mongolia, most people speak only their native language, and few people speak English. Consequently, they are unable to access available English language content, communication, and technologies, and achieving accessibility becomes a very slow process.

5.3.2 Capacity and Capability to Develop Accessible Technologies

The capacity to develop accessible and assistive technologies is impacted by multiple socioeconomic and cultural factors.

5.3.2.1 Affordability

By and large, the region falls short on both availability and affordability of assistive technologies, with even mainstream digital resources being inaccessible. There are some differences across countries; for example, what is affordable in one country may not be so in another. One factor affecting this is the extent of financial aid being provided in different countries to PWDs to purchase assistive technologies, which may determine whether they are affordable or not. The main factors affecting availability are twofold: The first factor is whether the technology is proprietary or open source. For example, software like the screen reader JAWS developed by Freedom Scientific in the USA is expensive, whereas open-source software like e-Speak is free. The second factor is whether the technology has been made available in the local language, such as Hindi in North India, whether people are comfortable with working in English, or whether it has been customized in the local language.

Overall, development of technologies for all accessibility requirements such as mobility and vision remains very low, and PWDs still look toward the global market for their products, which are expensive. It is not that solutions do not exist at all, as much as that they are very basic or in the form of workarounds to key problems, and even these are not as frequent as would be desirable. The workarounds arise from the need to address a specific problem, and could be as basic as tinkering with a device to make it work for someone in the family.

A good example of this is the modification of a regular cassette player to a four-track cassette player at very low cost in 1999 in India by Dinesh Kaushal, who needed to read audio books recorded in that format, by Recording for the Blind and Dyslexic. Dinesh describes this experience as follows:

At that time [1999] there were limited number of books available in India. I found out that recording for the blind and dyslexic ("RFBD"), an American library which is now known as Learning Ally, has thousands of books for various technical subjects such as computer science, mathematics, and engineering. These books were recorded in the format discussed above. But a cassette player to play back those cassettes cost $200 or INR 10,000 at the conversion rate of INR 50 per dollar at that time.

So, after some reflections and observation that a car stereo does not require one to take out the cassette in order to switch the side of the cassette, I realized that I could use play head of a car stereo to select the track of a cassette. I also figured, after reading a basic electronic book that a resistor can reduce the current of a circuit. So, I decided to change the resistor of the motor so that it rotated at half the speed. Being blind, I could not solder the connections, so my father learned soldering in order to help me. Later I learned that some blind persons can also solder, but I could not learn to do that.

The component cost of the modification was only $3. After modifying one cassette player, I had proven that I could do this, but I realized that I would not be able to modify cassette players for others. So, I contacted a willing electronic mechanic and explained to him how I modified a regular cassette player to a four-track cassette player. He was able to make this modification at the cost of $7. But in a few years, RFBD stopped sending the cassette in the four track format as they switched to DAISY.

Modification of cars for persons with physical and mobility issues are instances of customization of technologies at a local level, although these are done in a more structured and large-scale manner. However, these are insufficient, and most technological needs of PWDs remain unaddressed. The development of communication technologies poses an especially great challenge.

Developing successful communication technologies, again taking the example of text to speech (a screen reader), requires the partnership of many people: funding to support the work of developers, DPOs who give their input from start to finish, design, testing to completion, training for certification and beyond, deploying these technologies among PWDs and academia/industry, good knowledge of standards such as customization process, and working with the international partner to bring these changes and languages into the international build. There are also other complexities such as accent, which many render a technology useful for one but uncomfortable for another.

5.3.2.2 Geography

Geography and terrain also have a role to play in the progress of accessibility initiatives. Countries like Nepal, Afghanistan, and Bhutan have many remote and isolated regions, which affect their socioeconomic development and, consequently, accessibility. In Mauritius, both education and employment pose difficulties. Because of island geography, traveling and moving around is a problem. Hence, there is more of a logistical challenge—rather than financial or linguistic ones—impeding the education and employment of PWDs using ICT, in contrast to the case of India. Less training happens because road transport is expensive, railways are not operational, and traveling from one island to another is difficult.

5.3.2.3 Language

The linguistic complexity of the APAC region presents a huge communication challenge. Over 2,300 languages are spoken in this region, which has six times the population of Europe (Kiprop, 2018). Many of these languages have evolved independently and without a common base or script, which makes it more difficult to leverage solutions built for one region in another. It is not surprising that assistive technologies are not developed/deployed across countries. There is very low awareness and participation in international standards—only 29% of APAC countries follow ISO, W3C, Section 508, EN 301 549 ITU, or other international standards. Japan has developed its own industrial standard, JIS-8341, which lays out guidelines for older persons and PWDs related to information and communications equipment, software, and services.

5.3.2.4 Infrastructure

Lack of infrastructure is another limiting factor. In many smaller countries, because of the small size of the population with disabilities, the infrastructure for producing even basic non-technical assistive solutions does not exist (P. R. Verma, personal communication, July 2019). For instance, in countries like Sri Lanka, Bangladesh, and Nepal, some devices may need to be imported from other countries like India and China. In such cases, even these basic items need to be carefully guarded against breakage or loss, since replacement may take time if they are not in stock. Overall, technology production is negligible in this region, with only a few initiatives seen over the past years.

5.3.2.5 Case Study: Orbit and Braille Me

For persons with blindness, the emergence of Orbit and Braille Me and their manufacturing in India has made a huge difference in terms of price (D. Manocha, personal communication, July 2019).

The Orbit Reader 20 (n.d.) is a small, portable, refreshable Braille display with 20 eight-dot cells, which can help users read books and act as a simple note-taker in electronic Braille. This refreshable Braille display can be connected to any device, such as a computer or mobile phone with a screen reader, using USB or Bluetooth, and comes with rechargeable batteries. It also has a Braille keyboard and important keys to carry out typing functions. It is a stand-alone device which can be used by readers to read Braille content stored on an SD card that plugs into the Orbit Reader 20.

Braille Me (n.d.) is a six-dot, 20-cell refreshable Braille display, with a six-key Perkins-style keypad and 20 routing buttons for input. Students can access files from an SD card by browsing, editing, and searching content, and can create and edit files. It is small and portable and can be connected to any device with a screen reader such as a computer or mobile phone.

While Braille displays used to cost $2,500 and upwards in the market, Orbit and Braille Me are now available in the range of $400 to $500. Braille Me is a truly Indian product created by Innovision, a start-up incubated within the Indian Institute of Technology Mumbai, which has features similar to Orbit and is being sold worldwide. Such assistive products were earlier only made in Japan, Canada, Australia, South Korea, China, etc. Similarly, the cost of Digital Accessible Information System players, especially those connected to the internet and content in digital libraries, has come down to the range of $120 to $250 in India, from $600 a few years ago. This global trend and effort toward reducing the cost of assistive technology is having a trickle effect in some of the developing countries in this region, although still primarily in urban areas. DotBook is a Braille display/note-taker created by the Indian Institute of Technology Delhi and released by Saksham, and costs approximately $580 (Indian Institute of Technology Delhi, n.d.). Further information on accessibility in India can be found in Chapters 15 and 17, respectively.

In India, there have been a few efforts to develop assistive technologies in the premier engineering colleges and technical organizations such as the Indian Institute of Technology, Indian Institute of Science, and the Centre for Development of Advanced Computing. However, these technologies do not find their way into the market because there is no formal effective mechanism in place to ensure the transition of prototypes into commercially available products. However, these are just a handful of efforts toward assistive technology production and deployment in APAC. By and large, this region still remains a consumer of assistive technologies rather than a producer, although this trend will hopefully change soon. It is important that countries grow local skills and capabilities to address their own technology needs relevant to their affordability and linguistic requirements.

5.3.2.6 Urban and Rural Divide

There is a big divide between urban and rural awareness of ICT accessibility. Some countries such as Nepal, Bangladesh, Sri Lanka, Vietnam, Myanmar, and India have good awareness and are driving usage of some assistive technologies and smartphones. Nonetheless, this is mostly concentrated in major cities and urban areas. In rural and remote areas, the awareness and use of accessible technologies and availability of accessible content is still extremely low.

5.3.2.7 Availability of Assistive Technology

Due to a combination of the factors discussed earlier, the availability of assistive technology is also a concern in many countries. Let us consider the case of text-to-speech synthesizers, software which converts digital text into audio and reads it aloud. It can work with multiple devices and platforms like computers, tablets, and mobile phones, and in different languages to read different language content.

Text-to-speech synthesizers are still unavailable in many languages in this region. The past decade has seen progress in the localization of e-Speak, an open-source text-to-speech synthesizer engine for non-visual desktop access on Windows in some Asian languages such as Nepali, Bengali, and Gujarati. However, this is not very good quality, since the speaking voice is mechanical and the native language is spoken with an English accent. It takes getting used to, and not everyone is able to manage.

In many countries, the development of a text-to-speech synthesizers (TTS) is recent (P. R. Verma, personal communication, July 2019). For instance, in Mongolia, a TTS was developed only last year. Until then, blind people and PWDs in Mongolia had no other recourse, as few people in Mongolia speak English to be able to use English TTS. Similarly, in Bangladesh, people managed with Indian Bengali TTS, which is slightly different from the Bengali spoken in Bangladesh. In Malaysia, a TTS was developed last year; until then people used an Indonesian TTS. In many countries like Kazakhstan and Azerbaijan, there appear to be low awareness, availability, and deployment of accessible technologies. Most of these countries have limited Braille books and audio content. It is only now that they are starting to produce accessible textbooks.

5.3.2.8 Mobile Accessibility

Although smartphone penetration has been increasing rapidly in the region, with over half of the world's mobile subscribers living in the APAC region, accessibility of mobile phones is an area that has made very little progress, with only 26% of countries having an accessibility policy for mobile phones (GSM Association, 2019). Technology adoption is better in countries like India, Vietnam, Malaysia, and Indonesia. Among APAC countries, Thailand is slightly ahead with regard to the production of assistive technology. They have been producing their own devices for some time now, and have had a TTS for some years.

LIRNEasia, a regional thinktank based in Sri Lanka, carried out research to understand the level of awareness, access, and use of mobile phones and assistive technologies among PWDs in Nepal. Looking at the results for the 15–65 age group of the representative population, they found that close to 90% of PWDs lived in a household with a mobile phone. However, only 32% of PWDs actually owned a mobile phone. Moreover, of those who owned a phone, 53% used a basic phone and 41% had a smartphone. PWDs who didn't own a phone cited their disability and lack of knowledge in how to use them as the key reasons for not having a mobile phone, while PWDs who were basic and feature phone users thought smartphones were not needed (40%) and not affordable (35%). Many PWDs who owned mobile phones required others' help for all functions, with low use of disability-specific assistive features. The PWDs in Nepal lagged far behind in internet awareness and use compared to the rest of the population. These findings highlight the fact that good national mobile/technology penetration does not automatically extend to the population with disabilities in a country. Special attention needs to be given to ensure that they have access to technology and have the capability to use these technologies to have social, economic, and cultural independence.

In China, the situation is somewhat different—they were among the first to create accessible products available in the mainstream market. For instance, there have been Chinese mobile phones specifically targeting seniors and PWDs for over a decade. However, although assistive technologies exist, there is little clarity or information about the standards to which they conform. There is relatively little information available on accessible ICTs in English. The most widely used content production and storage format remains MP3 audio and Word, and the Digital Accessible Information System has not yet spread to here. Accessible Chinese products do not necessarily subscribe to international standards, and hence there may be interoperability issues. However, China is among the few countries which have been able to produce and deploy assistive technologies in their local languages, irrespective of whether these products are usable in the global market as yet.

5.3.3 Access to Content

Only around 50% of the countries in the APAC region have policies promoting the accessibility of e-books or digital documents. These include countries like Australia, Malaysia, the Russian Federation, India, Indonesia, New Zealand, and Singapore. When it comes to the implementation of these policies, the level of implementation on the ground is reported to be minimal or partial. Overall, just over half the countries in the APAC region have signed and adopted the copyright exception under the Marrakesh Treaty. As of July 2019, signatories include Afghanistan, Azerbaijan, Australia, China, the Democratic People's Republic of Korea, India, Indonesia, Japan, Kyrgyzstan, Mongolia, Nepal, Philippines, Republic of Korea, Russian Federation, Singapore, Sri Lanka, Tajikistan, and Thailand.

The World Intellectual Property Organization and Accessible Books Consortium have engaged in several capacity-building projects in these countries, which include training on creating accessible Digital Accessible Information System (DAISY)/EPUB content and use of accessible assistive technologies for reading.

Irrespective of whether countries have signed the Marrakesh Treaty, the general situation with respect to availability of accessible content and digital information is concerning. Accessible content is extremely limited, does not even cover all educational content, let alone content for leisure and general reading, and is not always in an accessible e-text format or one which enables sharing. Also, as is the case with technologies, this content is primarily only available to persons with print disabilities in cities and not to those in rural areas. In remote areas, even accessible school textbooks are a challenge, adding to the lack of an overall resource ecosystem to enable effective inclusive schooling. For instance, a common issue is the availability of qualified teachers who can support students with different disabilities. They are few in number and often need to travel long distances to support children in three or four villages. Again, availability of assistive technologies and resource centers are also equally problematic. In remote areas, it could take up to months for children to get their textbook in an accessible format like Braille.

In cities, however, more options are opening up for persons with print disabilities with respect to accessible content. For instance, the Bookshare initiative has a lot of potential to deliver more accessible content to the hands of readers with disabilities.

Bookshare provides the largest collection of e-books in customizable formats, including books in 34 languages across 83 countries that can be read using a variety of platforms such as smartphones, tablets, computers, MP3 players, and assistive technology devices (H. Mobedji, personal communication, September 2019). It has free or reduced fees for developing countries, based on World Bank income designations, and provides free and unlimited service to all South Asian countries such as India, Sri Lanka, Bangladesh, Nepal, Bhutan, Pakistan, and Myanmar. You can read more about Bookshare in Chapter 7: Digital Inclusion in the Global South.

5.4 Country Examples

5.4.1 Bangladesh

The disability law in Bangladesh is the PWDs Rights and Protection Act 2013, which covers ICT and accessibility. ICT accessibility is also covered under the Digital Bangladesh Strategic paper (V. Bhattacharjee, personal communication, July 29, 2019) as well as national policies on ICT. Accessibility of information is also mentioned in the Right to Information Act.

The Access to Information (A2I) program of the ICT Division of the government of Bangladesh has been working with various departments and ministries to implement web accessibility as per WCAG 2.0 and is also in the process of formulating a policy in this regard. As part of this initiative, the government is seeking to make 25,000 government websites accessible. The national content repository has also been made accessible with video, audio, animation, e-text, and DAISY books. The innovation fund of the Access to Information (A2I) program of the Bangladesh Government is supporting development of a device by which sign language users will be able to communicate with non-users. The fund is also supporting the organization Young Power in Social Action on a project to convert three accessible dictionaries—Bengali to English, English to Bengali, and Bengali to Bengali—which are available online and offline for all students, including persons with visual and print disabilities, learning disabilities, and information-disadvantaged groups. A project to create multimedia talking books is also being presently supported by the fund. The establishment of a Union Information Center (currently called Union Digital Center) by the government has made it easier for PWDs to get information. Government-run computer training centers help PWDs prepare for various job opportunities. A wide range of employment options are also being notified through community radio. Sign language is provided for national news broadcasts.

There is a basic open-source Bangla TTS that was developed in partnership with PWDs. Further work is, however, required to improve speech quality. School textbooks for grades 1 to 12 are available in DAISY text and multimedia formats. Accessible e-books and digital Braille copies are provided at no charge for students with disabilities. These accessible books are distributed to students with disabilities by the prime minister of Bangladesh on January 1 every year.

Informally, some young technology enthusiasts have been working on the development of currency indicator apps for Android phones. The Bangla Braille software (Duxbury Braille translator) has been developed and is commercially available in Bangladesh. Shahjalal University of Science & Technology has been working to develop a prototype Braille printer which will drastically bring down the cost of Braille printing in Bangladesh. Other than this, assistive devices such as Braille printers are still being imported from India and other countries.

Hence, various efforts are taking place to make Bangladesh inclusive for PWDs.

5.4.2 India

India has strong commitments to accessibility, with a disability rights law, a definition of accessibility that includes ICT, a definition of reasonable accommodation, and government bodies dedicated specifically to ICT and PWDs.

There is fairly good awareness and availability of technologies in urban areas. The DAISY Forum of India has members all over the country who are engaged in training PWDs to use digital technologies and content. There are also other organizations working with persons with different types of disabilities and supporting them to use technologies. There are instances of successful partnerships and collaboration between PWDs and developers/engineering institutions to develop technologies, devices, and applications which are suitable for specific disabilities.

There are several strong disability networks and coalitions and ICT accessibility experts who are engaged in advocacy for the rights of PWDs. Due to a strong exception clause for PWDs to convert and share copyrighted content in accessible formats, there is a movement to build up a national repository of accessible content on "Sugamya Pustakalaya" (n.d.), an online library for accessible books which was a collaborative effort on the part of the government (Ministry for Social Justice and Empowerment), DPOs (DAISY Forum of India), and industry (TCS, the company hosting the library platform).

India has witnessed a few large-scale campaigns with respect to digital accessibility. A very successful campaign, run by civil society, was the "Right to Read campaign," which was launched nationwide to advocate very strongly for copyright amendment and for India to support the Marrakesh Treaty. The Government of India also launched an "Accessible India campaign," which had some targets relating to accessibility of websites and documents. This was unfortunately not as much of a success, and targets remain unmet. There have been projects by organizations of the DAISY Forum of India (DFI) to provide devices coupled with training and subscription to the online library, which have been successful, but at a very small scale, since fundraising has been an issue. However, the model for connected devices with content is present in a limited way and needs to be scaled up. India presently also has two committees focusing on accessible ICT procurement and formulating a policy on accessible television, especially for the hearing impaired.

As is the case with most of the countries in this region, accessibility remains an exclusive issue which is given less priority. India really needs to focus on implementation of its policies and standards and scale up its advocacy and implementation of digital accessibility across programs. The sheer scale and variety of the population with disabilities in India makes implementation of ICT accessibility a challenge. Hence, specific attention and measures need to be taken to meet the goal of an accessible, inclusive digital India (Galpaya & Amarasinghe, 2018).

Further information on accessibility in India can be found in Chapters 15 and 17, respectively.

5.4.3 Myanmar

Myanmar has a legal framework in place to address disability rights which has a definition of reasonable accommodation, and a government body dedicated to

PWDs. However, progress needs to be made on the capacity and implementation fronts.

Disability has been included in the new Universal Service Strategy (2018–2022) of Myanmar, which specifically recognizes the role of the Government in facilitating access to ICTs for PWDs, who are mentioned in Program 3—Special Projects, including content, applications, pilots, and disability (Myanmar Ministry of Transport and Communications Posts and Telecommunications Department, 2018). One of the aims of these special projects is catering to the needs of PWDs.

The Myanmar Centre for Responsible Business has also been working to promote ICT accessibility among its members, working with industry, civil society, and government to do so. The Myanmar Centre for Responsible Business published a Handbook on Employing PWDs for Employers in Myanmar in 2018 and has also been advocating for promoting ICT accessibility for PWDs in collaboration with key partners (Myanmar Centre for Responsible Business [MCRB], Association for Aid and Relief Japan, & Japan International Cooperation Agency, 2018). It has also organized sessions on ICT accessibility to raise awareness in its various events such as the Digital Rights Forum in 2018 and 2019 and the second Multi-Stakeholder Forum on promoting employment for PWDs in March 2019 (MCRB, 2019).

5.4.4 Sri Lanka

Sri Lanka has a Disability Rights Act (1996), as well as government bodies dedicated to PWDs and for ICT. The Disability Act, however, does not cover ICT accessibility or reasonable accommodation. As with many other countries in the region, the implementation of accessibility in most of the technology areas is either non-existent or at a minimum/pilot project level.

The past year has seen good traction in the field of assistive technologies (A. B. Weerawardhana, personal communication, August 2019) and accessible content for persons with blindness and low vision in Sri Lanka. In terms of TTS packages, the Tamil people make use of technologies developed for India mainly by entities such as Microsoft and Nuance communications. There are also a few options for Sinhalese TTS. A TTS was developed by the Language Technology Research Laboratory of the University of Colombo. This Windows TTS, however, lacks good voice clarity and is hence not very popular. The Language Technology Research Laboratory is presently also developing a natural-sounding TTS for Sinhala. The second option is the most widely used TTS software, "Sinhala bhashakaya," which was developed by Mr. Ashoka Bandula Weerawardhana, also for Windows. This TTS was developed using a separate build of e-Speak. He has also developed an offline Sinhala optical character recognition system with about

98.2% accuracy, as well as Liblouis, a real-time Braille translation utility. A good TTS option on Android is the Google Sinhala TTS.

In terms of access to information, Sri Lanka is the 24th country to ratify the Marrakesh Treaty. So far, it has produced more than 4,500 titles in accessible formats for the print-impaired and is also part of Bookshare. Mr. Weerawardhana has also designed a mobile-phone-based DAISY telephony system using interactive voice response called "Books on Call," which enables a user to access audio books by dialing a short code (2525), which they can use irrespective of their geographical location.

Sri Lanka has a good level of literacy and free government education, and even among the blind, more children go to schools there than in India or Bangladesh.

5.5 Conclusion

Overall, the APAC as a region is committed to promoting the rights of PWDs, as can be seen from the various regional instruments and the initiatives which have been undertaken over the decades. However, as is often the case globally, it falls short on capacity and implementation as well as in terms of the involvement of DPO stakeholders in policy and implementation. Given the vast diversity of economic growth, geography, culture, language, and political situations, it has been difficult to share solutions and expertise across countries.

Priority, awareness, affordability, availability, localization, and training are key barriers affecting development and deployment of appropriate assistive technologies; often, lack of mainstream infrastructure such as broadband connectivity affects technology use in far-flung and remote areas. Internet and broadband connections may not be ubiquitous, leading to limited relevance of technologies for which connectivity is a prerequisite. Linguistic diversity does not always allow for technology sharing, except for a few countries; unaffordability and the fact that technologies are only available in English hinders their ready adoption and use in APAC countries. The usefulness and suitability of technologies vary from country to country, depending upon the economic and cultural context in which they are situated, making it difficult to assess the level of technology sharing and availability. There may often be the need for localization, which would need to be done by the local developer community, and the access, availability, and affordability of technologies for a country are driven by how active its developer community and its DPOs are in this area.

Where initiatives have been developed in partnership with the DPO community, policy and technology solutions have emerged, as is the case with TTS development in countries like India and Bangladesh. In many cases, these technologies have been developed privately and not by the governments. Hence engagement of the private sector and finding a way to make products commercially available at affordable prices are crucial.

The various factors of linguistic, cultural, geographic, and economic diversity make it difficult to characterize the development of accessibility and assistive technologies in the APAC region as a single entity. As described earlier, the region has countries with advanced accessibility awareness and technology availability as well as countries with very minimal access. However, it is relatively safe to say that overall there is a huge gap in the availability of assistive technologies and accessibility of mainstream technologies such as websites, television, and books. There is also a need for more policymaking in the space of ICT accessibility. A lot of ground needs to be covered in this area.

However, there are also some best practices that the rest of the world can learn from in the region. APAC has been home to some really great advocacy by DPOs. Whether it is the Right to Read Campaign or the localization of the TTS e-Speak in several Indian languages, or the inclusion of accessible ICTs in the Universal Service Fund, as in the case of Myanmar, there are important learnings which can be taken from this region. Most noteworthy is the fact that the examples reiterate the global acknowledgment that wherever there has been some success, whether in terms of technology or policy, it has always been strongly supported by the DPO community. Where there is little involvement of DPOs, there is little positive advancement. Countries are on the right track in terms of their commitment to promote ICT accessibility; however, this can only be effectively realized with the partnership of DPOs. A one-size-fits-all approach cannot be taken for all countries, and solutions will vary from country to country.

The key to accelerating accessibility in the region is to build local capacity to implement accessibility, so that it is sustainable and targeted to overcome local challenges. Given the large number of PWDs in APAC, marketing locally developed technologies, which are compliant with international standards, can offer huge opportunities for the Asian market, as can be seen by the example of Braille Me, the Braille display developed in India which is catering to a global market. This can serve as an incentive for involving the private sector. Governments can accelerate the process by adopting and implementing appropriate policies, like accessible ICT procurement.

Apart from private sector and government initiatives, there is also a need for greater global engagement on this issue and a concerted effort to build capability in the region. APAC is home to several accessibility visionaries who have the passion and skill to develop and customize innovative technologies; what the region lacks is sufficient funding to promote technology development. Grants from leading international foundations for accessibility innovation in the region could go a long way toward supporting the creation, adoption, and assimilation of accessible technologies to foster an inclusive world.

References

American Printing House for the Blind. (n.d.). *Orbit Reader 20: Product webpage—purchase.* https://www.aph.org/product/orbit-reader-20/333/#main

Disabled Peoples' International. *Homepage.* (n.d.). Retrieved from DPI: http://www.dpi.org

G3ict. (n.d.). Homepage. Retrieved from G3ict: http://www.g3ict.org

Galpaya, H. & Amarasinghe, T. (2018). *AfterAccess: ICT access and use in Nepal and the Global South* [Presentation]. LIRNE Asia. https://lirneasia.net/2018/10/afteraccess-ict-access-and-use-in-nepal-and-the-global-south/

GSM Association. (2019).*The mobile economy Asia Pacific* [Report]. https://www.gsma.com/mobileeconomy/asiapacific/

Indian Institute of Technology (IIT) Delhi. (n.d.). *DotBook: IIT Delhi launches India's first Braille laptop.* http://www.iitd.ac.in/content/dotbook-iit-delhi-launches-india%E2%80%99s-first-braille-laptop-1

Innovision. (n.d.). *Braille Me. About product.* Retrieved from https://innovisiontech.co/

Kiprop, V. (2018, July 19). *Major languages spoken in Asia.* World Atlas. https://www.worldatlas.com/articles/major-languages-spoken-in-asia.html

McKinsey Global Institute. (2018, September). *Outperformers maintaining ASEAN countries' exceptional growth* [Discussion paper—ASHX file]. https://www.mckinsey.com/~/media/McKinsey/Featured%20Insights/Asia%20Pacific/Outperformers%20Maintaining%20ASEAN%20countries%20exceptional%20growth/MGI-Outperformers-ASEAN-Discussion-paper.ashx

Myanmar Centre for Responsible Business (MCRB), Association for Aid and Relief, Japan (AAR Japan), & Japan International Cooperation Agency (JICA). (2018). *Employing persons with disabilities: A handbook for employers in Myanmar* [PDF]. MCRB. https://www.myanmar-responsiblebusiness.org/pdf/handbook-employing-persons-with-disabilities_en.pdf

Myanmar Centre for Responsible Business (MCRB). (2019, March 14). *Second multi-stakeholder workshop on promoting employment opportunities for persons with disabilities.* https://www.myanmar-responsiblebusiness.org/news/second-multi-stakeholder-workshop-people-with-disabilities.html

Myanmar Ministry of Transport and Communications (MOTC) Posts and Telecommunications Department. (2018, January). *Executive Summary: Universal Service Strategy for Myanmar (2018–2022)* [Report PDF]. https://www.motc.gov.mm/sites/default/files/Executive%20Summary%20of%20Universal%20Service%20Strategy%20%28English%29_0.pdf

Protection of the Rights of Persons with Disabilities Act, Sri Lanka (1996, October). http://www.hrcsl.lk/PFF/LIbrary_Domestic_Laws/Legislations_related_to_Employment/Protection%20of%20the%20Rights%20of%20Persons%20with%20Disabilities%20Act%20No%2028%20of%201996.pdf

Sugamya Pustakalaya (n.d.). *Accessible books online library—homepage.* https://library. daisyindia.org/NALP/welcomeLink.action

United Nations Economic and Social Commission for Asia and the Pacific (UNESCAP). (2011a, September 8). *Biwako millennium framework for action towards an inclusive, barrier-free and rights-based society for persons with disabilities in Asia and the Pacific.* https://www.unescap.org/resources/biwako-millennium-framework-action-towards-inclusive-barrier-free-and-rights-based-society

United Nations Economic and Social Commission for Asia and the Pacific (UNESCAP). (2011b, September 8). *Biwako plus five: Further efforts towards an inclusive, barrier-free and rights-based society for persons with disabilities in Asia and the Pacific.* https://www.unescap.org/resources/biwako-plus-five-further-efforts-towards-inclusive-barrier-free-and-rights-based-society

United Nations Economic and Social Commission for Asia and the Pacific (UNESCAP). (n.d.a). *Incheon strategy: 10 goals to "Make the Right Real" for persons with disabilities in Asia and the Pacific.* https://www.maketherightreal.net/incheon-strategy/introduction

United Nations Economic and Social Commission for Asia and the Pacific (UNESCAP). (n.d.b). *Incheon strategy strengthening the 2030 Agenda. Retrieved from Make the Right Real!* https://www.maketherightreal.net/incheon-strategy-strengthening-2030-agenda

United Nations Programme on Disability/Secretariat for the Convention on the Rights of Persons with Disabilities (CRPD). (2017, September 19). *CRPD and optional protocol signatures and ratifications* [Map of Signatories—JPEG]. https://www.un.org/disabilities/documents/maps/enablemap.jpg

World Intellectual Property Organization (WIPO). (2013, June 17). *Marrakesh Treaty to Facilitate Access to Published Works for Persons Who Are Blind, Visually Impaired or Otherwise Print Disabled.* https://www.wipo.int/treaties/en/ip/marrakesh/

6

The Role of Technical Standards in Improving Digital Accessibility in Under-Resourced Regions and Communities

Judy Brewer
World Wide Web Consortium (W3C), Massachusetts Institute of Technology
Computer Science and Artificial Intelligence Lab

Shadi Abou-Zahra
World Wide Web Consortium (W3C), European Research Consortium on
Informatics and Mathematics

6.1 Introduction

6.1.1 The Relevance of Digital Accessibility to Under-Resourced Regions

An estimated 4.4 billion people out of the world's estimated population of 7.7 billion are online, according to March 2019 data (Miniwatts Marketing Group, 2020). While the average worldwide online presence is 56.8%, regional differences in online access are striking. In Europe and North America, nearly 90% of the population is online. Across Latin America and the Caribbean, the Middle East, Oceania, and Australia, internet access averages in the high 60 percents. Across Asia, internet access is in the low 50 percents, and in Africa it is only in the high 30 percents.

In 2011, the World Report on Disability estimated that one billion people in the world have significant disabilities (World Health Organization & World Bank, 2011). Accessibility barriers still pervade every country, though to varying degrees, due to progress in disability rights in some countries in recent decades. In low-income countries and lower middle-income countries (World Bank Group, n.d.), people with disabilities sometimes face additional economic, physical, and communication barriers in comparison to their peers in other regions.

Accessible Technology and the Developing World. Michael Ashley Stein and Jonathan Lazar, Oxford University Press.
© Oxford University Press 2021. DOI: 10.1093/oso/9780198846413.003.0007

6.1.2 An International Human Rights Framework for Digital Accessibility

The United Nations (UN) Convention on the Rights of Persons with Disabilities (CRPD) (2006a) defines a broad set of rights that address barriers throughout many aspects of society, including barriers in digital technologies. Particularly important in the context of digital accessibility are Articles 9 and 21 of the CRPD (see Chapter 1: Development for All for additional detail), which continue to be essential and impactful concepts.

Article 9 (CRPD, 2006b) directs States Parties "to promote access for PWDs to new information and communications technologies (ICT) and systems, including the internet, and to promote the design, development, production and distribution of accessible information and communications technologies ICT and systems at an early stage, so that these technologies and systems become accessible at minimum cost." Article 21 (CRPD, 2006c) directs States Parties to ensure that PWDs can "exercise the right to freedom of expression and opinion, including the freedom to seek, receive and impart information and ideas on an equal basis with others and through all forms of communication of their choice" and outlines essential steps to achieving this, urging provision of information "in accessible formats and technologies appropriate to different kinds of disabilities in a timely manner and without additional cost."

6.1.3 Planning for Digital Accessibility Before Digital Technologies Are in Place

As new technologies emerge, will they be "born accessible" in order to meet human rights needs for the millions of people around the world with disabilities? Or, will the multi-year lags in digital accessibility that people with disabilities have experienced over the years continue in even worse forms, as these technologies become more prevalent in low-income and lower middle-income countries?

Many types of disabilities can be impacted by the presence or absence of accessible technologies. These disabilities include blindness and low vision, deafness and hardness of hearing, mobility and dexterity limitations, speech disabilities, and cognitive, language, learning, and neurological disabilities. Given the expanding scope of digital technologies—not only on personal computers and mobile but also within the web of things, digital publishing, telemedicine, autonomous vehicles, and more—timely development of digital accessibility is essential.

In some under-resourced regions, very few people have access to digital technology, though this can change rapidly, as in rural Kenya and other rural communities in Sub-Saharan Africa, and likewise in Papua New Guinea, where off-grid solar installations now generally come with mobile phone chargers (Ellsmoor, 2019).

In other regions, access to digital technologies may be inconsistent and unreliable. In these regions it may be tempting to think that planning ahead for digital accessibility should wait, given that people with disabilities are necessarily dealing with myriad survival needs. Given that retrofitting for accessibility is one of the least efficient approaches to making technology accessible, where possible it is valuable to anticipate and plan for future needs for digital accessibility. However, it is not always possible, and sometimes digital accessibility must wait until survival needs are met.

Colleagues in Venezuela emphasized this in recent years, noting that in the absence of functioning systems for education, healthcare, and food distribution and a general economy, there was no ability to promote awareness of applicable accessibility policies, let alone the means to monitor or enforce them. There was not a stable enough economy to develop accessibility specialties in web and mobile design and development businesses, despite the existence of free international resources providing accessibility guidance. Students with disabilities were cut off from schooling opportunities that supported their accessibility needs, and even from basic healthcare that would help them stay healthy enough to attend school, especially in rural areas of the country.

These circumstances highlighted the relevance of all the other basic human needs and civil rights in the CRPD as critical precursors for engaging meaningfully in advocating for digital accessibility.

6.1.4 Transformative Opportunities from Digital Accessibility

The ability to access online information and engage with others online can be radically transformative. But for people with disabilities in under-resourced regions and communities, age-old barriers in community infrastructure—as simple and pervasive as stairs at the entrances to computer training centers or a lack of interpreters for deaf students—can make it harder to even get in the door to learn these new technologies. These are barriers that we hear about from around the world, including from under-resourced communities within high-income countries.

Digital accessibility, or the lack thereof, can impact opportunities for people with disabilities across all aspects of life. It can facilitate or block opportunities to learn, to work, to engage in commerce, to participate in civic society, to obtain healthcare, to entertain and be entertained, to contribute to the community, to be informed about the world, and to inform the world about one's own experiences and perspectives.

When chalk and blackboard-based education suddenly leaps into the digital age, it can set people with disabilities further behind their peers. Internationally available "massive open online courses," if inaccessible, cannot be fixed at a local level. When new online banking options are inaccessible, entrepreneurs with disabilities can be shut out of local economies until accessibility barriers are resolved. When telemedicine is the only access to medical care, especially specialized providers needed for treatment of disabling conditions, if a telemedicine platform is not accessible, that can mean no access to medical care for patients who need to stay healthy enough to self-advocate.

But with advanced planning to ensure that new technologies are designed for inclusion, these sudden leaps into the digital future can be an equalizer rather than a new mechanism of exclusion.

6.1.5 Barriers in Digital Technologies

Every emerging technology brings the risk of new barriers for people with disabilities, just as in the physical world. Barriers may be introduced because of lack of awareness and training on the need for accessibility, or may stem from a lack of testing during design and development.

On websites, examples of barriers include images and videos without descriptions; text that does not reflow smoothly when enlarged; audio without captions, interactive web content lacking interoperability with assistive technologies; websites with inconsistent navigation menus and options; or blinking and flashing content that can cause distraction or seizures. Other barriers include e-books without chapter-by-chapter navigation, which is critical for people with print disabilities; and inaccessible interfaces to devices and sensors to control home or work environments.

6.1.6 Compounded Barriers in Under-Resourced Regions

For people in low-income and lower middle-income countries or in marginalized Indigenous communities, lower regional levels of online access often combine with other factors to exacerbate inaccessibility. A lack of training resources for users, designers, or developers can combine with inadequate local telecommunication infrastructures and an unreliable local power grid. Resources to evaluate and monitor accessibility, and to remediate inaccessibility, may be non-existent. There may be no policies that require accessibility for local government, businesses, or other organizations.

In Argentina, despite its upper middle-income economy and the presence of relevant policies, there is no ecosystem of awareness, including knowledgeable developers, skilled evaluation resources, and more, to drive web accessibility progress at the local or national levels.

People with disabilities often have less access to education, which includes the foreign language instruction needed to take advantage of international accessibility advocacy resources. This is the case in countries such as China and Mongolia, even though knowledge of English may be relatively high among the non-disabled population. Therefore, availability of technical and educational resources in local languages becomes especially important. Lags in localization of assistive technologies such as screen readers particularly impact language communities with small populations. For instance, although Slovenia is an upper middle-income country, local accessibility advocates note that the small population (2 million) sometimes means delays in updated translations of NVDA, a popular, free, volunteer-translated screen reader. This can interrupt the ability of individuals to use key software, websites, and browsers or authoring tools. Of the 60 languages where NVDA localizations are supported (GitHub, 2020), only 34 of those have up-to-date translations at the time of writing.

Even in the highest-income countries, under-resourcing of some communities can have significant impact on the technology needs of people with disabilities. In 2005, when Hurricane Katrina hit New Orleans in the United States, it particularly devastated low-income communities. At that time, the U.S. Federal Emergency Management System only supported one type of browser for emergency assistance applications, creating interoperability barriers for people who had lost access to their assistive technology when their neighborhoods were submerged by floods. Similarly, people with disabilities in many Indigenous communities in Canada and the United States must deal with little to no internet connectivity compared to other regions of these countries.

A combination of multiple barriers across infrastructure and social sectors can be insurmountable. To ensure that the benefits of digital accessibility are available to people with disabilities in under-resourced regions, it is important to ensure comprehensive barrier reduction throughout all aspects of the CRPD.

6.2 The Role of Technology Standards in Promoting Accessibility

Web and mobile accessibility can be keystone technologies for driving cross-sector accessibility progress and facilitating disability inclusion, which drives more progress. They can provide a platform over which people with disabilities can learn, share resources and strategies, and advocate for progress throughout their communities, helping to realize all parts of the CRPD.

Technology standards play a key role in promoting accessibility by defining feasible, consensus-backed approaches for addressing the needs of users with disabilities in a digital environment. Standards that have been accessibility-tested help counter myths about the feasibility of accessibility and demonstrate that the needs of different disability communities complement rather than contradict each other. They demonstrate that accessibility, rather than constraining innovation, instead drives innovation; that accessible interfaces can be vibrant and dynamic; and that accessible interfaces set the foundation for more inclusive interfaces for all.

6.2.1 The World Wide Web Consortium

The World Wide Web Consortium (n.d.) develops technologies for the web platform internationally, to ensure that the web remains open and interoperable. World Wide Web Consortium standards include HTML and hundreds of standards which support graphics, media, internationalization, payments, virtual reality, sensor networks, and automotive navigation.

Historically, the World Wide Web Consortium (W3C) has focused more on the needs of higher-income countries. W3C is jointly hosted in North America, Europe, Japan, and China. W3C brings together thousands of dedicated technologists representing more than 400 organizations. Outreach chapters include the low-income and lower middle-income countries India, Morocco, and Senegal. W3C's consensus process (Nevile, 2018), with multi-stakeholder participation, provides a setting in which user needs can inform and be considered alongside the interests of industry, research, education, and government, and within which any organization or individual from around the world can participate. Its commitment to a "Web for All" includes accessibility, internationalization (W3C, 2020), privacy, and security.

6.2.2 The Web Accessibility Initiative

The Web Accessibility Initiative (n.d.) develops accessibility guidelines that are considered the authoritative international standard for web content and applications, and standards for authoring tools and user agents for the web. It coordinates accessibility work within W3C and works with hundreds of organizations to make the web more accessible for people with disabilities and older users. Web Accessibility Initiative (WAI) work includes ensuring that all W3C technologies support accessibility, including emerging technologies on the web.

WAI has developed guidelines (Henry, 2019b) for accessible web content and applications (Henry, 2018b), browsers (Spellman et al., 2016), and authoring tools (Henry, 2015) that support production of accessible web content and improved interoperability of mainstream technologies with assistive technologies used by people with disabilities such as screen readers and speech recognition. WAI develops resources to support improved accessibility evaluation and to promote awareness and implementation of accessibility. It coordinates with accessibility research and development, and promotes harmonized uptake of international web accessibility standards. Recent WAI work, supported by the Ford Foundation, has enabled increased outreach to people in low-income and lower middle-income countries.

Uptake of WAI's Web Content Accessibility Guidelines into governmental policies increased following the signing of the CRPD. However, many countries still have little implementation planning, monitoring, or enforcement, and therefore progress on achieving web and mobile accessibility is often slow. This is the case even in countries such as India and Estonia—which perform extensive amounts of "in-sourced" accessibility work—because there is not yet a national infrastructure supporting digital accessibility and encompassing efforts in awareness, promotion, implementation support, and monitoring.

In contrast to this, in higher-income regions such as Hong Kong, where there is an excellent infrastructure for digital accessibility going back over two decades, the policy uptake has been good due to strong awareness and technical skills among the disability community, strong promotion efforts from government, and a robust training program from computer science programs. Similarly, in South Korea, a very high level of technical development, including work on mobile accessibility, has built upon multi-stakeholder interest and support among the disability community.

6.2.3   Web Accessibility Guidelines

Several complementary accessibility guidelines are needed to ensure accessibility of the digital environment, as described in "Essential Components of Web Accessibility" (Henry, 2018a) and below.

6.2.3.1   Web Content Accessibility Guidelines

W3C's Web Content Accessibility Guidelines (WCAG) 1.0, published in 1999, was largely focused on HTML, the primary web markup language at the time (Chisholm et al., 1999). However, this focus presented a problem for web developers, who used many other markup languages. Derivative versions of WCAG 1.0 emerged, then derivatives of those, each of which altered the meaning of the standard and created disincentives for organizations obligated to comply with conflicting versions of guidelines across different jurisdictions.

WCAG 2.0, published in 2008 (W3C, 2008), was organized around four key principles: that web content should be perceivable, operable, understandable, and robust. It was designed to be technology-neutral but also forward-compatible, and to apply equally well to HTML and to technologies that had not yet been developed. This was accomplished by having a general "normative" layer (the part to which conformance can be tested), and "informative" techniques which developers could use or even develop on their own. It was also designed to be more precisely testable, so that it would be easier to determine whether a website complied with WCAG 2.0.

In developing WCAG 2.0, W3C WAI reached out to a much broader community of accessibility experts and stakeholders around the world. For instance, WAI received feedback on integration of needs related to Chinese, Japanese, and Korean, all character-based languages. Improvements included addressing issues such as improved bidirectional character rendering and incorporating pronunciation annotations. The larger and more diverse volume of comments improved the standard, establishing a stronger international consensus.

The updated WCAG 2.1 adds more provisions for some accessibility needs, including low vision; cognitive, language, and learning disabilities; and mobile (Kirkpatrick & Cooper, 2018). The expansion of mobile accessibility is important for people with disabilities in low-income and lower middle-income countries that rely heavily on mobile phones. WCAG 2.2 will add further provisions for these areas. WAI is developing next-generation guidelines with a new structure and conformance model, and plainer language where possible for improved understandability and translatability to better serve smaller language communities.

6.2.4 Authoring Tool Accessibility Guidelines

The Authoring Tool Accessibility Guidelines address accessibility support in tools used by designers and developers of web content, and are an essential complement to WCAG. Authoring tools include markup editors, graphics editors, integrated development environments, content management systems, learning management systems, and social media platforms such as wikis and blogs, which are widely used to informally create content for the web. Authoring Tool Accessibility Guidelines (ATAG) 2.0, published in 2015, aligns with WCAG 2.0 (Richards et al., 2015a). Supporting materials include "Implementing ATAG 2.0" (Richards et al., 2015b), which provides support for designers and developers in specific contexts of web content authoring.

In low-resource regions, efficient production of accessible content is especially important. China, with a vibrant technical community and one of the largest language communities in the world, is starting steps to require ATAG 2.0 (Zhang, 2018, Trans.) within its own authoring tool market.

6.2.5 User Agent Accessibility Guidelines

User Agent refers to any software used to access, interact with, and render content, and includes browsers, media players, and assistive technologies. The rapid expansion of mobile applications as a means of accessing web content was one of the main motivations for updating the User Agent Accessibility Guidelines to the 2.0 version in 2015 (Spellman et al., 2016). Again, for low-resource regions heavily dependent on mobiles, effective built-in support for accessibility can help web designers and developers save time from working around the inconsistent feature support in browsers and assistive technologies that can particularly affect small language communities.

6.2.6 Accessible Rich Internet Applications

WAI Accessible Rich Internet Applications (Cooper, 2020) is a W3C (2018) technical specification that defines ways to make dynamic and interactive web and mobile applications more accessible to people with disabilities. WAI Accessible Rich Internet Applications (WAI-ARIA) addresses these barriers by, for instance, defining new ways for functionality to be provided to people using screen readers, or to people who cannot use a mouse or have limited use of gestures. Materials include WAI-ARIA Authoring Practices 1.1 (Fiers et al., 2019), which promote consistent implementations of ARIA, and a number of ARIA Application Programming Interfaces and ARIA Application Programming Interface Mappings.

6.2.7 Accessibility Conformance Testing Rules Format

Accessibility Conformance Testing Rules Format 1.0 (W3C, n.d.) defines a format for writing rules to test the accessibility of web content. These test rules encompass automated, semi-automated, and manual testing. Over time, these advances will increase automation of accessibility testing, thereby eventually supporting more automated and affordable conformance testing for under-resourced regions.

6.3 Localizing the Ecosystem for Promoting and Implementing Accessibility

The existence of accessibility guidelines, and supporting technical and educational materials, does not automatically lead to an accessible web even in regions of the world with extensive resources. The web still has major gaps and barriers in high-income countries, but in low- and middle-income countries (LMICs) and in Indigenous communities, gaps can be more severe.

Achieving an accessible web relies on a combination of many different elements: guidelines, conformance testing, technical specifications that work well with assistive technologies, and more. Since digital accessibility is still not well understood in many countries, it requires local efforts to promote awareness, training to support implementation, policies to help drive uptake, and integration of accessibility requirements into organizational processes—into the design, development, and maintenance of websites and mobile applications. To date, the needs of people in under-resourced regions have not been sufficiently reflected in this ecosystem.

Accessibility is best addressed from the beginning of the design process, as part of an inclusive design approach to maximizing inclusion for all users. While accessibility focuses on ensuring equal access for people with disabilities, it also benefits people who may not currently have disabilities. Additional benefits across other stakeholder groups are addressed in WAI's "Business Case for Digital Accessibility" (Henry, 2019a). Gathering diverse user scenarios, analyzing these, articulating the technical needs and accessibility requirements, and documenting these in standards, guidelines, and other supporting resources for designers and developers of websites, tools, and technologies requires the participation of a broad range of stakeholders from a representative global user base.

Often, local digital accessibility advocates explain that they need more background information before they can promote digital accessibility. Free online resource collections such as WAI's site (Henry, 2020a) can help, including "Accessibility Fundamentals" (Henry, 2019c); "Planning and Policies" (Henry, 2020b); "Designing and Developing" (Henry, 2020c); "Testing and Evaluating" (Henry, 2019b); "Teaching and Advocating" (Henry, 2020c); and "Standards and Guidelines".

Accessibility advocates in under-resourced regions report that video introductions can be an easier way to learn and to share information about digital accessibility. Contributing to the popularity of "WAI Perspectives" (Abou-Zahra, 2016) is a series of one-minute videos showing different aspects of the disability user experience as well as benefits of accessibility to other users. Additional video resources include the "Video Introduction to Web Accessibility and W3C Standards" (Abou-Zahra et al., 2019), and a TEDxMIT video, "Creating an Accessible Digital Future" (Brewer & TEDxMIT, 2019), which explains the impact that accessibility can have on people's lives, and why good implementation is so important. Other resources that can serve as easy on-ramps for accessibility include "How People with Disabilities Use the Web" (Abou-Zahra & Brewer, 2017), which provides a window into the user experience of people with disabilities; the massive open online course "Introduction to Web Accessibility" (Abou-Zahra 2019), which provides a free overview on web accessibility; and

"How to Meet WCAG (Quick Reference)" (Eggert & Abou-Zahra, 2019), an implementation guide for developers.

Several elements of the web accessibility ecosystem may be especially important for strengthening digital accessibility in under-resourced regions. These include translation of accessibility resources, closing stakeholder gaps, capacity building, and policy development, addressed in the next sections.

6.4 Scaling Up Translations

English has to date been the common language for most computer coding and digital standards development. However, progress on accessibility depends on many channels of uptake in addition to developer implementation of technical standards. Uptake of accessibility standards requires widespread awareness of the need for accessibility; awareness of relevant policies and of relevant technical standards and guidelines; and the ability to use technical support materials and accessibility conformance testing materials. For diverse stakeholders from different language communities, it is essential that these materials are available in languages other than English.

6.4.1 The Need for Localized Resources for Digital Accessibility

Disability advocates in low-resource regions often note that their efforts must be spread across a great many survival-level topics, including access to housing, transportation, local education, healthcare, employment opportunities, and civic participation. Among these survival-level priorities, the concept of digital accessibility can initially seem like a luxury. However, digital accessibility rapidly becomes a survival issue once local connectivity improves and critical information and interactions move online.

Disability advocates sometimes report not feeling confident about what to advocate for with regard to digital accessibility, or knowledgeable enough to educate other advocates on the topic. For disability advocates in under-resourced regions, their disability may mean less opportunity for advanced education and exposure to mainstream digital technologies and assistive technologies, and less opportunity to learn English compared to people from other stakeholder groups in their communities, making it harder to learn these materials on their own.

At an international gathering in Mongolia of disability advocates from countries in the Asia-Pacific region, we heard how important it is for advocates to have materials about web accessibility available in local languages throughout the region, to allow them to more easily pass these along to others in the advocacy community and to allies in other communities.

For some languages this requires technical improvements in the ability to render specific scripts or characters on the web. In other cases, it requires more specialized pronunciation rules. W3C has been working on both of these kinds of support. We have heard similar comments from developers in Argentina, who have said that they need key web accessibility materials available in Spanish, and in Brazil, where government promoters of web accessibility realized early on that Portuguese was an essential starting place for promoting interest within the developer community. In the advanced web industry in China, there is interest in an authorized translation of ARIA for accessibility of highly interactive websites, and an interest in the ATAG to support Chinese-language production tools for more efficient production of accessible websites.

6.4.2 Expanding Translations of Technical and Educational Materials

A volunteer community of translators has helped make a limited number of W3C standards available in languages other than the official English version over the years. Initially, translations of W3C materials were prepared without a formal review process, occasionally resulting in disagreements between different organizations within the same language community as to which was the authoritative translation. W3C's "Policy for Authorized W3C Translations" (Brewer & Herman, 2020) encourages reviews by different stakeholder groups using an established process for resolving discrepancies between translations. This policy, however, is primarily applied to standards, not to the many educational and implementation support materials needed for uptake of accessibility guidelines which require careful monitoring in order to ensure translations of a consistent quality.

To address the need for higher-quality translations, W3C is updating and extending the translation policy to be applicable to technical support and educational resources as well. To improve discoverability of translations, WAI has updated the infrastructure for posting translations on its website so that the available translations for a given resource can be directly accessed from that resource page, as well as from language-specific reference lists (WAI Expanding Access Project, n.d.). WAI also improved support for bidirectionality in page views and in navigation menus, so that Arabic and other languages would render properly.

WAI recruited volunteer translators to expand translations for an initial set of five priority WAI documents in each of the six major UN languages: Arabic, Chinese, English, French, Russian, and Spanish. Some WAI resources, such as the WCAG, have translations in more than 20 languages already, though most WAI resources have only a few translations. WAI will next expand the pool of translators to less widely spoken languages.

WAI is developing guidance to facilitate the work of translators. Because of the specialized technical vocabulary in WAI materials, we are exploring approaches such as translating glossaries of accessibility terminology that can be useful in supporting translations of multiple documents. This approach helps support development of consensus on appropriate translations of words referring to disability, since terminology relating to disabilities remains stigmatizing in many language communities, and neutral alternatives referring to different aspects of disability may not be widely known yet.

The complexity of the technical standards themselves also creates barriers to translation. WAI is therefore also developing guidance for writers of specifications and educational materials, to encourage the use of plain language in order to improve translatability of these resources.

6.5 Addressing Stakeholder Gaps

Thanks to efforts from people from many stakeholder groups, substantial progress has been made in digital accessibility standards over the past two decades. This effort has included contributions from industry, disability communities, developers, evaluators, research, government, education, and more. Nevertheless, there are gaps in who is represented in standards discussions, and who gets to have input into priorities and during the standards development process.

6.5.1 Who's at the Table When Developing Accessibility Standards?

People with disabilities have frequently spoken up about their needs with regard to digital accessibility, sometimes despite technology designers and developers being disinterested in their feedback, but this is starting to change. Technology companies have increased their hiring of people with disabilities, realizing that it can be a competitive advantage to have employees onboard who have knowledge and experience about human variation in the usage of digital technologies. Consequently, people with disabilities are having more direct opportunities to bring their voices into all stages of technology development.

Employment of people with disabilities in the technology field also allows contribution of their accessibility expertise to standards organizations such as W3C. Industry representatives, regardless of whether they have disabilities, become involved in accessibility standards development for a variety of reasons. Initially, much of the industry participation in accessibility work was on a volunteer basis, and most participation was from upper-income countries. W3C member organizations now more frequently cover their employees' work on accessibility standards. Motivations may include contributing to the quality of accessibility standards; learning more about accessibility so as to improve the customer experience; improving the work environment with regard to diversity and inclusion; or enhancing organizational standing on social responsibility indicators.

The quality of accessibility standards is enhanced by having different perspectives at the table. Researchers can delve into novel accessibility challenges. Web designers and developers can develop and refine reusable web components. Educators and content developers can develop tutorials that speak clearly to other developers.

6.5.2 Who's Missing from the Table?

Despite the increased diversity of stakeholders at the standards table, many communities' voices are still missing from the accessibility standards development conversation, and this includes voices from low-income and lower middle-income countries and Indigenous communities. The absence of people from under-resourced regions impacts prioritization of issues in standards development.

The Web Foundation's Web Index (Brudvig, 2020) describes the growing inequality of the web, including the extent to which worldwide economic differences underlie disparities in access to the web. These same disparities in access to the web are paralleled in differential engagement patterns in technology standards development. W3C's member organizations—of which there are over 450—are largely, though not entirely, drawn from higher-income countries. Participation in W3C working groups and task forces, and in community groups—the settings in which technical challenges are identified and approaches to address these are incubated—reflect this pattern of lower participation from low-income and lower middle-income countries. For people with disabilities in Indigenous communities, there may be barriers to establishing connectivity (National Congress of American Indians, 2016; U.S. Government Accountability Office, 2016) to the internet and the web, though attention to connectivity has begun to increase through the Internet Society's Indigenous Connectivity Summits (Internet Society, n.d.).

Uptake of WCAG 2.0 into national policy requirements shows similar disparities, with progress to date mainly in higher-income countries, as evidenced in W3C WAI's listing of "Web Accessibility Laws and Policies" (Mueller et al., 2018) and 3Play Media's map of WCAG uptake (Enamorado, 2019), which primarily show uptake in higher-income countries. In many countries, availability of authorized W3C translations of WCAG 2.0 and WCAG 2.1 in national languages is a necessary precursor to uptake. Yet beyond the six major UN languages, WAI resources are still more often translated into languages used in higher-income countries.

6.5.3 Creating a More Welcoming Table

Standards organizations are sometimes structured around who is already at the table and who already has a history of contributing, rather than around a deliberate effort to diversify stakeholder representation. Participation can become a self-reinforcing cycle.

It is important that standards organizations create a more welcoming table for people from around the world interested in helping to shape technology development and accessibility standardization. Some international standards organizations already have provisions to encourage more geographically diverse participation. W3C has always had a diversity of participation modes: email, collaborative web development, teleconferences, and face-to-face meetings. Efforts to diversify these further include developing easier-to-use review tools, geographically rotating meeting locations and times, and using captions which help participants who are less fluent in English, as well as those who are deaf or hard of hearing. For a limited number of people without other means, W3C waives registration fees and covers the costs of travel and/or accommodation. W3C's annual Technical Plenary meetings can be an excellent opportunity for people to meet standards developers working in many different working groups across W3C, yet can be expensive to get to and stay at. The expansion of virtual meetings due to COVID-19 restrictions have ironically leveled the playing field with regard to participation from under-resourced regions.

The opportunity to participate in making standards for the technology industry has not come easily for people with disabilities, just as it has not for women in the tech field, for people of color, and for people from lower-income regions. While there are have been improvements over the years, the needs of people with disabilities are often misunderstood, and have often been dismissed, and sometimes even mocked, in a standards setting. These issues have all been barriers to more diverse participation, and have eventually been taken up in direct conversations, leading to incremental improvements over time. More work is needed.

As people with disabilities working in the tech field, the authors can vouch that the lack of an accessible infrastructure across all dimensions—the built environment, transportation sector, communications—makes it difficult for people with disabilities to gather, to learn from each other, and to take action collectively on accessibility issues in the tech field.

These issues are often exacerbated in low-resource regions. For both authors, it has at times meant a lack of means to travel safely and accessibly, and even to access venues in which we have been invited to present about accessibility. This lack of an accessible infrastructure can similarly become a barrier for people from regions not currently represented in W3C WAI work, who are trying to join the international discussion; and it means a steep learning curve for people trying to join the discussion.

Nevertheless, the increasing use of online environments has the potential to broaden our ability to network across geographic regions. We have developed many more "easy on-ramp" resources from which to learn about accessibility, and opened more avenues to get involved in building an accessible digital future.

6.5.4 Updating Specifications and Educational Materials to Include More Relevant Examples

Sample use cases and user scenarios embedded in technical and educational resources are often not inclusive and may be biased toward situations from higher-income countries. Standards developers may design new standards and technologies with the assumption that these will be deployed in areas with high-bandwidth uninterrupted connectivity and a reliable electrical grid, whereas people with disabilities in under-resourced regions cannot take these qualities for granted.

Incorporating examples that reflect people's experiences from around the world can help educational materials be more relevant to people in more regions of the world. For instance, in some LMICs, access to personal computers remains limited, but everyone has a mobile phone. Therefore, including examples of mobile phone usage and/or examples that reflect the lives of people in more rural communities can increase the relevance of resources to local communities.

6.5.5 Remote Participation Options, Including Using Advanced Technologies

With the increase in remote work, including low-cost and no-cost teleconferencing and video conferencing, it has become easier to establish a sense of engagement and shared work community—instead of having to board a plane, which can be expensive and difficult for people from low-income and lower middle-income countries, as well as for people with disabilities. Given the difficulties, cost, and time associated with international travel, more accessible and reliable remote participation options could greatly expand participation in international standards work.

Often, though, these remote options also have accessibility barriers. Video conference controls may not work well for people who are blind or have low vision. It may be difficult to integrate streaming captions, or to display sign language interpreting next to shared content or slides.

Advanced technologies for telepresence and virtual gatherings could support transcending physical location for people with disabilities in low-income and lower middle-income countries. Telepresence devices enable a remote participant to visit a location and use a self-propelled device to "visit" an office or other location and move through space at the remote location under their direct control. These still have barriers, including bandwidth and reliability issues, and will only work for people with disabilities if these technologies are developed with accessibility in mind.

6.6   Building the Capacity to Create an Accessible Digital Future

In higher-income countries, a variety of accessibility skills and complementary roles evolved gradually over time: accessibility experts, tech-savvy disability advocates, web designers, developers and evaluators, software and app developers, policy makers, and more. But opportunities to develop a similar range of experts and organizations have been absent in other areas. Given the extent of technical and educational material that has already been developed, there is now an opportunity for local communities to deliberately plan to develop these skills at an accelerated pace, by drawing on international resources, without starting from scratch or recreating the wheel.

6.6.1   Training and Capacity Building

The saying "Nothing About Us Without Us" has long been popular in the disability rights movement, and applies to technology development and standards development as much as to other things that affect the lives of people with disabilities. In the case of standards development, it is vital to have users' voices involved in articulating use cases, user requirements, the suitability of potential solutions, and their design and development.

Co-training approaches that deliberately bring together participants from different stakeholder groups can be a good way to deliver local capacity-building training that is inclusive of people with disabilities. Combinations of stakeholder groups or roles can be deliberately sought; for instance, co-training initiatives could be focused on training designers, developers, and evaluators together. Policymakers and disability advocates could be trained together so as to build shared understanding, goals, and partnerships. If engineers train alongside people with disabilities, engineers get a direct view of the barriers that people with disabilities must contend with on a daily basis, including when trying to use today's technologies. Advocates and allies can learn how the design and development process works and get insight into gaps in developers' understanding of design requirements of people with disabilities.

The Inclusive Africa (2020) webinar and conference series organized by InABLE in Kenya, has highlighted the stories and experiences of people with disabilities and allies working on accessibility in industry, to build a community with shared advocacy and technical understanding. They have started with basics such as the importance of equal access in Africa; understanding digital accessibility; details of web accessibility; mobile accessibility; inclusive design; and more. This has elevated the work of digital accessibility proponents from different parts of Africa, strengthened ties with international industry representatives, and helped those representatives gain a better understanding of needs in some under-resourced areas of Africa.

6.7 Conclusion

6.7.1 Disability Rights, Procurement Legislations, New Markets, and Accessibility Standards

Human rights, including disability rights, have evolved since the early work of W3C WAI. In 2006, the CRPD was passed and quickly adopted by many countries around the world, driving increased interest in web accessibility guidelines as a means for addressing rights defined in the CRPD, especially those supporting access to information and information technologies.

Even before that, the market for accessible ICT had been impacted by a procurement legislation approach which started in 1986 with enactment of U.S. Section 508 of the Rehabilitation Act, and which was subsequently taken up in Europe. Section 508 required that U.S. federal agencies were responsible for ensuring that ICT that they developed, procured, maintained, and used must be accessible for people with disabilities so as not to exclude current or future employees from being able to become part of the federal workforce. It also included an obligation that agencies' public-facing technologies should be accessible, so that people with disabilities among the public would not be excluded from access to public information.

W3C's WCAG 1.0 were referenced in an update to Section 508 (U.S. Access Board, n.d.) in 1999. Section 508 modified some provisions of WCAG 1.0, creating some non-aligned requirements. Section 508 was updated again in 2017 (U.S. Access Board, 2018), harmonizing more precisely with W3C's WCAG 2.0. It was expanded to apply more broadly to ICT, including hardware and software such as printers and automated teller machines. W3C obtained endorsement of WCAG 2.0 as an International Organization for Standardization (ISO) standard 40500 (International Organization for Standardization, 2019) in 2012, which has led to additional international uptake of WCAG 2.0.

The European Union pursued a similar approach to accessibility procurement requirements, resulting in 2014 in the European Norm (EN) 301 549 (eSSEN-TIAL Accessibility, 2019), now aligned in scope with the updated U.S. Section 508 Technical Standard. Given that this adoption of WCAG 2.0 for non-web contexts had not initially been envisioned by W3C, some technical clarifications were required and were addressed during the standardization process in Europe. The most recent version of EN 301 549 was revised to align with WCAG 2.1, and to meet specific requirements of the "European Web Accessibility Directive" (European Commission Accessibility, Multilingualism, & Safer Internet Team, 2019).

6.7.2 National Laws, Policies and Regulations for Web Accessibility

Many countries have developed laws or policies that reference or adopt W3C's WCAG, either directly or by referencing ISO IEC 40500, or the broader EN 301 549. W3C maintains a reference listing of such laws and policies from around the world (Mueller et al., 2018), now including policies in Australia, Canada, China, Denmark, the European Union, Finland, France, Germany, India, Italy, Ireland, Israel, Japan, Netherlands, New Zealand, Norway, Republic of Korea, Taiwan, the United Kingdom, and the United States.

Some governments have developed local web accessibility standards. When these diverge technically from international standards, they can create barriers to sharing training and technical assistance materials across borders and difficulty maintaining local standards over time as technologies evolve. For organizations that must meet different guidelines in different countries, this fragmentation has the outcome of increasing compliance costs for web accessibility, instead of accelerating accessibility uptake by enabling sharing of training resources and authoring and evaluation tools. Feedback on existing accessibility standards and input into new guidelines under development are therefore important to ensure that future versions of accessibility guidelines address the needs of all communities, including those in under-resourced regions.

6.7.3 Policy Issues in Under-Resourced Regions

There are many potential barriers to policy uptake of digital accessibility in under-resourced regions and communities, and these parallel combinations of barriers are mentioned earlier in this chapter with regard to the implementation of web accessibility guidelines. Some barriers may be more common in under-resourced regions, and certain policy issues may need particular attention at a local level.

Uptake of accessibility guidelines into national and local policies is dependent on development of an understanding of the need and approaches among individuals who are positioned to affect policy changes. This usually takes multiple years, and this continuity can sometimes be more difficult to maintain in LMICs, leading to slower development of accessibility policies in under-resourced regions.

Abrupt changes of government can interrupt continuity, as the authors have, for instance, observed in Tunisia and Egypt in recent years, where WAI had been active in promoting awareness and uptake of WAI guidelines. In these circumstances, a deep network of people informed about digital accessibility across multiple stakeholder groups (government, disability community, industry, research, and education) can be especially important in maintaining a resilient network of accessibility proponents within different sectors of society. In some countries this has enabled progress on policy uptake to resume relatively quickly following social or governmental disruptions.

Even while international standards harmonization can serve as an accessibility accelerator, certain issues may, however, benefit from local policy development. To take the example of education, in meeting the goals of Article 24 of the CRPD (2006d), which defines educational rights for people with disabilities, it is important to consider the role of technology in educating students with disabilities—and to plan for that even before digital learning has entered local classrooms.

For instance, countries whose educational systems integrate disabled and non-disabled students together in the classroom sometimes maintain a practice of taking students with disabilities out of their regular classrooms for specialized instruction for a few hours each day. If this practice coincides with classes on science, technology, engineering, and math, students with disabilities may be deprived of learning the technical skills that would allow them to help design their digital future because of assumptions that they do not have the interest, aptitude, or need for the science, technology, engineering, and math curriculum.

From the perspective of local policies, it is therefore essential to ensure that science, technology, engineering, and math (STEM) curricula and classrooms are accessible for students with disabilities, and that adequate time and resources are allocated for making the curriculum accessible. When educational personnel are made aware of the need for an accessible curriculum in advance, including for STEM, and when training and resources are made available to ensure students with disabilities are accommodated, students can have equitable educational opportunities. Moreover, they can acquire skills that will enable them to build the accessibility solutions of the future. Ensuring opportunities for students with disabilities to learn about technology and accessibility can help build a generation of digital accessibility advocates. This can fuel a virtuous cycle of expanding capabilities and opportunities, rather than perpetuating a cycle of inaccessibility that perpetuates exclusion of people with disabilities. If we plan for digital inclusion now, we can help build an accessible digital future, sector by sector, including in under-resourced communities.

References

Abou-Zahra, S. (Ed.). (2016, September 15). *Web accessibility perspectives: Explore the impact and benefits for everyone.* World Wide Web Consortium (W3C) Web Accessibility Initiative (WAI). https://www.w3.org/WAI/perspective-videos/

Abou-Zahra, S. (2019, December 3). *Free online course "Introduction to web accessibility."* World Wide Web Consortium (W3C) mail, news, blogs, podcasts, and tutorials. https://www.w3.org/blog/2019/12/free-online-course-introduction-to-web-accessibility/

Abou-Zahra, S., & Brewer, J. (Eds.). (2017, May 15). *How people with disabilities use the web.* World Wide Web Consortium (W3C) Web Accessibility Initiative (WAI). https://www.w3.org/WAI/people-use-web/

Abou-Zahra, S., Henry, S. L., Brewer, J., & Eggert, E. (2019, February 21). *Video introduction to web accessibility and W3C standards.* World Wide Web Consortium (W3C) Web Accessibility Initiative (WAI). https://www.w3.org/WAI/videos/standards-and-benefits/

Brewer, J., & TEDxMIT. (2019, May). *Why we need a more accessible digital landscape.TED.* https://www.ted.com/talks/judy_brewer_why_we_need_a_more_accessible_digital_landscape

Brewer, J., & Herman, I. (2020, May 4). *Policy for authorized W3C translations.* World Wide Web Consortium (W3C). https://www.w3.org/2005/02/TranslationPolicy

Brudvig, I. (2020). *The web index.* World Wide Web Foundation. https://webfoundation.org/our-work/projects/the-web-index/

Chisholm, W., Vanderheiden, G., & Jacobs, I. (Eds.). (1999, May 5). *Web content accessibility guidelines (WCAG) 1.0.* World Wide Web Consortium (W3C). https://www.w3.org/TR/WCAG10/

Cooper, M. (2020). *WAI-ARIA Overview.* World Wide Web Consortium (W3C). https://www.w3.org/WAI/standards-guidelines/aria

Eggert, E., & Abou-Zahra, S. (2019, October 4). *How to meet WCAG (quick reference).* World Wide Web Consortium (W3C). https://www.w3.org/WAI/WCAG21/quickref/

Ellsmoor, J. (2019, September 8). *Mobile phones are driving a solar revolution in Papua New Guinea.* Forbes. https://www.forbes.com/sites/jamesellsmoor/2019/09/08/mobile-phones-are-driving-a-solar-revolution-in-papua-new-guinea/

Enamorado, S. (2019, June 3). *Countries that have adopted WCAG standards* [MAP]. 3Play Media. https://www.3playmedia.com/2017/08/22/countries-that-have-adopted-wcag-standards-map/

eSSENTIAL Accessibility. (2019, October 1). *EN 301 549: The European standard for digital accessibility.* https://www.essentialaccessibility.com/blog/en-301-549/

European Commission—Accessibility, Multilingualism & Safer Internet Team (Unit G.3). (2019, November 27). *Shaping Europe's digital future. Web accessibility.* https://ec.europa.eu/digital-single-market/en/web-accessibility

Fiers, W., Kraft, M., Mueller, M. J., & Abou-Zahra, S. (Eds.). (2019, October 31). *Accessibility Conformance Testing (ACT) rules format 1.0.* World Wide Web Consortium (W3C). https://www.w3.org/TR/act-rules-format/

GitHub. (2020, January 12). *NVDA translation and localization.* GitHub. https://github.com/nvaccess/nvda/wiki/Translating

Inclusive Africa Conference. (2020). https://www.inclusiveafrica.org

International Organization for Standardization (ISO). (2019). *ISO/IEC Standard 40,500.* https://www.iso.org/standard/58625.html

Internet Society. (n.d.). *Indigenous connectivity summit.* https://www.internetsociety.org/events/indigenous-connectivity-summit/

Kirkpatrick, A., & Cooper, M. (2018, June 5). *WCAG 2.1 is a W3C recommendation.* World Wide Web Consortium (W3C) Blog. https://www.w3.org/blog/2018/06/wcag21-rec/

Miniwatts Marketing Group. (2020). *World internet usage and population statistics.* Internet World Stats. https://www.internetworldstats.com/stats.htm

Mueller, M. J., Jolly, R., & Eggert, E. (Eds.). (2018, March 21). *Web accessibility laws & policies.* World Wide Web Consortium (W3C) Web Accessibility Initiative (WAI). https://www.w3.org/WAI/policies/

National Congress of American Indians (NCAI). (2016). *Telecommunications and technology.* http://www.ncai.org/policy-issues/economic-development-commerce/telecomm-and-tech

Nevile, C. M. (Ed.). (2018). *World wide web consortium process document.* World Wide Web Consortium (W3C). https://www.w3.org/2018/Process-20,180,201/

Nothing About Us Without Us. (n.d.). In *Wikipedia.* Retrieved November 2019, from https://en.wikipedia.org/wiki/Nothing_About_Us_Without_Us

Henry, S. L. (Ed.). (2015, September 24). *Authoring tool accessibility guidelines (ATAG) overview.* World Wide Web Consortium (W3C) Web Accessibility Initiative (WAI). https://www.w3.org/WAI/standards-guidelines/atag/

Henry, S. L. (Ed.). (2018a, February 27). *Essential components of web accessibility.* World Wide Web Consortium (W3C) Web Accessibility Initiative (WAI). https://www.w3.org/WAI/fundamentals/components/

Henry, S. L. (Ed.). (2018b, June 22). *Web content accessibility guidelines (WCAG) overview.* World Wide Web Consortium (W3C) Web Accessibility Initiative (WAI). https://www.w3.org/WAI/standards-guidelines/wcag/

Henry, S. L. (Ed.). (2019a, June 5). *Introduction to web accessibility*. World Wide Web Consortium (W3C) Web Accessibility Initiative (WAI). https://www.w3.org/WAI/fundamentals/accessibility-intro/

Henry, S. L. (Ed.). (2019b, March 13). *W3C accessibility standards overview*. World Wide Web Consortium (W3C) Web Accessibility Initiative (WAI). https://www.w3.org/WAI/standards-guidelines/

Henry, S. L. (Ed.). (2019c, September 20). *Design and develop overview*. World Wide Web Consortium (W3C) Web Accessibility Initiative (WAI). https://www.w3.org/WAI/design-develop/

Henry, S. L. (Ed.). (2020a, April 28). *Planning and policies overview*. World Wide Web Consortium (W3C) Web Accessibility Initiative (WAI). https://www.w3.org/WAI/planning/

Henry, S. L. (Ed.). (2020b, April 28). *Evaluating web accessibility overview*. World Wide Web Consortium (W3C) Web Accessibility Initiative (WAI). https://www.w3.org/WAI/test-evaluate/

Henry, S. L. (Ed.). (2020c, May 12). *Teach and advocate overview*. World Wide Web Consortium (W3C) Web Accessibility Initiative (WAI). https://www.w3.org/WAI/teach-advocate/

Henry, S. L., Eggert, E., Bakken, B., Miller, V. M., & Keen, L. (2018, November 9). *The business case for digital accessibility*. World Wide Web Consortium (W3C) Web Accessibility Initiative (WAI). https://www.w3.org/WAI/business-case/

Richards, J., Spellman, J., & Treviranus, J. (Eds.). (2015a, September 24). *Authoring tool accessibility guidelines (ATAG) 2.0*. World Wide Web Consortium (W3C). https://www.w3.org/TR/ATAG20/

Richards, J., Spellman, J., & Treviranus, J. (Eds.). (2015b, September 24). *Implementing ATAG 2.0—A guide to understanding and implementing authoring tool accessibility guidelines 2.0*. World Wide Web Consortium (W3C). https://www.w3.org/TR/IMPLEMENTING-ATAG20/

Spellman, J., Allan, J., & Henry, S. L. (Eds.). (2016, May). *User agent accessibility guidelines (UAAG) overview*. World Wide Web Consortium (W3C) Web Accessibility Initiative (WAI). https://www.w3.org/WAI/standards-guidelines/uaag/

United Nations Convention on the Rights of Persons with Disabilities (CRPD). (2006a). https://www.un.org/development/desa/disabilities/convention-on-the-rights-of-persons-with-disabilities/

United Nations Convention on the Rights of Persons with Disabilities (CRPD). (2006b). *Article 9–Accessibility*. https://www.un.org/development/desa/disabilities/convention-on-the-rights-of-persons-with-disabilities/article-9-accessibility.html

United Nations Convention on the Rights of Persons with Disabilities (CRPD). (2006c). *Article 21—Freedom of expression and opinion, and access to information*. https://www.un.org/development/desa/disabilities/convention-on-the-rights-of-persons-with-disabilities/article-21-freedom-of-expression-and-opinion-and-access-to-information.html

United Nations Convention on the Rights of Persons with Disabilities (CRPD). (2006d). *Article 24—Education.* https://www.un.org/development/desa/disabilities/convention-on-the-rights-of-persons-with-disabilities/article-24-education.html

U.S. Government Accountability Office. (2016, February 3). *Telecommunications: Additional coordination and performance measurement needed for high-speed internet access programs on tribal lands.* https://www.gao.gov/products/GAO-16-222

Web Accessibility Initiative (WAI). (n.d.). *Web Accessibility Initiative (WAI) home.* World Wide Web Consortium (W3C). http://www.w3.org/WAI/

Web Accessibility Initiative (WAI) Expanding Access Project. (n.d.). *All WAI translations.* World Wide Web Consortium (W3C) WAI. https://www.w3.org/WAI/translations/

World Bank Group. (n.d.). *World Bank country and lending groups.* (n.d.). Retrieved November 2019, from https://datahelpdesk.worldbank.org/knowledgebase/articles/906519

World Health Organization & World Bank. (2011, December 13). *World Report on Disability.* World Health Organization. https://www.who.int/publications/i/item/world-report-on-disability

World Wide Web Consortium (W3C). (n.d.). *Homepage.* http://www.w3.org/

World Wide Web Consortium (W3C). (2008, December 11). *Web standard defines accessibility for next generation web.* https://www.w3.org/2008/12/wcag20-pressrelease

World Wide Web Consortium (W3C). (2020). *World Wide Web Consortium (W3C) launches internationalization initiative.* https://www.w3.org/2018/07/pressrelease-i18n-initiative.html.en

Zhang, K. (Trans.). (2018). *Chinese translation of: Authoring tool accessibility guidelines (ATAG) 2.0.* World Wide Web Consortium (W3C). https://www.w3.org/Translations/ATAG20-zh/

7

Digital Inclusion in the Global South

Work Globally, Act Locally

Betsy Beaumon
Social Entrepreneur, Former CEO, Benetech

7.1 Introduction

There are some one billion people with disabilities in the world, with about 800 million of them living in developing countries (World Health Organization & World Bank, 2018). How can the innovations streaming out of the Global South incorporate the knowledge and benefit from progress born of decades of struggle from across the globe, South and North, without creating a new generation of inaccessible, exclusive products and services that further marginalize people with disabilities?

This chapter outlines the benefits of global initiatives in concert with locally driven content and tools for people with disabilities in the connected digital age, and highlights examples of relevant programs and collaborations that originate from or deeply engage those in the Global South. It first examines the key challenges to creating and implementing inclusive, sustainable digital products and services, and how these challenges manifest in the Global South. It then lays out how global projects and partnerships can help provide a pathway to success, especially when powered by open standards and open-source software developed and implemented in an inclusive, community-driven way. Finally, it is important to recognize the new challenges arising from new technologies and the role of global initiatives in addressing them.

7.2 Challenges: Inclusion, Implementation, and Sustainability

The fight for inclusion of people with disabilities in all aspects of society is ongoing, and inclusive access to digital products and services is a critical part of this battle. Despite significant gains, it is still notable when new products, services, and content are accessible, rather than this being the norm. The promise of information and communications technology (ICT) to help deliver huge benefits in the social sector is real, but not without challenges that parallel the challenges

Accessible Technology and the Developing World. Michael Ashley Stein and Jonathan Lazar, Oxford University Press.
© Oxford University Press 2021. DOI: 10.1093/oso/9780198846413.003.0008

in all social programs. From the global development perspective, just as food, healthcare, employment, and other initiatives succeed or fail based on the local relevance and quality of implementation, so too do digitally based interventions. The tech is not a panacea, and the social structures, policy supports, and, most importantly, the people at the center, matter—whether they're holding a bushel basket or a smartphone. Finally, funders and implementers tend to focus all of their resources on the upfront project when it comes to technology, forgetting that systems change requires sustainability to achieve lasting impact. Hardware breaks and software needs updating, either to take advantage of new needs or capabilities or due to changes driven by the larger environment, from security to communications infrastructure.

The solutions to each of these challenges facing ICT-based initiatives—inclusion, implementation, and sustainability—are works in progress. The challenges occur in every part of the world and will increasingly impact projects, companies, and state actors in the Global South as many locally driven, digital initiatives take off. It is useful to examine these challenges from the perspective of product development and their impacts on people with disabilities in the Global South.

7.2.1 Inclusion and Implementation Challenges

7.2.1.1 People with Disabilities Are Users, Too!

For a product developer, inclusion begins with knowing one's users. For a product to be of high quality, it must function efficiently and effectively in its target geographical context, be fit for purpose, and be designed with a high degree of usability. Modern software developers usually achieve as much of this as possible by employing Agile software development methodologies and user-centered design, whose core tenets involve working with users early and often throughout the design, development, and testing processes (The Agile Alliance, 2001).

However, they must first understand who their users are. Just as statisticians must be careful to design their samples to be representative and eliminate biases, so must software developers directly involve enough people with different characteristics to create meaningful personae, including the broad diversity among people with disabilities (World Wide Web Consortium, Web Accessibility Initiative, 2019). When it comes to people with disabilities, most products fail before one line of code is written. Designers and developers do not regularly consider these individuals among their target users. People with disabilities are only recently finding their way into user and tester groups for large tech companies, and it is still common for start-ups in the Global North to go backward, making the same mistakes as their peers in previous generations by thinking "we'll worry about people with disabilities later" (Wentz et al., 2011). The problem is, later often never comes, because retrofitting accessibility is far more expensive than doing it upfront (Ossmann et al., 2008).

Exacerbating this problem in the Global South are low levels of access to digital technologies by people with disabilities, likely due to lower socioeconomic status and related low employment rates, with 80–90% of working-age people with disabilities unemployed (United Nations (UN), 2017). That means people with disabilities are not seen as a market, as employees or as individuals. As noted by Watermeyer and Goggin (2019, p. 170), "In low-income countries and populations, the promises of and reliance upon the commercial encounter to provide the preconditions of digital citizenship is far more dubious." By all accounts, the digital revolution will further marginalize those who do not have core digital skills, similar to basic literacy and numeracy. Therefore, it is not enough to just introduce technology or let the local market take its course. Government and civil society must weigh in to provide basic education, explicit digital training, and access to a variety of devices and digital services for people with disabilities.

7.2.1.2 Inclusive Development and Implementation Requires a Knowledgeable Workforce

Even if people with disabilities are a recognized market, companies still need the wherewithal to do inclusive product development. That includes having the skillsets within the team to develop accessible products and having more diverse product teams that include people with disabilities. Both can be a catalyst for more focus on accessibility within product requirements and implementation. According to the U.S.-based Partnership on Employment and Accessible Technology, there is a wide and increasing gap between industry demand for accessible tech skills and the available pipeline of candidates. In a May 2018 survey in conjunction with Teach Access, they found that while over 84% of their industry respondents cited hiring developers and designers with accessible technology skills as important or very important, 60% said it was difficult or very difficult to find candidates with those skills (Partnership on Employment and Accessible Technology & Teach Access, 2018).

> **Spotlight:** Accessibility Skills as Core Technology Curriculum
>
> Teach Access (http://teachaccess.org/), made up of a number of progressive companies in league with academia and advocates, is working to close this gap by making the fundamentals of digital accessibility design principles and best practices a larger part of undergraduate education in programming, design, or digital technology-related courses. The companies include Verizon Media Group, Facebook, and Apple, with universities including Stanford and Michigan State.

There are a number of barriers to overcome before this type of program is widely feasible in the Global South—including increased numbers of post-secondary engineering and design programs and a louder call from local tech companies to meet accessibility needs. The good news: the curriculum is gaining interest as a solution in Nairobi. Irene Mbari-Kirika and InABLE, a 501(c) non-profit organization that establishes computer labs at schools for the blind in Kenya and studies the impact of assistive technology (InABLE, 2019), are working to tie together universities and fledgling tech companies in the region to begin this virtuous cycle of skills supply and demand using these global best practices (see Chapter 16: The Role of Ugandan Public Universities in Promoting Accessible ICT).

The Global South is Tomorrow's Workforce, and Education is Critical
Population distribution trends are clear, and global tech giants know that a very large number of their future employees will likely come from the developing world. In most of Sub-Saharan Africa, and in parts of Asia, Latin America, and the Caribbean, the population of working-age adults (25–64 years) is growing faster than other cohorts, creating an opportunity for accelerated economic growth. These same regions are also predicted to grow as a percentage of an increasing world population, as populations elsewhere age. For example, Africa's share of the global population is projected to rise from 17% (1.0 billion) in 2010 to 24% (2.2 billion) by 2050 (Canning et al., 2015, p. 52). Thus, there is a growing supply of potential workers, as long as education and skills keep pace. Companies such as SAP and Google (a Teach Access participant) have introduced over 1.8 million young Africans to coding skills through programs such as Africa Code Week, including some children with disabilities (South Africa: The Good News, 2018). It is imperative that such programs, and the next-level training and computer science degrees that follow, introduce *inclusive* design and coding practices as well, perhaps with curriculum from groups such as Teach Access (Putnam et al., 2016).

For people with disabilities to increase their numbers within the growing cohort of potential employees, it means now is the time to address education in these countries. Chapter 2 in this volume, by Charlotte McClain-Nhlapho and Deepti Raja, provides significant examples.

Benetech's inclusive education initiatives, including Bookshare and Imageshare, are examples of a path forward for reaching the educational goals outlined by the World Bank and United Nations Sustainable Development Goal #4, and those of many individual countries, by providing accessible educational content to people with disabilities so that all students can participate in school.

Spotlight: Bookshare

Bookshare (www.bookshare.org, a Benetech initiative) is the world's largest library of accessible materials for people with print disabilities, including people who are blind, cannot hold or manipulate books, or have learning disabilities that affect reading. (This definition is imposed by copyright law in many countries, and under the Marrakesh Treaty, as discussed by Paul Harpur and Michael Ashley Stein in Chapter 9: The Relevance of the CRPD and the Marrakesh Treaty to the Global South's Book Famine.) Books are available in Braille, audio, enlarged font, and combinations of formats, on devices that may be as affordable as an $8 MP3 player. Students in both inclusive and segregated school settings globally use Bookshare; however, it is currently serving only a fraction of the people who need it in the Global South. As development programs focus on scaling access to education, they must include services for students with disabilities that work within that local ecosystem. Bookshare is one example of a globally relevant solution for students with disabilities that can easily layer onto local and multi-national educational initiatives so that all students may be served.

Employ People with Disabilities in All Roles

Employing people with disabilities creates overall economic empowerment and helps accelerate awareness of their market potential. The path to improved employment outcomes not only requires inclusive education in primary and secondary school, but also extends to higher education and vocational training in job skills, including the overall digital skills outlined earlier. While this approach obviously takes a long-term commitment, it is important to pursue these goals in parallel with near-term employment opportunities *in all fields* for people with disabilities.

People in the Global South are tackling this challenge head-on by leveraging digital tools in order to boost employment for people with disabilities. For example, the Bookshare team in India has increased its focus on accessible digital books as a part of the employment value chain, rather than purely in education. The Bookshare team is collaborating with EnAble India to tackle employment of people with visual impairments, by incorporating Bookshare as a tool for both vocational skills building and developing some of the most important skills for getting hired and achieving success on the job—language skills and soft skills. As recognized by the Deshpande Educational Trust (2019) and others, it is important for job seekers to be able to engage with potential employers in India in conversational English. Individuals from rural villages, especially those with disabilities, may lack these conversational language skills (even if they have studied in English at a university level) and a broad range of life experience. This deficit greatly affects their confidence to engage at a social level. In addition to language and skills training, access to a wide range of books in multiple languages provides exposure and confidence.

Spotlight: #SeeAMillion

See A Million (http://www.enableacademy.org/seeamillion/) has a very specific goal: digital empowerment of one million persons with vision impairment by 2025. This mission aligns with Digital India, a Government of India initiative; through it, stakeholders form a value chain, from rehabilitation and education to skilling and mainstream inclusion, with the goal of enabling the livelihoods of persons with vision impairment through digital empowerment.

They are bringing together digital technologies and on-the-ground implementation. This campaign is supported by an ecosystem of individual and employer tools, including career awareness workshops, accessible cyber cafes, a job board, training on assistive technology, access to resources such as Bookshare, and ongoing support for program alumni.

In Kenya, a local social entrepreneur has also focused on connecting employers and job seekers with disabilities to implement the national employment policy, leveraging locally relevant digital tools to scale what he had been doing individually.

Spotlight: Riziki Source in Kenya

Riziki Source (http://www.rizikisource.org/) has a mission of *enabling access to job opportunities for persons with disabilities (PWDs) in Kenya*. Fredrick Ouki, a Kenyan with a disability, understands the issue directly: despite a constitutional protection around employment of people with disabilities, employment numbers in the formal sector are low. According to Mizunoya and Mitra (2013, pp. 18, 41, 43), 57% of people with disabilities in Kenya are employed, with 75% of those in the informal sector. As a social entrepreneur who understood the barriers, Fredrick came up with a locally relevant solution—he created a mobile-centric solution allowing any person with a disability looking for a job or internship in Kenya to send a text message (**KAZI**) to the short code **21499**. The application is preloaded with questions that a user responds to through a text message, and this information is aggregated into a profile of the client, including the type of disability, level of qualification, type of job being sought, and so on. Employers can then see how many people with disabilities are actively looking for employment, and from how many counties, and directly source employees through Riziki Source.

Inclusive employment, driven in part by the CRPD, is finally getting more attention globally. The 2018 UK Aid Connect disability inclusion awards from the

U.K. Department for International Development (since renamed the Foreign, Commonwealth and Development Office) focus on employment for people with disabilities and the premise of global/local partnerships to increase employment (UK Aid Connect, 2018). Global initiatives in the disability sector may be of additional help in introducing further gains in the South, including disability employment initiatives such as Disability:IN (https://disabilityin.org/) and the Valuable 500 (https://www.thevaluable500.com/), as most of the participants are global companies whose hiring practices can influence norms in all of the countries in which they operate.

7.2.2 Extra Implementation Challenges in the Global South

These solutions must operate in the context of significant challenges in the Global South, from unstable governments to underdeveloped roads and institutions, all of which have an amplified effect on people with disabilities and other vulnerable populations. **Legal frameworks**, especially security and privacy, have not yet caught up with the technology (Schia, 2018). **Digital infrastructure** is still lacking in many countries, and varies greatly between urban and rural areas and among people with disabilities, making "one size fits all" concepts extremely ineffective. It also makes software development more difficult in general. In a 2018 meeting about promoting tech start-ups in Africa, an African tech entrepreneur stated, "To do Africa and tech together is hard" (Social Capital Markets, 2018). For most software development, broadband is required, yet many developing countries have a second-class internet: slow, expensive, and rarely "always on," meaning it is regularly not working, leaving developers at a standstill.

7.2.3 Sustainability

Programs that are market driven, invest in the next generation, or can become the norm within society have the most potential to take hold long term and garner financial support. To be a critical part of meaningful systems change and truly impact the lives of large numbers of people with disabilities, ICT-driven solutions must either thrive as part of long-term funded programs, such as core services supported by the national government, have market-driven revenue sources, or find other paths to scale and sustainability.

In the social sector one of the biggest challenges around incorporating ICTs is the recognition of the need to invest in the digital technology over time, through one or more of these mechanisms. In the commercial sector, software is continually updated. The primary drivers in that investment are competition and customer demand. Companies know that without continual investment they will be overtaken. Other critical drivers are external, such as operating system updates that require applications to be updated in order to take advantage of new features, or even to function, and security patches required to close vulnerabilities. Unfortunately, in the social sector, software solutions are often seen as one-time investments. Funders are far more likely to fund new development than enhancements to existing tools, leaving many social sector software applications stuck at their first revision, often leading to early abandonment, or, in the best case scenario, limited adoption, which impacts perceived value and, thus, additional investment (Chang, 2018).

Many technologists, or those social sector players who believe digital tools are somehow different than other interventions, fail to enmesh ICTs within effective programming and adoption strategies, including outreach and training. To implement and sustain new tools, innovation is often required on supporting functions as well, to drive adoption and re-invent local systems. (Fruchterman, 2016).

Inclusion at all stages of this process can pose additional challenges and even more opportunities. While some additional investment may be required, providing solutions to a broader market, locally or globally, can create opportunities for sustainability based on larger market opportunities, and may spur government support by driving compliance with policies.

7.3 Global Solutions

Global initiatives can benefit projects in the Global South and help address some of these challenges. Truly global projects have the possibility of being more inclusive, more equitable, and more sustainable than a project developed in any one region; they take a local approach to need, context, users, and implementation, while benefiting from a formal, global consensus on both technology and policy levels. (To be truly "global" means real engagement from the Global South.)

Many modern developers build their tools to be global from the start. Ideally, that means they are considering the needs of the Global North and South at once, and working with teams from both, taking into account the approaches of the Agile, user-centered, inclusive design described above. With the rise of cloud computing, global distribution is immediately available, and with distributed development teams, a global perspective is increasingly the norm.

Two important leverage points in the quest for inclusive digital tools and content are global open standards and open-source software. With these building blocks, stronger and more inclusive development is possible wherever the project is based, and development can be less expensive and more sustainable. Leveraging global policy initiatives, such as the CRPD and the Marrakesh Treaty to Facilitate Access to Published Works for Persons Who Are Blind, Visually Impaired or Otherwise Print Disabled (Marrakesh Treaty) (see Chapter 9: The Relevance of the CRPD and the Marrakesh Treaty to the Global South's Book Famine), can dramatically increase the chances for impact, as projects take advantage of a global spotlight and often a new legal framework within a given country.

The disability sector can now more fully take advantage of ICTs to help countries implement the CRPD and ensure people with disabilities are not left behind in achieving the UN Sustainable Development Goals, leveraging shared data and experiences. Locally, nationally, and globally, disabled peoples' organizations (DPOs) can gain far more ground by collaborating to collect and share data and stories that amplify the needs, struggles, and gains in the rights of individuals with disabilities around the world.

7.3.1 Global Open Standards

Technical interoperability standards define the framework for interaction between systems. Without them, products from different companies would not likely work together, and you would have no electricity grid and no internet. The world wide web and EPUB content standards from the World Wide Web Consortium are a critical example of open technical standards developed within a global community, as noted in Chapter 6: The Role of Technical Standards in Improving Digital Accessibility in Under-Resourced Regions and Communities (Brewer & Abou-Zahra).

Open standards can instantiate best practices where laws are lacking or unenforced; however, these are not mandates. When standards bodies, industry players, governments, and civil society work together toward the same end, it can have global implications.

7.3.2 Accessible Digital Books: Ending the Global Accessible Book Famine

The past few years have seen stunning progress in the push to end what the World Blind Union describes as the global book famine for people with print disabilities (Rowland, 2008). This is a perfect example of the beneficial convergence of global policy and global technology standards, and those with the most to gain are in developing countries in the Global South.

On the policy side, the Marrakesh Treaty was adopted June 27, 2013, and has been ratified by 72 member states as of the end of 2018, including the 25 within the European Union and the Philippines (World Intellectual Property Organization, 2019). This treaty is covered extensively in Chapter 9.

Spotlight: The Open Standards that Built Marrakesh

The whole concept of an interchange of accessible books is only possible because of the longstanding work on global content standards that defines how to format and read these books, and it is arguable that without this work the support for a global treaty may not have happened at all.

The Digital Accessible Information System Consortium (https://daisy.org/) is a global organization of libraries for people who are blind or otherwise print disabled, and it supports organizations that make accessible reading tools. The development of the Digital Accessible Information System standard and implementation proof points over the years has included direct involvement of groups from the Global South, including the Digital Accessible Information System (DAISY) fora of India and Latin America. This involvement makes it far more likely that the resulting solutions can be relevant globally and can end the book famine where it is most acute.

The DAISY standard has merged with the global commercial digital publishing standard EPUB, to make accessible e-books available commercially and globally and allow a broader array of reading tools and learning platforms for people needing accessibility features. The EPUB 3.x standard represents that merge (Lazar & Stein, 2017). One way to think of this is that the rules defining webpages (such as how links work, how videos play, etc.) have a wrapper of book-related rules, such as how to represent chapters, tables of contents, and book metadata like author, title, and publisher. Given this tight linkage to the web, the International Digital Publishing Forum, who owned the EPUB standard, has turned over responsibility to the core web standards body, the World Wide Web Consortium. So, for the first time in history, there is a complete connection for accessibility standards across all traditional forms of digital content and media, and it is in concert with commercial players, enabling market-driven growth along with a social safety net. (This does not fully take into account new media, such as virtual reality, or other interactive applications where new standards work is definitely required [Accessible Books Consortium, 2019].)

Having the standards in place and a treaty adopted and going through ratifications is a major achievement. However, implementation of the Marrakesh Treaty is just beginning, and scaling this good will rely heavily on Global North/South collaboration in implementing technology solutions.

Spotlight: Implementation of the Marrakesh Treaty

To deliver on the promise of Marrakesh, the Bookshare model has been expanding to allow every country or region to have its own leading-edge digital library, leveraging the Bookshare technology platform with local front end and local content in local languages. National or regional organizations deliver and administer the system and handle training and outreach efforts. These organizations may already have a digital library to interface with Bookshare's back-end application programming interfaces, common in more industrialized nations, or they may leverage page templates to easily implement one. This model allows a national or regional organization to take responsibility for the development and management of front-end technology as their capacity allows and their own delivery model indicates, otherwise they may take ownership of the website content and leverage the entire Bookshare infrastructure.

This example illustrates the power of Global North/South collaboration. In a resource-constrained field, it importantly leverages what each group does best to serve individuals across the world, while improving all the players in the process.

Spotlight: Global Certified Accessible—The Sustainable Market Solution

The next generation of this work is to make all "born digital" content "born accessible" (https://benetech.org/our-work/born-accessible/). As publishers continue to migrate their product lines to e-books, they can and should make them accessible e-books. Merely advocating for this action is not enough to move the market, and the movement must happen while new production processes are being set. Thus, a global coalition is working with industry to make "born accessible" publishing a reality.

DAISY has defined a baseline definition of what it means to be accessible based on the standards (EPUB Accessibility 1.0 specification, as of this writing), and Benetech is leading an international coalition of DAISY members, publishers, and distributors in implementing a program to certify content that meets it— Global Certified Accessible (https://benetech.org/our-work/born-accessible/ certification/). This certification provides a third-party seal of accessibility, and distributors can use it as a mechanism to sell inclusive books. While the digital publishing revolution started with publishers in the Global North, this initiative is global from the start, not only based on global standards but including digital distribution partners in India and Africa.

As with digital publishing, there are other rigorous accessibility practices and products that were developed and honed in the Global North, whose countries have benefited from traditionally stronger national legal frameworks around accessibility and a headstart on technical leadership. Underlying operating systems and platforms started that way, but they are increasingly developed and delivered by global teams with increasingly global accessibility mandates.

In addition to leveraging global World Wide Web Consortium (W3C) standards to enforce corporate guidelines, Microsoft has also developed a set of how-to instructions for developers to make their own applications accessible when built on their platforms, such as the guidelines provided for Windows at https://docs.

microsoft.com/en-us/windows/uwp/design/accessibility/developing-inclusive-windows-apps. These types of tools and standards help local developers to build accessibly, as they focus on the right features and solutions for their users. In the example above, Riziki Source could innovate a Global South solution on top of a set of global standards and accessibility innovations that came with their chosen platforms. They did not have to invent implementations for text messaging or web accessibility; they could focus on inventing a useful employment solution for people with disabilities.

7.3.3 Open-Source Software

Open-source software is software with source code that anyone can inspect, modify, and enhance. Open-source projects, products, or initiatives embrace and celebrate principles of open exchange, collaborative participation, rapid prototyping, transparency, meritocracy, and community-oriented development (Opensource.com, 2019). Such projects can potentially serve many audiences and can be supported by volunteers as well as paid developers in communities across the world. They are often offered free of charge, in full or in part, which greatly decreases the cost of implementation using these products. All of these aspects together increase the chances for more sustainable software for social good.

> **Spotlight:** Open-Source Software for Accessibility
> Open-source software plays a big role in the accessibility field. Here are just a few examples: Liblouis is a multi-language, open source, Braille translator and back translator. It is used within the American Printing House Braille Blaster (https://www.aph.org/product/brailleblaster/), the DAISY Pipeline 2 EPUB conversion tool (http://daisy.github.io/pipeline/), and Bookshare (www.bookshare.org). NVDA software (https://www.nvaccess.org/) is an open-source screen reader used by over 70,000 people worldwide. DAISY (http://daisy.org) delivers all of its products as open source. Audiolux (https://www.cymaspace.org/audiolux/) is an open source "visual sound system."

Increasingly, software products are implemented using open application programming interfaces that allow other products to plug into each other, and they are especially useful in cloud-based products. This approach is easiest when a newer product is built with the existing application programming interfaces (API) in mind, but it is also quite achievable with two existing products when they are built to the same global standards. Developers of accessible tools from Brazil to Canada have used Bookshare's API to connect existing reading tools to Bookshare, while still others have developed tools for their specific audience because this interface existed. Open APIs regularly spur such innovation.

Local projects can take advantage of such architectures to enable a "best of both worlds" combination, with products developed in one region of the world using the API as a bridge to connect to another product developed elsewhere, as described above. Openness spurs innovation and inclusion, and will continue to benefit society immensely.

7.3.4 The Power of Collective Action and Shared Data

ICTs across global and national initiatives can play a critical role in promoting, implementing, and enforcing the CRPD and achieving the goals within the UN Sustainable Development Goals (SDGs) to give more individuals with disabilities in the Global South the rights and opportunities they deserve.

The disability sector has banded together to achieve major policy gains in the international arena. However, implementation could also benefit from a more collaborative approach. Disabled peoples' organizations in the Global South are often small organizations with limited reach and resources who know their beneficiaries and members well and provide solutions every day, but who do not always have the capacity or tools for collective action to monitor or press for policy implementation. What can international and national initiatives working toward the CRPD and SDGs bring to bear to assist in these shared objectives?

They can bring lessons learned from other policy implementations around the world, and they can bring tools and proven methodologies that will have an impact at the national and global level, informing the state-level CRPD reports and local authorities. This approach must go beyond the formal monitoring and evaluation data captured in international aid projects, where third-party organizations measure the direct outputs of a project, such as quantity delivered, and also attempt to measure the impact, often via statistical means. The voices of individuals often come in the form of stories that provide insights about needs and real life on the ground, before and after new laws or interventions.

7.3.4.1 Digital Tools for Amplifying the Voices of People with Disabilities

Digital technologies remain extremely underutilized in this effort and can provide an effective means of collaboration between organizations for amplification of individual voices. By using modern digital tools, multiple organizations can combine and share data to tell a much larger story than they could on their own, mapping the data to key global policy indicators while protecting individual privacy and achieving the goals of their own organizations. Digital tools can effectively propagate globally agreed-upon measurement approaches such as the Washington Group Questions (https://www.washingtongroup-disability.com/question-sets/), which allow disaggregation by disability status, and cross-nationally comparable population-based measures of disability, and help implement important frameworks such as the draft Human Rights Indicators for the CRPD (UNHCR, 2019).

Such efforts take more than technology—they require significant coordination, trust, and capacity building. However, it is worth the effort. When a community speaks in concert, especially in the quest for human rights, it is truly more powerful than one voice on its own.

Spotlight: Data for Inclusion

Bangladesh ratified the CRPD in 2007. Are individuals with disabilities getting the access they should be getting to education and employment? How can organizations such as Bangladesh Legal Aid and Services Trust get even more compelling information to include in a civil society shadow report to the CRPD Committee and thereby inform national efforts?

The needs of DPOs in Bangladesh and Kenya have driven the initial design and implementation of a new initiative, leveraging accessible, digital technologies. The goal is to help individuals with disabilities understand their national rights around education, employment, and transition, and report when they are denied these rights so that DPOs can then collaborate to drive enforcement of existing policies or implementation of new ones. An advisory group of global disability leaders helped to ensure the program was sufficiently comprehensive, learned from earlier efforts in other countries, and had a path to impact through the CRPD and national legislation.

Implementation specifics were locally driven as part of the iterative design process. The vision is one of a powerful community approach with collaboration across organizations, including agreements around goals, key questions, or topics, and desired outcomes and data sharing. If successful, local DPOs and individuals would be able to use this platform long term, creating relevant initiatives to fight for rights in these sectors or others.

This is an example of an early-stage product (or product suite) resulting from expressed needs across the world, developed in a Global South-first mode. It has leveraged global standards and accessibility frameworks, utilizing existing open-source software modules. Product development resources were located in both the Global North and the Global South and included people with disabilities.

7.4 New Technologies, New Challenges

Unfortunately, the very rise of digital technologies, which promises so many benefits, also brings a new set of parallel problems to overcome and can further marginalize the most vulnerable. Despite very real potential for good, hype can cloud the judgment of implementers who fail to think through the unintended consequences, and issues are not clear to users who adopt them with ever-increasing speed. International bodies, aid organizations, and global companies must understand these new challenges and how they impact digital projects, as well as how to affect change through their global reach.

7.4.1 Data and Privacy

The world is just beginning to grapple with the reality of our personal data as a product in the realm of social media and other free services. The evolution of advertising and personal data as the core commercial model of the internet, rather than fee-for-service or global public goods (Nautilus Institute, 2019), has set up a race to collect and sell our information. Every click, purchase, web search, map, song, social media post, or step count adds to the treasure trove of information about us. Meanwhile, our health records, financial data, and other sensitive information are also stored on the servers of private organizations, who, as we have seen increasingly in the past several years, are not always safeguarding it as we might hope, with data breaches up over 1,300 in 2017, compared to under 200 in 2005 (Marketwatch.com, 2019). This means people may not have a choice about disclosure of a disability if the information is allowed to leak. Clearly, privacy must be a paramount concern in all digital endeavors worldwide, protected by developers and by law.

Governments, notably in Europe with the General Data Protection Regulation, are beginning to act to protect the basic online privacy of citizens (https://gdpr-info.eu/), while others are looking to leverage these capabilities for increased control. As noted in *The Age of Digital Interdependence* report (UN Secretary-General's High-Level Panel of Digital Cooperation, 2019, Chapter 3), the right to privacy is contentious, as both governments and corporations exploit "vast new possibilities for surveillance, tracking and monitoring." Convenience and access to services are far outpacing the ability of governments and civil society to protect the rights of any of its citizens, let alone the most vulnerable. Authoritarian governments could, in fact, choose to use such data to further marginalize specific citizens, including those with disabilities.

7.4.2 Artificial Intelligence: Codifying Bias

The use of massive datasets is fueling the latest generation of technology, with extremely rapid adoption, amazing possibilities, and only a preliminary understanding of the risks. While artificial intelligence has been around in some form since the 1980s, based on concepts from decades earlier, it is now in increasingly regular use. Civil society groups and academics are beginning to sound the alarm that many artificial intelligence (AI)-driven products being used, for example, in hiring and employment, are discriminating against people with disabilities, among others (Ajunwa, 2019). Product vendors and employers frequently do not even realize it (Fruchterman & Mellea, 2018). While it is true that machines themselves are not biased, machine learning relies on training data, and algorithms rely on a certain level of "productive bias" to be able to extrapolate results in a meaningful way. Data is the focus of much attention, because if the data used to train systems are not inclusive, the algorithms will be non-inclusive. It's like the old saying, "You are what you eat," and machine learning eats data (Beaumon, 2019).

Why is this of concern in the Global South right now? With so many basic concerns, isn't this a problem for another day? The use of AI in all facets of life is a very current global phenomenon. Yet, systems created in the Global North, eating a diet of Global North data, do not always work elsewhere, even while the labor that tags data for some of the biggest players in AI often resides in the Global South.

With education and employment being such a critical part of the picture for advancing the state of people with disabilities in the Global South, it is important to understand that one of the areas of clear bias in AI is within the employment realm. Hiring among global companies includes AI-driven systems that screen resumes and others that screen individuals based on their voice or their interaction with a screen. If such an individual does not speak in a typical manner, or if they do not make eye contact with a figure on a screen, these systems will screen them out, and they will not advance to any personal interaction.

Spotlight: AI Bias—Potential Solutions

It is encouraging that people are recognizing the issues. So what are some of the solutions?

- Policies are beginning to emerge. Most notably, the European Union has adopted its Ethics Guidelines for Trustworthy AI (https://ec.europa.eu/ digital-single-market/en/news/ethics-guidelines-trustworthy-ai), including a set of seven key requirements that AI systems should meet in order to be deemed trustworthy, such as privacy and data governance, transparency, diversity, non-discrimination and fairness, and accountability (https:// ec.europa.eu/digital-single-market/en/news/ethics-guidelines- trustworthy-ai). The United States has a proposed bill, the Algorithmic Accountability Act (Wyden, 2019), which would require companies to report to the Federal Trade Commission and show how the system was built and designed, what kind of data was used to train it, and what it is being used for.
- Researchers are collaborating and exploring new models. The Association for Computing Machinery is convening a cross-disciplinary conference of

computer scientists, other researchers, and legal scholars annually on Fairness, Accountability, and Transparency (https://facctconference.org/). Other collaborations, such as the work at the Alan Turing Institute, are exploring concepts such as encrypting sensitive attributes, including gender, race, and disability, in order to allow data to be examined while maintaining privacy.

- Disability-related explorations are still emerging:
  - Dr. Jutta Treviranus, Director of the Inclusive Design Research Centre at OCAD University, has proposed an approach she calls the "Lawnmower of Justice" to increase the influence of the less common data points (such as data representing people with disabilities) near the edges of a data distribution by limiting the weight we put on data from the most represented groups (i.e., mowing off the top of the Gaussian curve) (Treviranus, 2019).
  - Bespoke job boards and other organizations, such as the Trace Center, are exploring solutions to augment the data about people with disabilities out near the edge of the distribution, which have huge gaps. Such gaps mean the algorithms will still be lacking in proper representation. I have referred to this complementary solution as the "Fertilizer of Justice." Simply put— just as within the CRPD Framework, we need to collect more data by and about people with disabilities that reflect a specific area of activity, such as employment or education. Data collection and analysis can raise the voices of people with disabilities not only for today's rights but for tomorrow's even more automated world.
- As noted in the rest of this chapter, engaging people from the community in all roles is a critical path toward meaningful solutions.

7.4.2.1   From Smart Cities to Inclusive Smart Cities

The current state of bias within AI does not bode well for smart city initiatives throughout the world, which are heavily AI-driven. There is a range of definitions of a smart city, but the consensus is that smart cities utilize Internet of Things sensors, actuators, and technology to connect components across the city (i.e., it is datacentric, with public and private participants). That means that data-driven automation will be touching the daily lives of urban residents and visitors, with the purpose of improving them. That can only happen if they are included in the data and the development of the all-important algorithms, and if there is enough transparency for relevant governments and civil society organizations to monitor. While there are already smart city initiatives in middle-income countries in the Global South, from Brazil to South Africa, the potential benefits, and a whole new host of challenges, multiply across the whole of Africa, Asia, and Latin America, where more than 90% of all future urban population growth through 2050 will occur, including the population of people with disabilities (Aggarwala et al., 2018).

This challenge argues for direct engagement by innovators in the Global South. The SDGs explicitly say to leave no one behind, and this directive definitely includes major urban environments, which means that no city can truly be a *smart* city, unless it is an *inclusive* city. Cities For All (http://www.cities4all.org/) is a leading example of a public–private partnership around urban development that focuses on including the global disability community in these initiatives.

Once again, global collaboration and transparency are critical. As the UN Panel notes, any of their envisioned solutions to today's serious global, digital challenges "would need to take special steps to ensure that they are broadly representative and develop specific mechanisms to ensure equitable participation of developing countries, women and other…groups who have often been denied a voice" (UN Secretary-General's High-Level Panel of Digital Cooperation, 2019, Chapter 4).

7.5 Conclusion

Developing countries are home to 80% of the world's population of people with disabilities (UNDP, 2019), and they are the best source of new digital content, products, and services to meet their needs. Their future depends on maximizing their success in these endeavors now and building toward equality in a world increasingly driven by information technology. By working in concert with each other and allies across the globe, they can take advantage of assets that will help overcome significant challenges to ensure more accessible, inclusive, sustainable digital initiatives.

The way forward toward this important goal includes the following key lessons:

(1) People with disabilities must be explicitly included as users and employees within the tech sector and across the board. This requires a purposeful change in attitudes, including positive, empowering media attention and strong training and skill-buillding programs, such as #SeeAMillion.

(2) Tomorrow's workforce will largely come from the Global South, so it is urgent to improve education for all. Inclusive education initiatives must actually include all students—not leave the kids with disabilities in a corner to fend for themselves. Train teachers (and families) locally to support students with disabilities, and implement proven assistive technology and services, such as Bookshare, as an element within global education programs.

(3) Accessible design and development skills are a must for all of tomorrow's tech workers, North and South. That means universities must teach accessible methods, and employers, who are their customers, must demand their graduates possess these skills. Emerging technology firms

from around the world can take a cue from large global players who are participating in programs such as Teach Access.

(4) Plan for sustainability in all programs—software does not update itself to meet the ever-changing infrastructure and demands placed upon it.

(5) Global solutions can bootstrap accessible digital projects in the Global South. Global open standards provide best practices, born out of multinational collaboration, for software and content. Open-source software products are often free and can be leveraged by developers worldwide to make extensions or even new products that best suit local users, as long as they are Born Accessible and global.

(6) Collaboration in the disability sphere occurs in policy contexts, but locally relevant collaboration is still largely absent in the implementation and data that are critical to implementing policies such as the CRPD. Data scientists and technology tools for data collaboration can greatly assist in fueling collaboration to amplify the voices of people with disabilities via share data.

(7) New technologies will always bring new challenges. Technologists and advocates, in concert with policymakers, data scientists, and most importantly, the community, can and must quickly address bias in data and algorithms. As in other areas, solutions must combine smart policy, transparency, and the will to make sure the data we use is less biased than society on the whole.

References

Accessible Books Consortium. (2019). *ABC global book service*. Retrieved August 19, 2019, from http://www.accessiblebooksconsortium.org/globalbooks/en/

Aggarwala, R., Hill, K., & Muggah, R. (2018). *Smart city experts should be looking to emerging markets. Here's why*. World Economic Forum. https://www.weforum.org/agenda/2018/10/how-the-developing-world-can-kickstart-the-smart-cities-revolution/

Agile Alliance. (2001). *The agile manifesto*. http://agilemanifesto.org/

Ajunwa, I. (2019, October 8). *Beware of automated hiring* [Opinion]. The New York Times. https://www.nytimes.com/2019/10/08/opinion/ai-hiring-discrimination.html

Beaumon, B. (2019, May 2019). *Aging and disability in the 21st century: How technology can help maintain health and quality of life* [Testimony before the United States Senate Special Committee on Aging]. https://www.aging.senate.gov/hearings/aging-and-disability-in-the-21st-century-how-technology-can-help-maintain-health-and-quality-of-life

Canning, D., Raja, S., & Yazbeck, A. S. (Eds.). (2015). *Africa's demographic transition: Dividend or disaster?* Africa Development Forum series. World Bank.

doi:10.1596/978-1-4648-0489-2. License: Creative Commons Attribution CC BY 3.0 IGO.

Chang, A. M. (2018, October). *Lean impact: How to innovate for radically greater social good*. Wiley.

Deshpande Educational Trust. (2019). [Homepage]. https://www.detedu.org/

Fruchterman, J., & Mellea, J. (2018). *Expanding employment success for people with disabilities*. Benetech. Retrieved October 9, 2019 from https://benetech.org/about/ resources/expanding-employment-success-for-people-with-disabilities/

Fruchterman, J. (2016). Using data for action and for impact. *Stanford Social Innovation Review*, Summer 2016. https://ssir.org/articles/entry/using_data_for_ action_and_for_impact#

GDPR.eu. (2019). The General Data Protection Regulation (GDPR) is the toughest privacy and security law in the world. Retrieved June 2021 from https://gdpr.eu/ what-is-gdpr/

Lazar, J. & Stein M. A. (2017). *Disability, human rights and information technology*. University of Pennsylvania Press. Part III, Chapter 9, p. 153.

InABLE Accessible Technology. (2019). [Homepage]. https://inable.org/

Marketwatch.com. (2019). *How the number of data breaches is soaring—in one chart.* https://www.marketwatch.com/story/how-the-number-of-data-breaches-is-soaring- in-one-chart-2018-02-26

Mizunoya, S. & Mitra, S. (2013). Is there a disability gap in employment rates in developing countries? *World Development, Elsevier*, vol. 42(C), 18, 41, 43.

Nautilus Institute. (2019). *What are global public goods?* https://nautilus.org/gps/ applied-gps/global-public-goods/what-are-global-public-goods/

Opensource.com. (2019). *What is open source?* https://opensource.com/resources/ what-open-source/

Ossmann, R., Miesenberger, K., & Archambault, D. (2008). A computer game designed for all. In *Proceedings of International Conference on Computers Helping People with Special Needs 2008* (pp. 585–592).

Partnership on Employment and Accessible Technology (PEAT) and Teach Access. (2018). *Accessible technology skills gap report.* https://www.peatworks.org/ skillsgap/report

Putnam, C., Dahman, M., Rose, E., Cheng, J., & Bradford, G. (2016). Best practices for teaching accessibility in university classrooms: Cultivating awareness, understanding, and appreciation for diverse users. *ACM Trans. Access. Comput.* 8, 4 (Mar. 2016), 13:1–26. https://doi.org/10.1145/2831424

Rowland, W., President. (2008, April 23). *Campaign launch: Global right to read* [Press conference]. World Blind Union (WBU). Accessed October 20, 2016, from https:// view.officeapps.live.com/op/view.aspx?src=http%3A%2F%2Fwww.worldblindun- ion.org%2Fenglish%2Fgeneral-assembly%2F7th%2520General%2520Assemb ly%2520Documents%2Fprogram%2520Women%27s%2520Forum.doc/

Schia, N. (2018). The cyber frontier and digital pitfalls in the Global South. *Third World Quarterly*, *39*(5), 821–837. 10.1080/01436597.2017.1408403

Social Capital Markets (SoCap). (2018, October 25). *Beyond Silicon Valley: Frontiers of tech in emerging markets.*

South Africa: The Good News. (2018). *Africa Code Week 2018 helps hearing-impaired children in Mozambique.* https://www.sagoodnews.co.za/africa-code-week-2018-helps -hearing-impaired-children-in-mozambique/

Treviranus, J. (2019). *Sidewalk Toronto and why smarter is not better.* Data Driven Investor. October 30, 2018. Retrieved October 9, 2019, from https://medium.com/datadrivenin-vestor/sidewalk-toronto-and-why-smarter-is-not-better-b233058d01c8/

UK Aid Connect. (2018). *UK Aid Connect: Consortia grants formally announced.* https://assets.publishing.service.gov.uk/media/5c128861ed915d0b7268eeb4/ Consortia-Grants-formally-announced-April19.pdf/

United Nations DESA/Population Division. (2019). *World Population Prospects 2019: Highlights, June 2019.* https://population.un.org/wpp/

United Nations Development Programme (UNDP). (2019). *Disability-inclusive deveopment.* https://www.undp.org/content/undp/en/home/2030-agenda-for-sustainable-development/peace/governance/disability-inclusive-development.html

United Nations High Commissioner for Refugees (UNHCR). (2019). *EU and OHCHR project Bridging the Gap I.* https://www.ohchr.org/EN/Issues/Disability/Pages/ EUAndOHCHRProjectBridgingGapI.aspx

United Nations. (2017). *Disability and Employment Factsheet 1.* Retrieved December 10, 2018 from https://www.un.org/development/desa/disabilities/resources/ factsheet-on-persons-with-disabilities/disability-and-employment.html

United Nations Secretary-General's High-Level Panel of Digital Cooperation. (2019). *The age of digital interdependence* [Report]. https://www.un.org/en/pdfs/ DigitalCooperation-report-for%20web.pdf

Watermeyer B., & Goggin, G. (2019). Digital citizenship in the Global South: Cool stuff for other people? In B. Watermeyer, J. McKenzie, & L. Swartz (Eds.), *The Palgrave Handbook of Disability and Citizenship in the Global South.* Palgrave Macmillan (pp. 167–181). https://link.springer.com/chapter/10.1007/978-3-319-74675-3_12

Wentz, B., Jaeger, P. T., & Lazar, J. (2011). Retrofitting accessibility: The legal inequality of after-the-fact online access for persons with disabilities in the United States. *First Monday*, *16*(11). https://journals.uic.edu/ojs/index.php/fm/article/view/3666/3077

World Bank. (2016). *World development report 2016* (pp. 212–213). http://documents. worldbank.org/curated/en/896971468194972881/pdf/102725-PUB-Replacement-PUBLIC.pdf

World Health Organization (WHO) & World Bank. (2018). *World report on disabilities.* WHO Press.

World Intellectual Property Organization (WIPO). (2019). *WIPO-administered treaties.* https://www.wipo.int/treaties/en/ShowResults.jsp?lang=en&treaty_id=843

World Wide Web Consortium (W3C) Web Accessibility Initiative (WAI). (2019, January). *Planning and managing accessibility: Involving users in web projects for better, easier accessibility.* https://www.w3.org/WAI/planning/involving-users/#diverse

Wyden, R. (2019). A bill introduced to the 116th Congress of the United States: "Algorithmic Accountability Act of 2019." Retrieved October 9, 2019, from https://www.wyden.senate.gov/imo/media/doc/Algorithmic%20Accountability%20Act%20of%202019%20Bill%20Text.pdf

8

Digital Accessibility and Intersectional Discrimination

G. Anthony Giannoumis and Rannveig A. Skjerve
Department of Computer Science, Oslo Metropolitan University

8.1 Introduction

This chapter explores the relationship between intersectionality and the accessibility of information and communications technology (ICT) for persons with disabilities (PWDs). Research in universal design and ICT accessibility has begun to examine access and use of ICT from an intersectional perspective (Giannoumis & Stein, 2019; Skjerve et al., 2016). Nonetheless, policies, programs, and technical standards that recognize intersectionality have yet to emerge for designing technology. Because accessible ICT can benefit everyone, an intersectional perspective provides a new tool for reducing the digital divide.

This chapter asks, "What barriers do persons who hold intersectional identities experience when accessing and using ICT?" Data from 11 in-depth interviews with nine persons from the Global North and two persons from the Global South have shown three key findings. *Cost and affordability* affect an individual's access to ICT and assistive technology—that is, to technology that enables an individual to increase, maintain, or improve their abilities. *Online aggression* results in exclusion from social media and the use of the web. *Digital skills* influence the decision to adopt, access, or use ICT. Although these findings are not unique, this chapter is one of the first attempts to examine these barriers from an intersectional perspective (Gagliardone et al., 2015; Hellawell, 2001; van Deursen & van Dijk, 2014) and to consider the implications of removing these encumbrances for sustainable development.

The Committee on the Rights of Persons with Disabilities (CRPD Committee) has not yet fully considered intersectionality with respect to ICT accessibility. The United Nations (UN) Convention on the Rights of Persons with Disabilities (CRPD) obligates States Parties to ensure access to ICT for PWDs (Article 9). The Committee recognizes that intersectionality can promote substantive equality and address socioeconomic disadvantages (CRPD, 2018). Intersectionality therefore provides a useful rationale for considering ICT accessibility for PWDs who also hold other socially disadvantaged identities.

Accessible Technology and the Developing World. Michael Ashley Stein and Jonathan Lazar, Oxford University Press.
© Oxford University Press 2021. DOI: 10.1093/oso/9780198846413.003.0009

The relationship between disability and socioeconomic status provides a proxy for considering the experiences of PWDs in the Global South. The intimate linkages between poverty and disability are well-documented and longstanding challenges in the Global South (Groce et al., 2011; Moyo & Ferguson, 2009), including strong correlations between socioeconomic status, disability, and poverty in health and education (American Psychological Association, 2003; Grech, 2015; Groce et al., 2011). The UN is attempting to remediate these challenges by integrating ICT accessibility into sustainable development schemes (Habitat III, 2017; IGF, 2018; WSIS, 2003), and research is developing that considers the impact of those efforts. Contributing to that evolving academic assessment, this chapter provides early evidence on the interrelated factors that connect ICT accessibility and sustainable development in relation to three areas of social disadvantage: disability, gender, and socioeconomic status.

The chapter is structured in five sections. First, it provides the background for examining the experiences of persons with intersectional identities. Second, the chapter presents the empirical methods, data collection, and analysis utilized. Third, it analyses the data according to three overarching themes: cost and affordability, exclusion and online aggression, and learning digital skills. Fourth, the chapter summarizes the implications and good practices for accessibility and intersectional discrimination. Fifth and finally, it concludes by synthesizing the overall contribution that the analyses make in terms of recommendations for future research, policy, and practice.

8.2 Background

Digital divides separate those with access to ICT from those without (Ching et al., 2005; Datta et al., 2019; Goggin, 2016; Jackson et al., 2008; Latimer, 2001; Mossberger et al., 2003; Wong et al., 2015). Ragnedda and Muschert (2017b) argue that bridging or closing the digital divide must take into account the social barriers that limit or prevent people from accessing or using ICT. Similarly, according to Ragnedda and Muschert (2017a), the digital divide consists of access to the internet; ICT use, skills, motivation, and purposes of use; opportunities that ICT provides for participating in and influencing social life.

Digital divides correlate with many forms of social disadvantage, including race, color, sexual orientation, language, religion, political or other opinion, national, ethnic, Indigenous or social origin, property, birth, disability, and age. It is beyond the scope of this chapter to examine all forms of social disadvantage. Thus, this chapter focuses on disability, gender, and socioeconomic status in order to illustrate the overlapping forms of social disadvantage that people experience when accessing and using ICT and to provide potential explanations for those disadvantages.

In a separate area of research, Christensen and Jensen (2012) argue that intersectionality constitutes both macro-level institutions and micro-level experiences. From this perspective, intersectionality recognizes the complex and dynamic interactions between a person's social identities and systems of oppression and power (Collins & Bilge, 2016; Crenshaw, 1991). Therefore, intersectionality relates to both the personal experience of accessing and using ICT and broader systems of privilege and disadvantage. Essentially, this means that accessing and using ICT is not only related to a single identity, and the process of creating ICT can support existing and create new systems of inequality and injustice.

Notably, research on ICT has only begun to take an intersectional approach (Alper et al., 2016; Kvasny et al., 2009; Leurs & Ponzanesi, 2014; Skjerve et al., 2016). Kvasny et al., (2009) examined the relationship between gender, race, and class identity. Low-income ICT users and professionals experienced various forms of oppression in education and employment, such as being provided with low or inferior resources. In another study, Leurs and Ponzanesi (2014) examined the experiences of Moroccan Dutch migrant youths in using ICT. Intersectionality helped the authors frame the interdependent aspects of social identity and their relationship with a person's behavior online and offline. Skjerve et al., (2016) provided one of the first examples of research on intersectionality and ICT accessibility. They argued that oppressive content has led to self-exclusion. In a similar study, Alper et al., (2016) investigated the intersectional experiences of young people using new media. The authors found that intersectionality connected the use of new media to broader concerns for the rights of children.

Research on the universal design of ICT aligns with an intersectional approach to ICT accessibility. According to Article 2 of the CRPD, universal design refers to "the design of [ICT] to be usable by all people, to the greatest extent possible, without the need for adaptation or specialized design." The Convention goes on to state that universal design requires consideration of assistive technologies. Scholars have argued that universal design provides a basis for considering human diversity in the design of ICT and that a universal design approach should take into account other forms of social disadvantage, including, but not limited to, disability (Fuglerud, 2009; Giannoumis, 2016; Giannoumis & Stein, 2019; Hamraie, 2013; Lid, 2013, 2014; Winance, 2014). Universal design relates to scholarship on ICT accessibility. Research on ICT accessibility has focused on disability-related usability and accessibility barriers as well as the provision of assistive technology (Ferati & Sulejmani, 2016; Juárez-Ramírez, 2017; Lazar et al., 2010; Petrie & Kheir, 2007).

8.3 Methods, Data Collection, and Analysis

This chapter uses a qualitative approach to examine the complex and multi-dimensional barriers that are created at the nexus of disability and other disadvantaged identities (Creswell, 2003, 2007; King et al., 1994; Maxwell, 1996; Miles & Huberman, 1994; Schiellerup, 2008; Srivastava, 2009; Weiss, 1994). Researchers have used qualitative methods to investigate complex social interactions such as systems of disadvantage, social identity, and ICT access and use. This chapter uses in-depth qualitative interviews with participants from the Global South (Mongolia, India) and Global North (Poland, Italy, United States, United Kingdom, Norway, and Russia).

This chapter acknowledges that participants from the Global North are over-represented, compared to participants from the Global South. The two participants from the Global South were selected to provide useful and illustrative examples of the potential barriers experienced in the Global South. Their experiences do not represent the broader challenges experienced in the Global South. Taken together, however, the experiences of all participants provides an initial base of evidence for understanding intersectional identities in various social, economic, and cultural contexts.

Eleven interview participants were purposively selected based on whether they self-identified as having two or more oppressed identities. Table 8.1 provides a breakdown of their social identities.

Five PWDs had sensory disabilities, including blindness, partial sight, deafness, and hardness of hearing; four had physical disabilities, including paralysis and muteness; and one had a psychosocial disability. The participants were selected to provide initial, not representative, evidence on intersectionality and ICT accessibility. Participants could have been selected based on any combination of two or more social identities. This chapter primarily focuses on the experiences of PWDs, and so the participants were selected first based on their disability status and then on whether they held at least one additional socially disadvantaged identity. The one participant without a disability was selected to illustrate that barriers to ICT access and use barriers also relate to their experiences.

Table 8.1 Participants' Social Identities

Social identity	Number
Disability	10 (GC, OB, LB, CP, MG, SN, SI, FO, TA, KC)
Transnational migrant	6 (GC, OB, LB, CP, SN, KC)
Women	8 (OB, LB, CP, MG, SN, SI, TA, KC)
Low socioeconomic background	4 (GC, OB, MG, KC)
Transgender	1 (FO)

The interview results provided an initial base of evidence and point of reflection for considering the potential barriers that may exist for persons that hold intersectional identities. The results of this chapter should be interpreted with discretion, as this chapter is an exploratory study on the substantive issues related to intersectionality and ICT accessibility. The authors acknowledge that an individual's multiple social identities are complex, and as such, the analysis has limitations. Nonetheless, the data highlight the need for further investigation and the rich benefits that systematic research on the topic could provide. Further research is necessary in order to continue to develop the insights from this chapter, in particular as they pertain to considerations of accessibility and intersectionality in the Global South.

8.4 Intersectional Experiences of Technology Accessibility

Overall, the interviews showed the complex, intersectional experiences of PWDs. Three themes emerged. First, cost and affordability are critical factors for accessing and using ICT for PWDs from low socioeconomic backgrounds. Second, the exclusionary effects of online aggression can impact the opportunities of girls and women with disabilities to access and use the web. Third, acquiring digital skills can act as a barrier to accessing ICT—or an opportunity—for women with disabilities and PWDs from low socioeconomic backgrounds.

The participants' experiences of accessing and using ICT were not separated or isolated based on a single social identity. For example, OB described her experience of accessing public transportation:

> Just imagine...public buses, they don't have accessible platforms...and, some drivers, most drivers, they are not playing the audio announcing [the stops]... and I'm not [an] economically privileged person to take a taxi all the time, and plus, it's dangerous, especially for a woman who is blind. (OB)

OB suggests that disability, gender, and socioeconomic status intersect and interact with her experience of accessing public transportation in a way that a person from a high socioeconomic status, a man with a disability, or a person without a disability would not experience. From an intersectional perspective, OB illustrates the barriers that someone may experience when accessing public transportation, affording alternatives to public transportation, and ensuring their personal safety and security. Addressing one of these barriers may help bridge the digital divide for some persons, but for a woman with disability who is from a low socioeconomic area, the barriers may remain.

8.4.1 Cost and Affordability

Cost and affordability affect the ability of a person that holds an intersectional identity to access ICT. One participant, GC, from the Global South, considered their experiences using assistive technology, and in particular highlighted the cost of licensing fees for assistive technology: "I was using this [demo] software...because I could not afford to [buy it]...[it] is pretty expensive to have" (GC). GC stated that workarounds, using demonstration or "demo" versions of software programs, can provide a useful mechanism for accessing assistive technology. However, the affordability of assistive technology, in particular for persons in the Global South, can act as a barrier to accessing and using ICT.

Another participant, MG, similarly considered cost in accessing assistive technology:

I have...had the benefit of insurance when it comes to technology...in terms of fixing things, particularly disability related...for example my insurance company...originally paid for my wheelchair. (MG)

KC, also discussed the affordability of assistive technology:

My hearing loss is bordering on deafness...my education and professional background require exceptionally good hearing aids so that I can work with people in different languages and actually understand speech...so for me these expensive hearing aids are the best but there is a limit [for] insurance compensation. (KC)

From a sustainable development perspective, the relationships between accessibility, usability, and intersectionality also revolve around cost and affordability of assistive technology. Cost and affordability disproportionately affect PWDs from low socioeconomic backgrounds in ways that persons with and without disabilities from high socioeconomic ones would not experience.

Another participant, OB, highlighted the necessity of knowing English in order to access ICT.

The youth and young generation, they want to study English...[the] language center fee is very high, and language is not something that they can learn and acquire good skills...after they have completed...one course. (OB)

OB argued that cost, both in terms of the financial affordability of language courses and the time spent acquiring English, affects access to, and use of, ICT. Cost and affordability relates not only to the direct cost of purchasing goods or services, but also personal costs in terms of time and attention.

8.4.2 Exclusion and Online Aggression

Online aggression excludes women with disabilities from the web. Essentially, the participants discussed different forms of abuse, including harassment and hate speech, and the effect that these have on their use of the web. One participant, MG, provided an example:

> There was some hate speech…that resulted in somebody…hating their own life…that really escalated and made somebody…[feel] pretty worthless…having to…navigate that was hard and…pretty scary. (MG)

MG pointed out that her experience of hate speech on social media influenced her use of the web. These experiences have led to different strategies for self-exclusion.

For example, some women and trans persons hesitate to participate in online discussions due to fear of harassment. They described various strategies to avoid this. One participant, CP, discussed her approach.

> I can't use a picture…on [social media]…I notice a difference between using a picture of mine and using a random picture…[hate speech] usually happen[s] when you comment on things, because…they see you…they insult you because they don't agree with something you said, and then it's…sexual right away. (CP)

CP adopted strategies for coping with and avoiding harassment and hate speech. Specifically, CP used an anonymous avatar for her public profile. Another participant, SI, further discussed online interactions:

> You go…10,000 rounds with yourself, even if you have an opinion and you want to express it…because you don't want to deal with the backlash, if you are disabled, or disabled and queer, or if you are disabled and Muslim or woman, or man…or have an immigrant background…it doesn't help. (SI)

According to SI, interacting online is a process of self-reflection where they weigh the costs of "backlash" against the benefits of expressing "an opinion." This self-reflection process takes into account their social identities and the effect of their expression. This self-moderating behavior was also described by other participants. Some participants (SI, CP, FO, and AB) participated in online discussions regardless of the potential for harassment. However, they would be more active in online discussions if the potential for harassment was lower.

SI further described the role of digital skills in using the web. SI, who works in a civil society organization for youth empowerment, described their experience:

> I often speak to…youths…about writing [blogs and social media] posts…they write posts, but then suddenly they don't want them published…because they are scared…when it gets out in the public, what reactions they will get. (SI)

According to SI, digital skills are important for empowering disadvantaged youths. The opportunities for those individuals to use those skills to express themselves freely are limited. Digital skills are not sufficient for coping with the fear of online harassment and hate speech. Several participants went into more detail and described more severe consequences, which included suicidal thoughts and self-exclusion from the web.

8.4.3 Learning Digital Skills

Learning digital skills is a necessary precursor to using ICT. In particular, knowledge of and fluency in a common language is an advantage in learning digital skills. This was particularly the case for the participants from the Global South. One participant, GC, detailed his experience:

> I had a prime advantage…compared to other people…who were predominantly relying on…local or regional languages…for instance what is meant by a menu…I could visualize, and it came…naturally…but for my colleagues and friends…they were like "what is a menu?" (GC)

SI, mentioned previously, similarly described the need to get "help from their children, or wife, or neighbors that speak [language of resident country] better." The participants both argued that language fluency provides a mechanism for learning digital skills.

The participants did not explain the role of English fluency in learning digital skills. However, they suggested that language barriers may prevent persons living in the Global South from accessing and using ICT. This particularly affects communities whose first language is a minority language, because language support is not available for different software. Language fluency affects the use of ICT and, according to one participant, caused them to rely on others for support.

Learning digital skills also occurred through observation and peer mentoring. One participant, LB, considered her experience of learning to use a hearing aid.

> I want to observe the person…I've got photographic memory…most of time, for technology, I don't get it, I don't understand, and my first kind of reaction [is] show me, that's what makes sense. (LB)

LB argued that seeing someone show them how to use technology is more effective than verbal instructions. LB also argued that when instructions were confusing, it impacted their confidence, a particularly salient finding for older PWDs.

Other participants, all of whom identify as women, discussed their hesitancy to learn or experiment with new ICT due to their lack of self-confidence.

> For people who are hard of hearing, who have lost the easy way of doing things like communicating on the phone for example...that's another layer of feeling being excluded, and that's how you lose your confidence. (LB)

Other participants described a strategy of asking friends or family members, in particular, men, when they experienced a barrier. One participant, CP, illustrated this experience:

> I grew up thinking I didn't understand technology...because I had some male friends that are very good at technology like programming or fixing problems...I always called them when I had problems. (CP)

Other participants brought up the relationship between their and their parents' use of technology. Their mothers had lower levels of digital skills than their fathers, and in some cases did not use technology at all. They argued that their mothers had a lack of interest or need, or that they were scared or worried about using technology. According to LB, "My mom still refuses. She still will not adopt . . . technology . . . my dad on the other hand, he was the gadget man, he loves gadgets." LB suggests that girls with disabilities may experience barriers to adopting ICT due to their mothers' attitudes and behaviors toward technology.

One participant, KC, provided a parallel example regarding assistive technology:

> My mom put a hearing aid on me when I was two [years old] and thanks to this, she could start teaching me to speak...that's the most important example of how technology affected my life. (KC)

Some of the participants (KC, SI, MG, LB) argued that encouragement and support from their family or community helped encourage them to use technology. One participant, SI, described their experience playing video games:

> Everybody encouraged me to do it...we played together, I played together with my cousin...I learned to play together with my uncle too...Dad started encouraging me to, because he was so busy at work. (SI)

Participant MG, mentioned previously, described similar family support. For instance, she related that:

> [Our] first computer was...a hand-me-down a...clunky old computer, we had [computers] at school, so...my family wanted to...[have] that at home as well...to use at home, and mostly used for educational purposes. (MG)

Finally, participant OB, referred to earlier in the chapter, described the role that disabled peoples' organizations (DPOs) play in promoting digital skills. "My friend; he is also totally blind; we joined [a DPO]...he was taught several times how to use screen reader" (OB). OB argued that the DPO gave him the skills and the ongoing support to learn to use a screen reader.

8.5 Implications and Good Practices for Accessibility and Intersectional Discrimination

An intersectional perspective extends notions of ICT accessibility from a disability-specific consideration to an issue of discrimination that affects all socially disadvantaged groups. Barriers to ICT usability and accessibility affect everyone. However, these barriers can prevent PWDs from participating in society and create further disadvantages for persons who hold intersectional identities.

Creating low-cost and affordable ICT and assistive technology can help promote access to ICT for PWDs from low socioeconomic backgrounds. The Global South is often an importer of ICT from the Global North, and the imported ICT is not traditionally designed for markets in the Global South. This is beginning to change with higher internet penetration and more affordable mobile phones. However, as GC and OB suggest, for PWDs in the Global South, who often live in poverty, the affordability of ICT can act as a significant barrier to access and use. OB argued that they also must consider the cost of time and effort to learn English or other common language as a prerequisite for using ICT. For persons from low socioeconomic backgrounds who speak Indigenous languages, having access to free or low-cost English language courses can promote access to ICT.

Fighting online harassment and hate speech can improve access to the web for women with disabilities. This means creating policies and guidelines that can reduce or eliminate harassment and hate speech on social media. It also means promoting awareness about harassment and hate speech and, as MG and SI allude to, providing skills for women with disabilities to cope with online aggression. These programs must be accessible for PWDs and must specifically account for the needs and preferences of persons with cognitive and psychosocial disabilities.

In addition, creating accessible tools for filtering harmful content can also prevent exposure to online aggression, promote inclusion, and reduce self-exclusion.

Promoting digital skills can improve access and use of ICT for women with disabilities and PWDs from low socioeconomic backgrounds. Specifically, awareness and training for using and maintaining all ICT, and in particular assistive technology, can help improve adoption and use. Integrating English language courses in schools and providing training through DPOs can provide an additional mechanism for improving access and use. Designing ICT that is not dependent on one or a few dominant languages can also improve access for persons in the Global South. In addition, providing digital skills training for older persons, in particular older women, as LB, KC, SI, and MG point out, can help ensure that younger women with disabilities have family role models and support in adopting and using ICT.

The perspectives of OB and KC further illustrate the complex lived experiences of persons that hold intersectional identities. Their identities cover the disadvantages of disability, gender, socioeconomic status, race, and transnational migration and show that their participation in many aspects of social life is fluid. As OB pointed out, the personal mobility of a woman with a disability from a low socioeconomic area is affected not only by the accessibility of public transportation, but also the safety and security of public and other forms of transportation. Both OB and KC showed that their relationships with their communities and their friends and family members boosted their access and use of ICT. Their experiences show that ICT accessibility is not only an issue of a person's interactions with technology. It is an issue of power relationships and systems of social disadvantage that are embodied in broader public infrastructures, market demands, and human relations.

Universal design may provide a framework for creating ICT that acknowledges intersectional identities and remediates disadvantages. Universal design must explicitly consider the intersectional experiences of PWDs and the complex power structures and systems of oppression that frame ICT design and development. An intersectional approach to universal design can help decision makers to consider broader aspects of social disadvantage, and it provides an impetus for considering broader social issues. Decision makers that adopt universal design in policy or practice therefore must take into account cost and affordability, exclusion and online aggression, and digital skills.

The experiences of persons who hold intersectional identities are complex and situated in specific social and cultural contexts. These multi-dimensional barriers are difficult to illustrate in a way that informs decision making. Researchers can help concertize these experiences and accelerate knowledge in this field through in-depth ethnographic examinations of intersectionality and universal design. This would provide new perspectives and support the development of new models and theories that can help guide decision makers.

Policymakers should consider intersectionality in policies and programs aimed at closing the digital divide and promoting universal design. Policies and programs to close the digital divide have yet to fully consider the intersectional experiences of PWDs. Future efforts must explicitly recognize and attempt to eliminate the barriers that persons who hold intersectional identities experience. This includes providing ICT that is low cost and affordable, creating policies and programs that mitigate the exclusionary effects of online aggression, and creating programs and opportunities for learning digital skills. In addition, universal design and ICT accessibility law and policy should explicitly recognize intersectionality and prohibit intersectional forms of discrimination that limit or prevent access to ICT. These provisions should be mainstreamed in all human rights and sustainable development laws, policies, and programs.

8.6 Conclusion

This chapter has shown that cost and affordability, exclusion and online aggression, and learning digital skills shape the experiences of PWDs in a way that cuts across disability, gender, and socioeconomic status. On the one hand, these experiences show that accessing and using ICT can empower and create opportunities for PWDs to participate in society. On the other hand, the results help expose systems of power and oppression evident in the barriers that persons who hold intersectional identities experience. Future research could continue to investigate the complex lived experiences of women with disabilities and PWDs from low socioeconomic areas, and extend this research by additionally examining the role of age, language, cultural identity, race, or sexual orientation in accessing and using technology across a wider spectrum of physical, sensory, and psychosocial disabilities.

Numerous high-level UN and international development initiatives have attempted to bridge the digital divide by deploying high-speed and mobile broadband connectivity. These efforts have increased internet penetration rates in the Global South. However, cost and affordability of ICT are not typically taken into consideration. For countries in the Global South, growing income inequalities may make ICT and specifically assistive technologies more unaffordable and further out of financial reach for PWDs.

The potential for harassment and hate speech has caused self-exclusion and self-moderated behavior and has mediated ICT use. When women with disabilities are excluded from or rendered invisible in online interactions, they cannot fully participate in and influence social life. In the context of the Global South, this is particularly pertinent, as women are less likely to have access to and use ICT. Online aggression can exacerbate this disparity by further reducing a woman's interest in accessing and using ICT.

The UN has promoted the acquisition of digital skills because it is a key consideration for sustainable development and bridging the digital gender divide (EQUALS, 2018, 2019). Digital skills are acquired, in part, through family and peer relationships. For women with disabilities, the gendered aspects of those relationships can either enhance or reduce their chances to acquire digital skills. For PWDs in the Global South, learning English can support the acquisition of digital skills and use of assistive technologies.

References

Alper, M., Katz, V. S., & Clark, L. S. (2016). Researching children, intersectionality, and diversity in the digital age. *Journal of Children and Media*, *10*(1), 107–114.

American Psychological Association. (2003). Resolution on poverty and socioeconomic status. *Roeper Review*, *25*(3), 103–105.

Ching, C. C., Basham, J. D., & Jang, E. (2005). The legacy of the digital divide: Gender, socioeconomic status, and early exposure as predictors of full-spectrum technology use among young adults. *Urban Education*, *40*(4), 394–411.

Christensen, A.-D., & Jensen, S. Q. (2012). Doing intersectional analysis: Methodological implications for qualitative research. *NORA-Nordic Journal of Feminist and Gender Research*, *20*(2), 109–125.

Collins, P. H., & Bilge, S. (2016). *Intersectionality*. John Wiley & Sons.

Crenshaw, K. (1991). Mapping the margins: Intersectionality, identity politics, and violence against women of color. *Stanford Law Review*, 1241–1299.

Creswell, J. W. (2003). *Research design: Qualitative, quantitative, and mixed methods approaches*. Sage Publications.

Creswell, J. W. (2007). *Qualitative inquiry & research design: Choosing among five approaches*. Sage Publications.

CRPD Committee. (2018). *General comment No. 6: On equality and non-discrimination*. Retrieved from https://digitallibrary.un.org/record/1626976?ln=en

Datta, A., Bhatia, V., Noll, J., & Dixit, S. (2019). Bridging the digital divide: Challenges in opening the digital world to the elderly, poor, and digitally illiterate. *IEEE Consumer Electronics Magazine*, *8*(1), 78–81.

EQUALS. (2018). *Resources to help you advance digital gender equality*. Retrieved from https://www.equals.org/resources

EQUALS. (2019). *Global partnership to bridge the digital gender divide*. Retrieved from https://www.equals.org/

Ferati, M., & Sulejmani, L. (2016). *Automatic adaptation techniques to increase the web accessibility for blind users* [Paper presentation]. International Conference on Human-Computer Interaction, Toronto.

Fuglerud, K. (2009). Universal design in ICT services. In T. Vavik (Ed.), *Inclusive buildings, products & services: Challenges in universal design*. Tapir Academic Press.

Gagliardone, I., Gal, D., Alves, T., & Martinez, G. (2015). *Countering online hate speech*. United Nations Educational, Scientific and Cultural Organization (UNESCO).

Giannoumis, G. A. (2016). Framing the universal design of information and communication technology: An interdisciplinary model for research and practice. *Studies in Health Technology and Informatics, 229*, 492–505.

Giannoumis, G. A., & Stein, M. (2019). Conceptualizing universal design for the information society through a universal human rights lens. *International Human Rights Law Review, 8*(1), 38–66.

Goggin, G. (2016). Disability and digital inequalities: Rethinking digital divides with disability theory. In M. Ragnedda & G. W. Muschert (Eds.), *Theorizing digital divides* (pp. 69–80). Routledge.

Grech, S. (2015). *Disability and poverty in the Global South: Renegotiating development in Guatemala*. Palgrave Macmillan UK.

Groce, N., Kett, M., Lang, R., & Trani, J.-F. (2011). Disability and poverty: The need for a more nuanced understanding of implications for development policy and practice. *Third World Quarterly, 32*(8), 1493–1513.

Habitat III. (2017). *New urban agenda*. https://habitat3.org/the-new-urban-agenda/

Hamraie, A. (2013). Designing collective access: A feminist disability theory of universal design. *Disability Studies Quarterly, 33*(4). https://dsq-sds.org/article/view/3871/3411

Hellawell, S. (2001). *Beyond access: ICT and social inclusion* (Vol. 54). Fabian Society.

Hickel, J. (2017). *The divide: A brief guide to global inequality and its solutions*. Random House.

IGF. (2018). *Dynamic coalition on accessibility and disability*. http://www.intgovforum.org/multilingual/content/dynamic-coalition-on-accessibility-and-disability

Jackson, L. A., Zhao, Y., Kolenic III, A., Fitzgerald, H. E., Harold, R., & Von Eye, A. (2008). Race, gender, and information technology use: The new digital divide. *CyberPsychology & Behavior, 11*(4), 437–442.

Juárez-Ramírez, R. (2017). User-centered design and adaptive systems: Toward improving usability and accessibility. *Universal Access in the Information Society, 16*, 361–363.

King, G., Keohane, R. O., & Verba, S. (1994). *Designing social inquiry scientific inference in qualitative research*. Retrieved from http://site.ebrary.com/id/10035824

Kvasny, L., Trauth, E. M., & Morgan, A. J. (2009). Power relations in IT education and work: the intersectionality of gender, race, and class. *Journal of Information, Communication and Ethics in Society, 7*(2/3), 96–118.

Latimer, C. P. (2001). *The digital divide: Understanding and addressing the challenge*. New York State Forum for Information Resource Management, New York.

Lazar, J., Beavan, P., Brown, J., Coffey, D., Nolf, B., Poole, R., Turk, R., Waith, V., Wall, T., Weber, K., & Wenger, B. (2010). Investigating the accessibility of state government web sites in Maryland. In P. M. Langdon, P. J. Clarkson, & P. Robinson, *Designing inclusive interactions* (pp. 69–78). Springer.

Leurs, K., & Ponzanesi, S. (2014). Intersectionality, digital identities and migrant youths. In C. Carter, L. Steiner, & L. McLaughlin, *The Routledge companion to media and gender* (pp. 632–642). Routledge.

Lid, I. M. (2013). Developing the theoretical content in Universal Design. *Scandinavian Journal of Disability Research*, 15(3), 203–215.

Lid, I. M. (2014). Universal design and disability: An interdisciplinary perspective. *Disability and Rehabilitation*, 36(16), 1344–1349.

Maxwell, J. A. (1996). *Qualitative research design: An iterative approach*. Sage Publications.

Miles, M. B., & Huberman, A. M. (1994). *Qualitative data analysis: An expanded sourcebook*. Sage Publications.

Mossberger, K., Tolbert, C. J., & Stansbury, M. (2003). *Virtual inequality: Beyond the digital divide*. Georgetown University Press.

Moyo, D., & Ferguson, N. (2009). *Dead Aid: Why aid is not working and how there is a better way for Africa*. Farrar, Straus and Giroux.

Petrie, H., & Kheir, O. (2007). The relationship between accessibility and usability of websites. In *Proceedings of the Special Interest Group on Computer-Human Interaction 2007 Conference on Human Factors* (pp. 397–406). https://doi.org/10.1145/1240624.1240688

Ragnedda, M., & Muschert, G. W. (2017a). Introduction. In M. Ragnedda & G. W. Muschert (Eds.), *Theorizing digital divides*. Taylor & Francis.

Ragnedda, M., & Muschert, G. W. (2017b). *Theorizing digital divides*. Taylor & Francis.

Schiellerup, P. (2008). Stop making sense: the trials and tribulations of qualitative data analysis. *Area*, 40(2), 163–171. https://doi.org/10.1111/j.1475-4762.2008.00817.x

Skjerve, R., Giannoumis, G. A., & Naseem, S. (2016). An intersectional perspective on web accessibility. In P. Langdon, J. Lazar, A. Heylighen, & H. Dong, *Designing around people* (pp. 13–22). Springer.

Srivastava, P. (2009). A practical iterative framework for qualitative data analysis. *International Journal of Qualitative Methods*, 8, 76–84.

van Deursen, A. J. A. M., & van Dijk, J. A. G. M. (2014). *Digital skills: Unlocking the information society*. Palgrave Macmillan US.

Weiss, R. S. (1994). *Learning from strangers: The art and method of qualitative interview studies*. Free Press; Maxwell Macmillan Canada; Maxwell Macmillan International.

Winance, M. (2014). Universal design and the challenge of diversity: Reflections on the principles of UD, based on empirical research of people's mobility. *Disability and Rehabilitation*, 36(16), 1334–1343.

Wong, Y. C., Ho, K. M., Chen, H., Gu, D., & Zeng, Q. (2015). Digital divide challenges of children in low-income families: The case of Shanghai. *Journal of Technology in Human Services*, *33*(1), 53–71.

World Summit on the Information Society (WSIS). (2003). *Declaration of principles. Building the information society: A global challenge in the new millennium.* Document WSIS-03/GENEVA/DOC/4-E. WSIS.

LEGAL FRAMEWORKS, DESIGN APPROACHES, AND APPLICATIONS

9

The Relevance of the CRPD and the Marrakesh Treaty to the Global South's Book Famine

Paul Harpur
University of Queensland

*Michael Ashley Stein**
Harvard Law School

9.1 Introduction

Historically, the print disabled (individuals who cannot read standard print)[1] were not easily able to purchase or acquire books in accessible formats. Instead, persons with print disabilities relied upon themselves, family or friends, disabled peoples' organizations (DPOs), and charities and non-governmental organizations (Fruchterman, 2017; Harpur, 2017, pp. 4–5). By contrast, technology has made it possible for everyone to access the written word; notably, e-books are now technologically capable of being published in modes that can be consumed by the print disabled without modification (Harpur & Suzor, 2014).

Nevertheless, an inability to access information continues to profoundly impact the lives of the world's 235 million persons with print disabilities, 90% of whom live in the Global South (Accessible Books Consortium, 2019a). Indeed, this abject lack of access to books has been categorized as a "book famine" (Accessible Books Consortium, 2019b, pp. 12–16). Specifically, despite some 129 million books existing worldwide, persons with print disabilities in the Global North can obtain less than 10% of these books in accessible formats (Harpur & Suzor, 2013; Taycher, 2010). The Global South situation is even more dire. In India, for instance, a mere 0.5% of published books are converted into formats that print-disabled persons can use (Hely, 2010). Scarcity is caused because books are not technologically accessible (access failure); are technologically accessible but not available for

---

* The authors thank Jocelyn Bosse and Riannah Burns for research assistance; Maria Dolhare for Spanish translations; Janet E. Lord for Persian translations; and Betsy Beaumon for terrific feedback.

[1] We broadly define "print disabled" to include "all impairments that restrict consumption of print" and without regard to whether such impairment "is categorised as sensory, physical or cognitive."

Accessible Technology and the Developing World. Michael Ashley Stein and Jonathan Lazar, Oxford University Press.

use (market failure); or are technologically accessible but readers lack capacity to consume them (instrumental failure).[2]

National copyright laws likewise contribute to the book famine by creating legal and normative barriers to accessible e-books. Copyright protections can restrict the conversion of books into accessible formats (Danielsen et al., 2011). For example, many persons with print disabilities use screen readers similar to those found on smartphones to consume digital content (Goggin, 2009). However, the digital content itself is often designed to restrict entry and thus preserve intellectual property rights (Harpur, 2017, pp. 106–119). A prime example is ring-fenced protections whereby Amazon requires reading on a Kindle or Kindle app. Consuming e-books likewise involves the use of information and communications technologies (ICT) that, due to copyright constraints and industry practices that embrace such constraints, may be designed specifically to prevent usage (World Intellectual Property Organization Copyright Treaty, 1996; World Intellectual Property Organization Performances & Phonograms Treaty, 1996).

Divergent cultural norms regarding intellectual property can also impede the wider sharing of copyright-protected works (Okediji, 2003). Notably, the Global North and Global South have different perspectives, experiences, resources, and commitments to intellectual property laws (Okediji, 2014; Vaughan, 1996). Indeed, some States even restrict the sharing of work to instances where the exporting body has direct knowledge of who will access the specific work, to what purpose, and within what jurisdiction (Helfer et al., 2017). Along the same lines, substantial differences between Chinese and Western culture can result in vastly divergent responses to respecting and protecting copyright interests (Alford, 1995; Yu, 2015). Relatedly, many African nations view intellectual property laws and their restrictions as a modern-day avatar of colonial control and repression, and therefore eschew compliance (Okediji, 2003, 2016).

Fortunately, two ground-breaking international human rights instruments— the United Nations (UN) Convention on the Rights of Persons with Disabilities (CRPD), and the Marrakesh Treaty to Facilitate Access to Published Works for Persons Who Are Blind, Visually Impaired or Otherwise Print Disabled (Marrakesh Treaty)—have transformed the international legal framework (Lazar & Stein, 2017). Read in unison, these instruments empower persons with print disabilities to exercise their right to read by mandating that States both provide them access to and facilitate the sharing of accessible materials (Helfer et al., 2017). Yet the extent to which these instruments will genuinely precipitate such access or

---

[2] Beyond the scope of this chapter, but highly relevant, are practical and instrumental concerns. Notably, granting a print-disabled person the right to log into an e-library and read a book will not address the book famine unless that individual can access a computer or equivalent; obtain internet access; appropriate adaptive technology; secure training to use that technology; and exists within an environment that permits them to exercise the right to read, including being able to afford the e-book when fees attach.

support their cross-border exchange—either North to South, South to South, or South to North—remains to be seen. For the foreseeable future, much of the heavy lifting will continue to be borne by charities and non-governmental organizations whose mandates include the dissemination of e-books. (Parenthetically, we note the efforts of the Global Certified Accessible initiative to instigate a publisher-driven model.) This is especially true regarding the Marrakesh Treaty's potential impact in the Global South, where some 212 million persons with print disabilities live (World Health Organization & World Bank, 2018).

Part 1 of this chapter sets out the implications of the CRPD and Marrakesh Treaty on international copyright regimes brought about by rebutting their traditional exclusionary approach. Next, Part 2 considers the potential impact of the international sharing framework established by the Marrakesh Treaty. We conclude with a few thoughts as to future prospects for ending the book famine.

9.2 Transforming International Legal Norms on Reading Equality

The CRPD and Marrakesh Treaty together have transformed the established print-disabled exception to copyright from a limited usage based on fair use to a general obligation to make the written word accessible (Harpur & Suzor, 2013). The CRPD altered the longstanding status quo by prioritizing the human right to read by persons with print disabilities over the property right of copyright holders (Harpur, 2017, pp. 50–62). The Marrakesh Treaty reinforces the CRPD's obligation on States to provide exceptions to absolute copyright regimes for the print disabled, and also provides guidance on operationalizing reading equality by encouraging the formation and exchange of works in accessible e-book formats.

9.2.1 Copyright Before and After the CRPD

Copyright laws construe concepts, information, and knowledge as protected goods (Elkin-Koren & Salzberger, 2013). Specifically, they create an intangible property interest when works are created (Knight, 2013) and grant authors an exclusive right to control how their work is used (Landes & Posner, 1989). Standard books and e-books attract copyright protection under the category of authorial works, meaning literary, dramatic, artistic, and musical works (Copyright Act, 1968; Copyright Act, 1985; Copyright, Designs and Patents Act, 1988).

The global copyright regime was established in 1878 by a group of authors, and manifested in 1886 as the Berne Convention, under which the right of authors became paramount (Burger, 1998). Creative works are automatically protected without being asserted or declared, and so authors need not "register" or "apply for" a copyright in every jurisdiction. The Berne Convention likewise provides that authors of works have an exclusive right to reproduce, translate, and create derivative works (Berne Convention for the Protection of Literary and Artistic Works, 1886). In its original manifestation, the Berne Convention did not permit any variation of a work without express permission from the rights-holder.

Persons with print disabilities need to modify standard books to consume them, and many e-books are not fully usable without some form of non-authorized modification (Harpur & Suzor, 2014). Australia, Canada, the United Kingdom, and the United States have each introduced statutory licenses which authorize derivative copies of books to be made in formats accessible to the print disabled (Copyright Act, 1968; Copyright Act, 1985; Copyright, Designs and Patents Act, 1988; Copyright Act of 1976). These exceptions contain tight restrictions on who can convert the works into an accessible format and who can ultimately use the converted work. Without these statutory licenses, the print disabled and entities supporting them would need to contact rights-holders and obtain a limited license before they could start creating derivative accessible works. While these exemptions have reduced the barriers to the written word, they fall a long way short of achieving equality, because they do not require that works be rendered accessible or be shared (Harpur, 2017, pp. 124–134).

Exemptions for non-standard uses to copyright regimes in national-level laws of the Global North have enabled charities and non-governmental organizations to create millions of Braille and embossed books, as well as digital source files which enable such formats to be printed.[3] In contrast to the Global North, copyright laws in the Global South contain less support for persons with print disabilities (Li, 2014). China, Brazil, Indonesia, and Cameroon, for instance, have only a sentence in their copyright laws relating to access for persons with print disabilities to the written word (Copyright, 2002; Copyright and Neighbouring Rights, 1998; Copyright Law of the People's Republic of China, 1991). The laws in other countries in the Global South—such as Afghanistan, Albania, Cambodia, Egypt, Gambia, and Zambia—do not make reference to persons with print disabilities or contain exemptions to enable their right to read (Copyright Act, 2004; Copyright and Performance Rights [Amendment] Act, 2010; Copyright and Related Rights, 2005; Protection of Intellectual Property Rights, 2002; Law on Copyright and Related Rights, 2003; Law Supporting the Rights of Authors, Composers, Artists and Researchers, 2008).

---

[3] Among these organizations: American Printing House for the Blind, Australian Braille Authority, Blind Foundation Library (New Zealand), Braille Institute of America, Canadian Federation of the Blind, Canadian National Institute for the Blind, Cardiff Institute for the Blind (U.K.), National Council for the Blind Ireland Library, National Federation of the Blind (U.S.), National Library for the Blind (U.K.), Royal National Institute of Blind People (U.K.), Vision Australia, and the Washington Talking Book and Braille Library (U.S.).

Human rights instruments have largely supported this approach of treating human rights as secondary to copyrighted property interests. The International Covenant on Economic, Social and Cultural Rights (1966), for example, recognizes that all people have a right to take part in cultural life and to enjoy the benefits of scientific progress and its applications. Yet the right of people to access scientific, literary, or artistic production is limited by a right of individuals "to benefit from the protection of the moral and material interests resulting from" creating such works (International Covenant on Economic, Social and Cultural Rights, 1966, Article 15 [1] [c]). In consequence, and as far as the International Covenant on Economic, Social and Cultural Rights is concerned, copyright interests can limit persons' rights to access scientific, literary, or artistic productions.

In stark contrast to this traditional approach, the CRPD clearly recognizes the right of PWDs to access otherwise inaccessible information in at least three of its provisions. CRPD Article 9 requires States to ensure that PWDs can access information and communications, including ICT and systems, on an equal basis with others. This includes hardware and software that enables persons with print disabilities to consume digital content such as e-books. In addition, Article 21 requires States to "take all appropriate measures to ensure that PWDs can exercise the right to freedom of expression and opinion," utilizing "accessible formats and technologies appropriate to different kinds of disabilities in a timely manner and without additional cost," and to urge "private entities that provide services to the general public, including through the Internet, to provide information and services in accessible and usable formats for PWDs" (Article 21 [a] and [c] CRPD). Moreover, Article 30 which entitles PWDs to equal participation in culture, requires States parties "to ensure that PWDs . . . [e]njoy access to cultural materials in accessible formats" (Article 30 [1] CRPD, 2017). Perhaps most trenchantly, Article 30 (3) specifically requires States "to ensure that laws protecting intellectual property rights do not constitute an unreasonable or discriminatory barrier to access by PWDs to cultural materials."

The primacy of CRPD was in turn bolstered by the independent body of experts tasked with implementing the CRPD—the Committee on the Rights of PWDs. General Comment 2, which interprets the CRPD's mandate of accessibility, elaborates the connection between Article 30 and the Marrakesh Treaty (CRPD Committee, 2014). General Comment 2 further observes that CRPD Article 9 interacts with the Marrakesh Treaty to "ensure access to cultural material without unreasonable or discriminatory barriers for PWDs . . . especially those facing challenges accessing classic print materials" (CRPD Committee, 2014, p. 13).

In sum, the CRPD reverses the traditional hierarchy of copyright law and instead makes the right of copyright holders to exclude subservient to the right by the print disabled to access materials, including e-books. Put another way, the CRPD provides that copyright exclusions can exist up to the point where they create conflict with PWDs' access to cultural material. This is a clear, significant, and ground-breaking mandate that intellectual property laws must yield to the right of print-disabled persons to read. In effect, the CRPD created a "ceiling" on international intellectual property law in the service of disability access rights, and in direct contravention of the conventional dynamic asserted by the Berne Convention and other related agreements (Grosse Ruse-Khan, 2009; Harpur, 2017, p. 91).

9.2.2 The Marrakesh Treaty and E-Books

The Marrakesh Treaty provides operational avenues for implementing the CRPD's paradigm shift that mandates equal access to materials as a human right. Most pertinently, Articles 4, 5, and 6 of the Marrakesh Treaty require States to adopt copyright exceptions for the non-profit creation and distribution of accessible versions of works for the benefit of persons with print disabilities and, further, by permitting the cross-border exchange of certain works. Had the Marrakesh Treaty required "only" the removal of barriers to exchanging accessible works within a country, that would have been a substantial achievement; the additional implications arising from the duty to promote an international exchange of accessible works can have global impact for persons with print disabilities. Whether it is a K-12 child at school in Massachusetts or Malawi, or a worker in Australia or Afghanistan, the Marrakesh Treaty's international exchange provisions can positively impel access to works.

The Marrakesh Treaty has broad application to most copyright-protected works, even expanding the already expansive United States regime, as well as enhancing capacity in other States such as Australia and Canada. This is because Article 2 defines "works" broadly to include literary and artistic works within the meaning of Article 2 (1) of the Berne Convention, which itself includes all works, regardless of their mode of expression. Although sharing driven by the Marrakesh Treaty is likely to occur most extensively with e-books, it likewise applies to accessible format copies of books; pamphlets and other writings; lectures, addresses and sermons; choreographic works; musical compositions with or without words; cinematographic works; works of drawing, painting, architecture, sculpture, engraving, and lithography; illustrations, maps, and three-dimensional works relative to geography, topography, architecture, or science. In addition to the traditional means of consuming content, the internet is creating obligations upon States to at least consider making web-based content accessible (Bantekas et al., 2017, p. 882). The combination of the CRPD and the wide reading of works could even create obligations to caption certain web-based cultural materials

(Blanck, 2017). That the Marrakesh Treaty was adopted by World Intellectual Property Organization—an entity created to enforce the very copyright laws that have restricted access to protected works—is remarkable, illustrating the far-reaching impact of the CRPD's disability rights agenda.

Notably, the growth in e-books can revolutionize reading for the print disabled across the globe. The Worldreader project explains that "[d]igital reading has the power to reach billions" (Worldreader, 2019). Moreover, as noted by e-book pioneer Jim Fruchterman, the "economics and convenience of digital delivery of E-Books are dramatic for people without disabilities" (2017, pp. 152–153). Because e-books are cheaper to mass produce and cost practically nothing to distribute across the globe, they have become a powerful tool in efforts to enable reading in the Global South (Fruchterman, 2017). If these e-books are distributed in formats that the print disabled can consume, then reading equality will be achieved for the first time in history.

At present, however, the vast majority of books are not currently available as e-books, due to both the prevailing production model and a general lack of social acceptance. This begs the question: Can the Marrakesh Treaty precipitate the distribution of e-books in formats that are usable for the print disabled, and do so globally, including in the lesser-resourced Global South?

9.2.3 Adoption and Ratification of the Marrakesh Treaty

There is no doubt that the CRPD and the Marrakesh Treaty are causing a legally based human rights paradigm shift, and that technology is increasing the capacity for realizing reading equality (Harpur & Loudoun, 2011). But will persons with print disabilities in the Global South actually benefit?

A review of signatures, ratifications, and commencements as of August 2019 suggests that States from the Global South are more committed to the Marrakesh Treaty than States from the Global North. A substantial number of States from the Global North have signed but not yet ratified the Marrakesh Treaty. These include Austria, Belgium, Denmark, Finland, France, Germany, Greece, Holy See, Ireland, Lithuania, Luxembourg, Norway, Poland, Switzerland, and the United Kingdom (WIPO, 2019a). In contrast to this long list of States who have signed but not yet ratified the Treaty, only Australia, Canada, the European Union, the Russian Federation, and the United States have signed and ratified, and are bound by, the Marrakesh Treaty (WIPO, 2019a).

Arguably Global North States are comparatively well resourced and have mechanisms in place to share works between themselves. HathiTrust, discussed below, is an e-book library based predominately in the Global North, sharing works within the United States. Consequently, Global North States may evince less enthusiasm for ratifying the Marrakesh Treaty than under-resourced Global

South States that recognize the potential of e-books to address the book famine. This conjecture would seem to be supported by the far greater interest in the Marrakesh Treaty from the Global South, where a significant number of States from Africa,[4] Asia,[5] the Persian Gulf,[6] and Latin America[7] have signed and ratified, and are now bound by, the Marrakesh Treaty. And yet the picture is more nuanced than a North–South duality. Whereas India may have the resources, inexpensive labor, and population to generate substantial numbers of accessible works, other Global South States, such as Ghana, Guatemala, and Kenya, are differently positioned. There, insufficient resources exist to provide access to copyrighted works to the wider population, making it reasonable to assume the print disabled will not fare as well.

In addition to voluntary ascension and compliance to the Marrakesh Treaty, the Committee on the Rights of Persons with Disabilities (CRPD Committee) has a role in combating the book famine by issuing Concluding Observations regarding the intersection of the CRPD and the Marrakesh Treaty as part of its review of State CRPD compliance reports (Harpur & Stein, 2018). We therefore examined all 70 Concluding Observations handed down after the Marrakesh Treaty was adopted on June 27, 2013, through August 1, 2019. The CRPD Committee raised the Marrakesh Treaty in 64 of 70 Concluding Observations (COs), invoking the highest legal authority on international disability norms, and thereby a powerful advocacy tool. The CRPD Committee noting a State's failure to ratify the Marrakesh Treaty seems to have provoked action from Ecuador (2014), Mexico (2014), the Republic of Korea (2014), the Dominican Republic (2015), Brazil (2015), Uganda (2016), Jordan (2017), and the Republic of Moldova (2017). Within three years of CO-based criticism, each of these States ratified and became bound by the Marrakesh Treaty (WIPO, 2019a). To date, each of the 2019 COs called for Marrakesh Treaty ratification, with the exception of Saudi Arabia, which had already acceded to the Treaty (2019a). Thus, it appears the CRPD Committee has recognized the value of calling for States to ratify the Marrakesh Treaty.

Where the CRPD Committee commented on Marrakesh Treaty enforcement, those COs focused on different articles of the CRPD. The CO on the Russian Federation (2018) concerned the Article 21 right to freedom of expression and

---

[4] The African States who have signed and ratified, and are bound by the Marrakesh Treaty are: Botswana, Burkina Faso, Ghana, Kenya, Lesotho, Liberia, Malawi, Mali, Nigeria, Tunisia, and Uganda.

[5] The Asian States who have signed and ratified, and are bound by the Marrakesh Treaty are: Afghanistan, Democratic People's Republic of Korea, India, Kyrgyzstan, Mongolia, Philippines, Republic of Korea, Singapore, Sri Lanka, and Thailand.

[6] Qatar, Saudi Arabia, and the United Arab Emirates.

[7] The Latin American and Caribbean States who have signed and ratified, and are bound by the Marrakesh Treaty are: Argentina, Brazil, Bolivia, Chile, Costa Rica, Dominican Republic, Ecuador, El Salvador, Guatemala, Honduras, Mexico, Panama, Paraguay, Peru, Saint Vincent and the Grenadines, and Uruguay.

opinion and access to information, while the COs on Guatemala (2016) and Panama (2017) concerned Article 30's right to participate in cultural life, recreation, leisure, and sport. In other COs, the Marrakesh Treaty was not mentioned, even though the State party could have been complimented for their action. Providing negative and constructive detail on Marrakesh Treaty implementation is likely to sharpen advocacy and political attention; the variation in focus between COs may also hinder efforts to create international benchmarks of good practice.

9.3 Sharing Across Borders

The Marrakesh Treaty is shifting copyright norms to permit more cross-jurisdiction sharing of accessible works, and thereby benefiting the Global South. This is because the Marrakesh Treaty requires States to allow the creation of accessible works and also requires promotion of their international distribution and sharing. Four well-resourced Global North States—the U.S., U.K., Australia, and Canada—have responded to the Marrakesh Treaty and demonstrated good international sharing practices. Prior to the Marrakesh Treaty there were a number of disability-specific charity and non-governmental organization (collectively, NGO) e-book libraries which were sharing accessible works. While such bodies will continue to do the heavy lifting in combating the book famine, Marrakesh Treaty adoption has motivated domestic law reforms which in turn support international sharing and also result in direct support to the World Intellectual Property Organization to likewise facilitate sharing.

9.3.1 The Marrakesh Treaty and Sharing Accessible Works

Disabled peoples' organizations—groups of PWDs advocating for disability-based rights—have reported for decades that copyright laws hindered the international exchange of books and publications in formats accessible to the print disabled (Vezzoso, 2014). Considering the cost that is involved in creating an accessible copy, preventing the cross-border sharing of such material has represented an added and significant barrier to accessing the written word (WIPO, 2007). The Marrakesh Treaty directly addresses copyright frameworks acting as barriers to exchanging accessible works across borders. The Marrakesh Treaty requires States to permit the exchange of such works if they were lawfully created. Moreover, to create efficiencies and enhance access, it requires States to allow lawfully made accessible copies to be distributed by organizations assisting persons with print disabilities to similar organizations in other States, or directly to individuals with disabilities in those States. The Marrakesh Treaty further requires States to actively foster the cross-border exchange of accessible format copies.

To derive a sense of whether and to what extent the CRPD in combination with the Marrakesh Treaty will impact the Global South, we analyze how four well-resourced Global North States are enabling the cross-border sharing of works. Australia, Canada, and the United States have ratified and are bound by the Marrakesh Treaty, while the United Kingdom has signed but not yet ratified the Treaty. Out of these jurisdictions, only Canada had expressly permitted cross-border sharing of materials at the time of the adoption of the Marrakesh Treaty (Copyright Act, 1985). Copyright regimes in Australia and the United Kingdom remain silent on whether or not derivative works created under the statutory license can be transferred to people with print disabilities in other countries.[8] These latter two regimes permit distribution of derivative works to persons with print disabilities; however, there is no indication whether those recipients must be in the country where the derivative work was created under the statutory license. Thus, even though it might be possible to transfer files internationally, in general, many authorized entities are reluctant to share files between nations. This was one of the reasons the World Blind Union and other rights advocates successfully fought to have the capacity to share files internationally enshrined in the treaty (Helfer et al., 2017).

The tangible impact of the Marrakesh Treaty was clearly demonstrated, and the dream of reading equality became one step closer, when the U.S. Congress passed the Marrakesh Treaty Implementation Act (U.S. Marrakesh Act) in 2018. Introducing Section 121A to the U.S. Copyright Act, the Marrakesh Treaty Implementation Act powerfully permits export of accessible works from the United States to authorized entities or eligible people in States that are parties to the Marrakesh Treaty. Because the United States has the largest numbers of e-books available in the world, the U.S. Marrakesh Act will open considerable resources to the Global South.

9.3.2 WIPO, E-Book NGOs, and Libraries

The tangible impact of the Marrakesh Treaty extends beyond law reform to operational measures. The Treaty requires the World Intellectual Property Organization (WIPO) to establish information-sharing procedures to enhance cooperation between Member States by establishing a voluntary register of institutions assisting persons with print disabilities (Marrakesh Treaty, Article 9).

---

[8] Even though Australia has enacted reforms to implement the Marrakesh Treaty, international sharing was not included in the Copyright Amendment (Disability Access and Other Measures) Act 2017.

These provisions reduce the marked inefficiency of the "modernized" Berne Convention that required institutions in each country to digitize their own accessible copies of each work. The voluntary register also hastens the development of economies of scale by enabling institutions in some countries to share works directly to beneficiaries in other countries, particularly when they have common languages. Importantly, as illustrated by the analysis of the Accessible Book Consortium in the next section, this scheme is already having a strong positive effect by allowing comparatively well-funded NGOs working on e-book access in some countries to support print-disabled people in the Global South with fewer resources or institutional support, via sharing. This is so even if Bookshare is currently the only accessible library specifically funded to create materials for international libraries.

9.3.3 WIPO and the Accessible Books Consortium

The Accessible Books Consortium is a public–private partnership made possible by the Marrakesh Treaty. The Accessible Books Consortium is led by WIPO, operationalizing the international sharing of disability-accessible works via a single-exchange contract, and thereby negating the need for separate agreements with each member (ABC, 2019b). It has partnered with an impressive list of DPOs and groups representing publishers, including the: World Blind Union; DAISY Consortium; International Authors Forum; and International Federation of Library Associations and Institutions (ABC, 2019b). The Accessible Books Consortium (ABC) aims to facilitate the cross-border exchange of books in formats that the print disabled can consume. To realize this vision, ABC operates an international "library-to-library" service, where participating members can search the catalogs of participating libraries. If an individual desires a book held by another ABC-participating library, their own ABC library requests an interlibrary loan (if available in Braille, large print, or on cassette tape) or to have the digital text file provided (if it is an e-book) (ABC, 2019b). Importantly, ABC is not creating new resources, but instead is facilitating the exchange of existing resources across the globe between member libraries. This is an enormously positive development, even if it means that the burden for creating accessible works remains primarily with NGOs.

The ABC's capacity-building and coordination efforts have increased the capacity of people with print disabilities in the Global South to fulfill their human right to read (ABC, 2019b). The ABC is promoting the availability of accessible works in the Global South by significant capacity-building efforts in "developing and least-developed countries" (ABC, 2019c). These projects are providing training and technical assistance to DPOs, government agencies, and commercial publishers on how to produce works in EPUB 3, DAISY, and Braille (both electronic and embossed paper) formats. These capacity-building efforts also support the building of collaborations between all stakeholders in the process, to both enhance awareness of accessible publishing and ensure the sustainability of accessible book production in the medium and long term. As part of this strategic approach, the ABC capacity-building efforts requires those who become partners to commit to gain support in their countries for ratification and implementation of the Marrakesh Treaty (ABC, 2019c). Accordingly, the ABC capacity-building efforts are not just trying to support the production of accessible books; they are in fact helping change societies and regulatory structures to normalize reading diversity.

Although the Marrakesh Treaty's mandate to WIPO is only a couple of years old, there are already indications that ABC facilitation is having a positive impact for Global South States. In Nepal, for example, Action on Disability Rights and Development (2015) reports that they are creating capacity to leverage the ABC model of international cooperation. Similarly, the South African Library for the Blind (2019) indicates that it is supporting members to obtain accessible works through international interlibrary loans and by offering discounted assistance to participate in the U.S.-based Bookshare. Likewise, the DAISY Forum of India (2006), which has a large collection of accessible books in Hindi, Tamil, and other languages, has partnered with the ABC and Bookshare to provide members "with accessible books from all over the world," and as a result, the Marrakesh Treaty is having a practical impact on the lives of Indians with print disabilities (WIPO, 2015).

The Marrakesh Treaty also creates the possibility of sharing from the Global North to the Global South in non-English speaking languages (ABC, 2019b). While there is currently no participating library based in Spain, more resources exist in French. This is unfortunate, as there are numerous ABC-participating libraries in the Spanish-speaking Global South that could benefit from accessing Spanish resources, including: Asociación Civil Tiflonexos from Argentina (2019a), Biblioteca Central para Ciegos from Chile (2019), Instituto Nacional para Ciegos from Colombia (2019), Discapacitados Visuales IAP (2019) from Mexico, and Fundación Braille de Uruguay from Uruguay (2019). Pointedly, ABC-participating libraries from Argentina, Colombia, Mexico, and Uruguay do not mention sourcing accessible books from Spanish sources. Bookshare, however, provides print-disabled persons in these States with access to resources, some of which are in Spanish (Asociación Civil Tiflonexos, 2019b; Bookshare, 2019b; Coalición México, 2019; Fundación Braille de Uruguay, 2019; INCI, 2019; Spanish Unión Nacional de Ciegos del Uruguay, 2019). Bookshare likewise provides Spanish e-books to U.S. and other international users, although only a fraction of the total published in Spanish across the globe (Bookshare, 2019c).

The Marrakesh Treaty is having a relatively greater impact on the cross-border exchange of accessible works in the French-speaking world, and in particular on the two Francophone Global South countries who have ABC-participating libraries in them: Burkina Faso, with the Union Nationale Des Associations Burkinabé pour la Promotion des Aveugles et Malvoyants (UN-ABPAM) (2019), and Vietnam, with its Sao Mai Vocational and Assistive Technology Center for the Blind (2019). Users in these Francophone Global South countries can source accessible works from the three French ABC-participating libraries (Association Valentin Haüy, 2019; Braillenet, 2019; Groupement des Intellectuels Aveugles ou Amblyopes, 2019) and three members from Canada (Bibliothèque et Archives Nationale du Quebéc, 2019; Canadian National Institute for the Blind, 2019; Centre for Equitable Library Access, 2019), where French is the second national language and the laws, discussed below, facilitate cross-border sharing of files. The fact that there are Global North and South libraries prepared to share accessible works across borders should be promoted and resources provided to Global South libraries to be able to take advantage of these resources from all ABC-participating members.

Strikingly, the relevance of the Marrakesh Treaty to the Global South goes beyond the provision of books from any one country and region to another. The purpose of the Marrakesh Treaty is to enhance a global exchange of accessible works. The focus should not be upon region or resource, but where the work is created. Many Global South accessible works are created in languages that are also spoken widely in other Global South countries, and to a lesser extent in the Global North. For example, libraries from Argentina, Bangladesh, Sri Lanka, and Vietnam are not seeking support from libraries in the Global North and are instead producing, in local languages, books for the consumption of people with print disabilities who speak these languages (Asociación Civil Tiflonexos, 2019b; DAISY Lanka Foundation, 2018; Sao Mai Vocational and Assistive Technology Center for the Blind, 2018; Young Power in Social Action, 2018). Hence, print-disabled persons can benefit from other forms of exchanges, such as South to South (for example, from Sri Lanka to Tamil Nadu in India) or South to North (e.g., from Vietnam to the four million members of the Vietnamese diaspora, many of whom are in the Global North [Valverde, 2012]). Notably, the Argentinian Tiflolibros Association creates works and obtains them directly from publishers, and distributes these accessible works to Association members across Argentina, as well as across the Global North and South, including Andorra, Algeria, Austria, Brazil, Bolivia, Canada, Chile, Colombia, Costa Rica, Croatia, Cuba, Denmark, Ecuador, Egypt, El Salvador, Germany, Spain, and the United States (Tiflonexos, 2018).

9.3.4 Disability E-Book NGOs: Bookshare

Some of the libraries that have joined ABC are enhancing their international sharing activities through the giant global NGO, Bookshare (Fruchterman, 2017,

p. 147). In addition to scanning copies, Bookshare approaches publishing houses and obtains copies of books in digital formats and a license to use those files in particular ways (Bookshare, 2019a). The majority of e-books on Bookshare are now provided directly from publishers, and thus the Marrakesh Treaty and the U.S. Marrakesh Act do not directly impact on the provision of these e-books. Arguably, the normative pressure created by these transformational international law developments may encourage publishers to be more generous with access for the print disabled (Geisinger & Stein, 2016).

Bookshare has a long history of improving access for persons with print disabilities in the United States (Beaumon, 2011) and, due to the Marrakesh Treaty and the U.S. Marrakesh Act, now has greater capacity to operate internationally (Harpur et al., 2008). Strikingly, as part of its global empowerment of the print disabled, Bookshare has extended its operation to the developing world. It now serves over 80 countries, including States across Africa, Asia, and South America, and offers an average of 500,000 titles to people in those jurisdictions (Bookshare, 2018, 2019c). The precise numbers of books available in each country will vary according to permissions by publishers. For example, as of this writing, persons with print disabilities in Afghanistan can access over 448,000 titles, those in India some 405,000 titles, their peers in South Africa over 404,000 titles, and those in Australia about 380,000 titles (Bookshare, 2018).

Bookshare also has a powerful role in supporting South-to-South sharing. Many Global South countries are not set up to exchange works between each other and to ensure users in other Global South libraries meet the disability criteria. This is where Bookshare can provide assistance. Global South libraries can upload their works to Bookshare. Once the works are included in the Bookshare library, they become internationally available. To facilitate this, Bookshare has launched "local language interfaces in Africa and South Asia, including India, Bangladesh, Nepal, Sri Lanka, the Philippines and Kenya" (Investor Base, 2019).

Almost certainly, the Marrakesh Treaty is enabling Bookshare to offer more titles through its global sharing initiatives. As of August 2018, Bookshare offered an average of 400,000 titles to non-U.S. members, and some 680,675 titles to members living in the United States (Bookshare, 2019a). Relevant for the Global South, Bookshare publishes works, such as Braille Ready Format digital files, in 29 languages, including languages spoken across the Global South such as Afrikaans, Arabic, Bengali, Chinese, Dutch, French, Gujarati, Hindi, Mararthi, Panjabi, Portuguese, Spanish, Telugu, and Turkish (Bookshare, 2019d). It is highly probable that legal reforms driven by the Marrakesh Treaty, and in particular the U.S. Marrakesh Act, will result in a global pooling of resources. In the case of Bookshare, this means that print-disabled users in the Global South will be able to access many titles which were previously restricted to U.S. citizens, resulting in approximately 200,000 more titles being offered. This is enormously positive, but if one recalls that there are over 120 million books in the world, it means that

Bookshare effectively provides persons with print disabilities access to about 0.4% of published books.

9.3.5 The Mass-Digitization of Books: The HathiTrust

Where Bookshare has relied increasingly upon obtaining works from publishers, the HathiTrust has grown its e-library through a mass-digitization project. The HathiTrust is a not-for-profit consortium controlled by a collective of university libraries across the globe (York, 2010). It operates the HathiTrust universal library, which enables staff and students associated with partner libraries to search and download digital copies of works (HathiTrust 2018a). And, in return for access to millions of hard-copy books from partner libraries, Google has provided the results of the mass-digitization project to the HathiTrust (Sookman, 2014). As of October 10, 2018, the day the U.S. Marrakesh Treaty Implementation Act was signed into law, HathiTrust had 16,759,501 total volumes or 5,865,825,350 pages scanned and available to the print disabled, associated with member institutions primarily within the United States (HathiTrust, 2018b).

The U.S. Marrakesh Act increases the legal capacity of HathiTrust to share works with institutions internationally, something it already does to a limited degree. Of the 157 individual members (additional universities will gain access to the HathiTrust through Consortia/State Systems membership), 11 are international universities, with three from Australia (Monash University, University of Sydney, and the University of Queensland), five from Canada (McGill University, McMaster University, Queen's University, University of British Columbia, and University of Calgary), one from New Zealand (University of Auckland), one from Spain (Universidad the Complutense de Madrid), and only one from the Global South (American University of Beirut, Lebanon) (HathiTrust Digital Library, 2019). It is hoped that more institutions across the globe become members of HathiTrust and, even more ambitiously, that universities worldwide will provide access to their works. The HathiTrust is already having a tangible impact on access for many print-disabled university staff and students across the Global North. Even though the numbers of university academics and students with print disabilities in the Global South are comparatively lower than those in the Global North, guest access to libraries can be used to enable access to these vast resources. There is also the hope that the massive increase in accessible works may open up new opportunities for further education and work to those in the Global South.

9.4 Conclusion

For most of human history the written word has been inaccessible to the print disabled, or only available to them as an exception to the norm. While hard-copy

Braille has improved access, it remains expensive to create and difficult to transport (Bhickel, 1988), and thus does not provide anything approaching equality. The advent of e-books and improvement in scanning technology have created the possibility of addressing the worst excesses of the book famine. At the same time, publishing accessible material is becoming more complex due to the growth of web-based self-publishing, which can increase the numbers and type of bodies who are publishing material that is of interest to people with print disabilities.

Hence, reversing the book famine will require dramatic political/policy/technological reforms. An expedient solution would be to ensure that all future (and some number of extant) books appear in accessible formats. Such an occurrence currently appears Utopian,[9] although we note with approbation the CRPD Committee's censure of the nearly complete lack of access for individuals with intellectual and developmental disabilities (CRPD Committee, 2014). A policy approach would increase awareness regarding the economic efficiency and development synergies related to literacy for the print disabled (Blanck, 2014, pp. 29–31). Equal reading access for the group advances State-based compliance with the CRPD and Marrakesh Treaty, as well as the UN Sustainable Development Goals, in a manner that has an empirically verified impact on poverty alleviation, health, and social inclusion (UN, 2019).

The change in copyright norms which enabled the Marrakesh Treaty to be adopted should open up other discussions on what these changes might mean for information access for the print disabled in the Global South. Creative Commons licenses encourage fairness and promote human development (Suzor, 2014, p. 185). The role that Creative Commons licenses have in enhancing access to information more broadly, and the legislative reforms to support this in the United States and elsewhere, have been analyzed (Harpur, 2017, p. 137). Future research should examine how the Creative Commons movement and open-source publishing can be utilized to increase the availability of accessible works generally and in the Global South.

Finally, and far exceeding the scope of this chapter, a technological solution ultimately might employ artificial intelligence tools that improve the mechanisms for converting books.[10] Besides more efficient scanning, artificial intelligence can identify paintings and photos in books and appropriately tag them; take text presented in complex and unformatted tables and present it in more usable formats; insert page numbers where they do not correctly align between e-book and print book; and other related tasks (AbilityNet, 2019). The scope of solutions is only limited by our political will and imagination.

---

[9]   See Chapter 7 in this volume, "Digital Inclusion in the Global South."
[10] See Chapter 7 in this volume, "Digital Inclusion in the Global South."

References

AbilityNet. (2019). *5 ways AI could transform digital accessibility.* Retrieved October 5, 2019, from https://www.abilitynet.org.uk/news-blogs/5-ways-ai-could-transform -digital-accessibility

Accessible Books Consortium (ABC). (2019a). *Accessible books consortium announces winners of 2019 ABC international excellence award.* Retrieved October 5, 2019, from https://www.accessiblebooksconsortium.org/news/en/2019/news_0004.html

Accessible Books Consortium (ABC). (2019b). *ABC global book service.* Retrieved October 5, 2019, from http://www.accessiblebooksconsortium.org/globalbooks/en/

Accessible Books Consortium (ABC). (2019c). *Capacity building.* Retrieved October 5, 2019, from https://www.accessiblebooksconsortium.org/capacity_building/en/

Action on Disability Rights and Development (ADRAD). (2015). *About Us: About ADRAD.* Retrieved November 15, 2018, from http://www.adradnepal.org/

Alford, W. P. (1995). *To steal a book is an elegant offense: Intellectual property law in Chinese civilization.* Stanford University Press.

Asociación Civil Tiflonexos. (2019a). *Inicio.* Retrieved October 5, 2019, from https:// tiflonexos.org/

Asociación Civil Tiflonexos. (2019b). *¿Quiénes somos?* Retrieved October 5, 2019, from https://tiflonexos.org/libros-accesibles

Association Valentin Haüy. (2019). *Accueil.* Retrieved October 5, 2019, from https:// www.avh.asso.fr/fr

Bantekas, I., Chow, P., Karapapa, S., & Polymenopoulou, E. (2017). Article 30: Participation in cultural life, recreation, leisure, and sport. In J. Lazar & M. A. Stein, *Disability, human rights and information technology* (pp. 863–923). University of Pennsylvania Press.

Beaumon, B. (2011). *Bookshare: Making accessible materials available worldwide* [Conference session]. World Library and Information Congress: 77th IFLA General Conference and Assembly. San Juan, Puerto Rico.

Berne Convention for the Protection of Literary and Artistic Works, September 9, 1886, S. Treaty Doc. No. 99–27.

Bhickel, L. (1988). *Triumph over darkness: The life of Louis Braille.* Allen & Unwin Books.

Biblioteca Central para Ciegos (BCC). (2019). *Chile: Biblioteca Central para Ciegos (BCC).* Retrieved October 5, 2019, from http://www.bibliociegos.cl/

Bibliothèque et Archives Nationale du Quebéc (BanQ). (2019). *Accueil.* Retrieved October 5, 2019, from http://www.banq.qc.ca/

Blanck, P. (2014). e-*Quality: The struggle for web accessibility by persons with cognitive disabilities.* Cambridge University Press.

Blanck, P. (2017). eQuality: The right to the web. In P. Blanck & E. Flynn (Eds.), *Routledge handbook of disability law and human rights* (pp. 166–194). Routledge.

Bookshare. (2018, August 2). *Books by country.* Retrieved October 5, 2019, from https://www.bookshare.org/cms/get-started/how-find-books/books-country

Bookshare. (2019a). *An accessible online library for people with print disabilities.* Retrieved October 5, 2019, from https://www.bookshare.org/cms/

Bookshare. (2019b). *Bookshare without borders.* Retrieved October 5, 2019, from https://www.bookshare.org/cms/help-center/bookshare-without-borders

Bookshare. (2019c). *Join Bookshare around the world.* Retrieved October 5, 2019, from https://www.bookshare.org/cms/international

Bookshare. (2019d). *What languages are available in BRF?* Retrieved October 5, 2019, from https://www.bookshare.org/cms/help-center/what-languages-are-available-brf

Braillenet. (2019). *L'accessibilité numérique au service d'une société plus inclusive.* Retrieved November 15, 2018, from http://www.braillenet.org/

Burger, P. (1998). The Berne Convention: Its history and its key role in the future. *Journal of Law and Technology, 3*(1), 5.

Canadian National Institute for the Blind (CNIB). (2019). *CNIB Foundation.* Retrieved October 5, 2019, from http://www.cnib.ca/

Centre for Equitable Library Access (CELA). (2019). *Public library services for patrons with print disabilities: CELA.* Retrieved October 5, 2019, from https://mlb.libguides.com/printdisabilities/CELA

Coalición México. (2019). *¿Quiénes Integran Coamex?* Retrieved October 5, 2019, from http://www.coalicionmexico.org.mx/quienes-integran-coamex.html

Convention on the Rights of Persons with Disabilities (CRPD). March 30, 2007, 2515 United Nations Treaty Series 3.

Copyright Act 1968 (Cth) s 10.

Copyright Act 1968 (Cth) Part VB, Division 3.

Copyright Act of 1976 2003 (USA) § 121.

Copyright Act RSC 1985, c C-42, s 5, s 32.

Copyright Act, 2004 (Gambia).

Copyright Amendment (Disability Access and Other Measures) Act 2017 (Commonwealth of Australia).

Copyright and Neighbouring Rights (Brazil) of 19 February 1998, Law No. 9610.

Copyright and Performance Rights (Amendment) Act, 2010 (Act No. 25 of 2010) (Zambia).

Copyright and Related Rights (Albania) of 28 April 2005, Law No. 9380.

Copyright, Designs and Patents Act 1988, (United Kingdom) s 1(1), s 31, 1B.

Copyright (Indonesia) of 29 July 2002, Law No. 19.

Copyright Law of the People's Republic of China of 1 June 1991.

CRPD Committee, Concluding Observations on the Initial Report of Brazil, UN Doc CRPD/C/BRA/CO/1 (September 29, 2015). United Nations.

CRPD Committee, Concluding Observations on the Initial Report of Ecuador, UN Doc CRPD/C/ECU/CO/1 (October 26, 2014). United Nations.

CRPD Committee, Concluding Observations on the Initial Report of Jordan, UN Doc CRPD/C/JOR/CO/1 (May 15, 2017). United Nations.

CRPD Committee, Concluding Observations on the Initial Report of Mexico, UN Doc CRPD/C/MEX/CO/1 (October 26, 2014). United Nations.

CRPD Committee, Concluding Observations on the Initial Report of Panama, UN Doc CRPD/C/PAN/CO/1 (September 29, 2017). United Nations.

CRPD Committee, Concluding Observations on the Initial Report of the Republic of Korea, CRPD/C/KOR/CO/1 (October 20, 2014). United Nations.

CRPD Committee, Concluding Observations in Relation to the Initial Report of the Russian Federation, CRPD/C/RUS/CO/1 (April 9, 2018). United Nations.

CRPD Committee, Concluding Observations on the Initial Report of Saudi Arabia, CRPD/C/SAU/CO/1 (May 13, 2019). United Nations.

CRPD Committee, Concluding Observations on the Initial Report of the Dominican Republic, CRPD/C/DOM/CO/1 (May 8, 2015). United Nations.

CRPD Committee, Concluding Observations on the Initial Report of the Republic of Moldova, UN Doc CRPD/C/MDA/CO/1 (May 18, 2017). United Nations.

CRPD Committee, Concluding Observations on the Initial Report of Uganda, CRPD/C/UGA/CO/1 (May 12, 2016). United Nations.

CRPD Committee, Concluding Observations on Uruguay, CRPD/C/URY/CO/1 (September 30, 2016). United Nations.

CRPD Committee, "General Comment No 2 on Article 9: Accessibility" UN Doc CRPD/C/GC/2 (May 22, 2014). United Nations.

DAISY Forum of India. (2016). *Sugamya Pustakalaya.* Retrieved October 5, 2019, from https://library.daisyindia.org/NALP/welcomeLink.action

DAISY Lanka Foundation. (2018). Member Detail: DAISY Lanka Foundation (DLF). Retrieved October 5, 2019, from https://www.wipo.int/cooperation/en/funds_in_trust/australia_fitip/news/2017/news_0003.html

Danielsen, C., Taylor, A., & Majerus, W. (2011). Design and public policy considerations for accessible e-book readers. *Interactions, 18*(1), 67–70.

Discapacitados Visuales I.A.P. Mexico. http://www.coalicionmexico.org.mx/quienes-integran-coamex.html

Elkin-Koren, N., & Salzberger, E. M. (2013). *The law and economics of intellectual property in the digital age.* Routledge.

Fruchterman, J. (2017). E-books and human rights. In J. Lazar & M. A. Stein, *Disability, human rights and information technology* (pp. 144–145). University of Pennsylvania Press.

Fundación Braille de Uruguay. (2019). *Bienvenidos a la fundación Braille del Uruguay.* Retrieved October 5, 2019, from http://www.fbu.edu.uy/

Geisinger, A. C., & Stein, M. A. (2016). Expressive law and the Americans with Disabilities Act. *Michigan Law Review, 114*(6), 1061.

Goggin, G. (2009). Adapting the mobile phone: The iPhone and its consumption. *Continuum: Journal of Media & Cultural Studies, 23*(2), 231.

Grosse Ruse-Khan, H. (2009). Time for a paradigm shift: Exploring maximum standards in international intellectual property protection. *Trade, Law and Development, 1*(1), 56–62.

Groupement des Intellectuels Aveugles ou Amblyopes. (2019). *À la Une.* Retrieved October 5, 2019, from https://www.giaa.org/

Harpur, P. (2017). *Discrimination, copyright and equality: Opening the e-book for the print disabled.* Cambridge University Press.

Harpur, P., & Loudoun, R. (2011). The barrier of the written word: Analysing universities' policies to include students with print disabilities and calls for reforms. *Journal of Higher Education Policy and Management, 33*(2), 153.

Harpur, P., & Stein, M. A. (2018). Indigenous persons with disabilities and the convention on the rights of persons with disabilities: An identity without a home? *International Human Rights Law Review, 7*(2), 165–200.

Harpur, P., & Suzor, N. (2013). Copyright protections and disability rights: Turning the page to a new international paradigm. *University of New South Wales Law Journal, 36*(3), 745.

Harpur, P., & Suzor, N. (2014). The paradigm shift in realising the right to read: How ebook libraries are enabling in the university sector. *Disability and Society, 29*(10), 1658–1671.

Harpur, P., Suzor, N., & Thampapillai, D. (2008). Digital copyright and disability discrimination: From Braille books to Bookshare. *Media and Arts Law Review, 13*(1), 1–17.

HathiTrust. (2018a). *About HathiTrust.* Retrieved October 5, 2019, from https://www.hathitrust.org/about

HathiTrust. (2018b). *Collections.* Retrieved October 5, 2019, from http://babel.hathitrust.org/cgi/mb?colltype=updated

HathiTrust Digital Library. (2019). *Member community.* Retrieved October 5, 2019, from https://www.hathitrust.org/community

Helfer, L. R., Land, M. K., Okediji, R. L., & Reichman, J. H. (2017). *The World Blind Union guide to the Marrakesh Treaty: Facilitating access to books for print-disabled individuals.* Oxford University Press.

Hely, P. (2010). A model copyright exemption to serve the visually impaired: An alternative to the treaty proposals before WIPO. *Vanderbilt Journal of Transactional Law, 43*, 1369–1375.

Instituto Nacional para Ciegos (INCI). (2019). *Nuestros servicios.* Retrieved October 5, 2019, from http://www.inci.gov.co/

International Covenant on Economic, Social and Cultural Rights, opened for signature 19 December 1966, 993 United Nations Treaty Series 3 (entered into force January 3, 1976).

Investor Base. (2019). *Benetech announces major international expansion to help end the global book famine*. Retrieved October 5, 2019, from https://finance.yahoo.com/news/benetech-announces-major-international-expansion-120000549.html

Knight, P. (2013). *Copyright: The laws of Australia*. Thomson Reuters.

Landes, W. M., & Posner, R. A. (1989). An economic analysis of copyright law. *Journal of Legal Studies, 18*, 325–326.

Law on Copyright and Related Rights 2003 (Cambodia).

Law Supporting the Rights of Authors, Composers, Artists and Researchers 2008 (Afghanistan).

Lazar, J., & Stein, M. A. (2017). Introduction. In J. Lazar, & M. A. Stein, *Disability, human rights, and information technology* (p. 158). University of Pennsylvania Press.

Li, J. (2014). Copyright exemptions to facilitate access to published works for the print disabled—the gap between national laws and the standards required by the Marrakesh Treaty. *IIC, 45*(7), 740–742.

Okediji, R. L. (2003). The international relations of intellectual property: Narratives of developing country participation in the global intellectual property system. *Singapore Journal of International & Comparative Law, 7*(2), 315.

Okediji, R. L. (2014). Chapter 13: The role of WIPO in access to medicines. In R. C. Dreyfuss & C. Rodriguez-Garavito, *Balancing wealth and health: The battle over intellectual property and access to medicines in Latin America*. Oxford University Press.

Okediji, R. L. (2016). Government as owner of intellectual property? Considerations for public welfare in the era of big data. *Vanderbilt Journal of Entertainment and Technology Law, 18*(2), 331, 354–355.

Marrakesh Treaty Implementation Act 2018 (United States of America).

Protection of Intellectual Property Rights 2002, Law No. 82 0f 2002 (Egypt).

Sao Mai Vocational and Assistive Technology Center for the Blind. (2018). *Homepage*. Retrieved October 5, 2019, from http://www.saomaicenter.org/en

Sookman, B. (2014). The Google Book Project: Is it fair use? *Journal of the Copyright Society USA, 61*, 485.

South African Library for the Blind. (2019). *Catalogues*. Retrieved October 5, 2019, from http://salb.org.za/catalogue/

Spanish Unión Nacional de Ciegos del Uruguay. (2019). *Bienvenidos a la UNCU*. Retrieved from http://www.uncu.org.uy/

Suzor, N. (2014). Free-riding, cooperation, and peaceful revolutions in copyright. *Harvard Journal of Law and Technology, 28*, 137.

Taycher, L. (2010, August 5). *Books of the world, stand up and be counted! All 129,864,880 of you*. Retrieved from Booksearch: http://booksearch.blogspot.com.au/2010/08/books-of-world-stand-up-and-be-counted.html

Tiflonexos. (2018). *Asociación civil Tiflonexos*. Retrieved December 27, 2018, from https://tiflonexos.org/libros-accesibles

Union Nationale Des Associations Burkinabé pour la Promotion des Aveugles et Malvoyants. (2019). *Bienvenue sur le site de l'UN-A.B.P.A.M.* Retrieved October 5, 2019, from http://www.un-abpam.bf/

United Nations. (2019). *Sustainable Development Goal 10.* Retrieved October 5, 2019, from https://sustainabledevelopment.un.org/sdg10

Valverde, K. L. C. (2012). *Transnationalizing Viet Nam: Community, culture, and politics in the diaspora.* Temple University Press.

Vaughan, R. E. (1996). Defining terms in the intellectual property protection debate: Are the North and South arguing past each other when we say "property"? A Lockean, Confucian, and Islamic comparison. *Journal of International & Comparative Law*, 2(2), 307, 316.

Vezzoso, S. (2014). The Marrakesh spirit: A ghost in three steps? *IIC*, 45(7), 796, 807–808.

World Blind Union. (2018). *Marrakesh Treaty ratification and implementation campaign.* Retrieved October 5, 2019, from https://worldblindunion.org/programs/marrakesh-treaty/

World Health Organization (WHO) & World Bank. (2018). *World report on disabilities.* WHO Press.

World Intellectual Property Organization (WIPO). Copyright Treaty, December 20, 1996, 2186 United Nations Treaty Series 121.

World Intellectual Property Organization (WIPO). Performances and Phonograms Treaty, December 20, 1996, 2186 United Nations Treaty Series 203.

World Intellectual Property Organization (WIPO). Standing Committee on Copyright and Related Rights, 15th sess. (February 20, 2007). *Study on copyright limitations and exceptions for the visually impaired*, WIPO Doc SCCR/15/7 47.

World Intellectual Property Organization (WIPO). (2015). *Bringing accessible textbooks to visually impaired students in India.* Retrieved August 19, 2019, from YouTube: https://www.youtube.com/watch?v=r5K5YZ5ziaI

World Intellectual Property Organization (WIPO). (2019a, August). *Contracting Parties to the Marrakesh VIP Treaty.* Retrieved August 15, 2019, from https://www.wipo.int/treaties/en/ShowResults.jsp?treaty_id=843

Worldreader. (2019). *Creating a world where everyone can be a reader.* Retrieved from https://www.worldreader.org/

York, J. (2010, June 7). *Building a future by preserving our past: The preservation infrastructure of HathiTrust digital library.* Retrieved October 5, 2019, from HathiTrust Digital Library: https://www.hathitrust.org/documents/hathitrust-ifla-201,008.pdf

Young Power in Social Action. (2018). *Accessible book.* Retrieved October 5, 2019, from http://ypsa.org/accessible-book/

Yu, P. K. (2015). Intellectual property and Confucianism. In I. Calboli, & S. Ragavan, *Diversity in intellectual property: Identities, interests and intersections* (pp. 247–270). Cambridge University Press.

10

The Patent System, Assistive Technologies, and the Developing World

Amy L. Landers
Drexel University

10.1 Introduction

This chapter considers the positives and negatives of the patent system as an incentive for the development of new assistive technologies in the developing world. Stated broadly, the patent system began in Europe and has been developed in the Global North. Initially, the law was imposed on the Global South through colonialism. Today, the patent system has been adopted in nearly every nation through accession in the World Trade Organization's Agreement on Trade-Related Aspects of Intellectual Property Rights. This Agreement requires all member states to employ minimal levels of patent protection and enforcement in all fields of technology. Currently, the prevailing version of patent law assumes that an inventor can access sufficient funding for an infrastructure that is capable of iterating advances in complex technology. If this research results in a patent, its owner can restrict the output of products that incorporate the invention, and thereby possesses the ability to charge higher prices. These prices allow the owner to recoup their research and development costs.

At the same time, there is a critical need for broadly distributed, low-cost assistive technology, particularly throughout the Global South. According to the World Health Organization, only one in ten individuals who need assistive devices have access (World Health Organization, 2018). The reasons are several, including "a lack of awareness of this need, discrimination and stigma, a weak enabling environment, lack of political prioritization, limited investment and market barriers on the demand and supply side" (ATScale, 2019). In resource-constrained countries, the problem is worse (World Health Organization, 2018).

In the Global South, some inventions in this field meet conventional patent law standards. Others, relying on design principles that do not meet the Global North's model, do not. Often, the ingenuity of the latter group centers on the end user's physical and economic needs. Designed for wide diffusion in the resource-constrained regions, their primary goals include cost reduction and broad access.

Accessible Technology and the Developing World. Michael Ashley Stein and Jonathan Lazar, Oxford University Press.
© Oxford University Press 2021. DOI: 10.1093/oso/9780198846413.003.0011

Their designs adapt solutions to local conditions, use less expensive materials, or remove superfluous features to lower costs. Principles that drive these forms of assistive design, including appropriate technology, frugal innovation, and systems thinking, have little traction to meet the traditional patentability criteria.

The Agreement on Trade-Related Aspects of Intellectual Property Rights has an open structure that permits individual nations to set standards to advance their national priorities. Members that wish to increase patenting activity might consider whether to adopt changes to encourage domestic innovators to obtain more patents. Yet such action should be undertaken cautiously, as patents can create both positive and negative impacts in developing countries. Where nations conclude that, on balance, the patent system is not optimal for their overall economies, they might consider alternative incentive systems to foster new solutions for assistive technology.

10.2 Assistive Technologies and the Patent System

The role of the patent system in the creation of assistive technologies raises questions about appropriate incentive systems to encourage broad availability in resource-constrained regions. As originally conceived, the patent system has long been intended to incentivize investment in research (Shavell & Van Ypersele, 2001). In the Global North, there is comparatively ready access to venture and grant funding. There are numerous universities with a scientific infrastructure and a path toward spin-off commercialization. The law contemplates that patentees will sell products that incorporate their new inventions at the highest price that the market will bear. This allows patentees to recoup a return for their research investment. The system works best in economic systems with sufficient resources for customers to pay these supra-competitive prices. As one source observes, "richer countries specialize in goods requiring infrastructure, institutions, and human and physical capital" (Hildago et al., 2007).

A number of international agreements require nations to facilitate access as a worldwide priority. The United Nations (UN) promulgated the 2030 Agenda for Sustainable Development that includes the goal of "a just, equitable, tolerant, open and socially inclusive world in which the needs of the most vulnerable are met" (UN, 2015, Para. 8). In this same vein, the UN Convention on the Rights of Persons with Disabilities (CRPD) reaffirms the rights and freedoms of those with disabilities, including the importance of access to the full range of social, educational, informational, and cultural environment at a minimal cost (Preamble [v] + Article 9 CRPD). Consistent with the purpose of the CRPD, the World Intellectual Property Organization's Marrakesh Treaty to Facilitate Access to Published Works for Persons Who Are Blind, Visually Impaired or Otherwise Print Disabled (Marrakesh Treaty) is targeted to alleviate the problem that only a small fraction of works protected by copyright are accessible (Helfer et al., 2017).

Together, the UN Sustainable Development Goals (SDGs), the CRPD, and the Marrakesh Treaty consider market interventions to ensure human dignity, education, and informational choice. These treaties encourage wide distribution to ensure that those in need can obtain access. In contrast, the point of patent law is the grant of market exclusivity to provide the inventor with the ability to narrow output and thereby command higher profits. As one stakeholder described, "inventors and investors often require the protection of a period of exclusivity to assume the substantial risk of investing the significant resources needed to bring a product to the public" (The State of Patent Eligibility in America, Part II., 2019). Moreover, the patent seeks to advance a technological state of the art in a technology-neutral manner. The legal standards do not vary based on end-user needs. Public policy concerns, including widely shared support for the human rights of the disabled, are not part of the patentability inquiry.

In general, assistive technology in developing nations derives from smaller producers that are "manufacturing products of varying degrees of quality and frequently at a limited price range" (MacLachlan et al., 2018). Consistent with this statement, the clear majority of patenting activity for assistive technology occurs in developed countries, including the United States, Japan, Germany, South Korea, and France (Solomon, 2015). By comparison, developing countries have generated "very small" patent numbers (ibid.). Although patent counts are not always reliable indicators of innovative activity, these statistics raise concerns.

There are reasons to encourage patenting beyond conferring the ability to charge supra-competitive pricing. Patented inventions can be sold for a low or no cost and selectively enforced. Therefore, it is conceivable that a patent owner can permit manufacturing and sales for non-profit or low-profit uses. Moreover, patents can be an important bridge to distribute products globally, including for buyers in wealthier regions. In some contexts, patents perform a signaling function that the work performed under a grant is valuable. Further, the promise that a particular project is or will be patented can be used to attract collaborators or start-up funding. Patents can be an important component of technology transfer agreements and an effective means to control sub-licensors, manufacturers, and distributors. For example, a patent owner can authorize the sale and distribution of products subject to low price restrictions or a requirement that distribution is channeled to non-profit entities. These examples demonstrate how patents can add value beyond their conventional purpose.

10.3 Bridging the Gap

In the absence of robust financial support, how does new assistive technology creation occur in resource-constrained regions? In some cases, *collaboration* with entities that provide financial assistance or technological infrastructure works. For example, two students at IIT-Delhi developed the DotBook, a computing tablet with a refreshable Braille display (IIT-Delhi, n.d.). IIT-Delhi's Department of Science & Technology provided development support (ibid.). As the project moved to the prototype stage, a grant from the United Kingdom's Wellcome Trust provided financial support (ibid.). Together, these organizations enabled the creation of low-cost devices, leading to communicative freedom for those who cannot afford full-featured alternatives.

Beyond this, *market shaping* can be strategically employed (MacLachlan et al., 2018). Generally, the free market sets supply and demand on different axes. In contrast, market shaping considers different interventions and relationships between supply and demand to shape the market to achieve larger goals (United States AID, 2014). Market shaping aided the development of the Orbit Reader, an inexpensive refreshable Braille display that was far more affordable than full-featured versions. The design was driven by Transforming Braille, an organization that coordinated several international collaborators, including the United States National Federation of the Blind and India's Sightsavers, for testing and feedback (Transforming Braille, 2016). In the Global North, a full-featured Braille reader costs thousands of dollars, which is "out of range for many users" (ibid.). Instead of a technological breakthrough, the Orbit Reader's design process prioritized simplification, usability, and a dramatically lower cost. Its design placed the end users' physical and economic needs at the center. In another example of market shaping, the Kilimanjaro Blind Trust relied on donations to give away Orbit Readers throughout Africa (Kaluyu, 2018). The Orbit Reader grants transformative access to many, even though "[i]t is not the sleekest, most elegant, smallest, or most feature-laden device available" (Transforming Braille, 2016). As this example shows, the market shaping used to create and distribute the Orbit Reader is distinct from the profit and price competition that are the mainstays of the free market.

Systems thinking is another form of creativity that leads to low-cost, effective solutions. (Hossain & Yeasin, 2013). This design methodology considers the end user within their entire environment to determine the advantages and limitations of particular solutions. The MBraille app is one example. Applying systems thinking, this product was developed by a team of researchers from Malaysia and Bangladesh to help students conveniently learn Braille (Nahar et al., 2015). This smartphone app, which runs on widely available Android phones, is designed for use in schools in remote regions of Bangladesh (ibid.). These schools do not have adequately trained specialists or learning tools for students with visual challenges (ibid.). The MBraille's design was driven by discussions with educators who described problems that their students encountered with then-existing solutions, which typically required a slate and stylus (ibid.). The MBraille designers considered a range of factors that considered the students' learning environment. These included keeping the cost low, the solution small and lightweight, and developing a design that permitted students to work independently. This MBraille design fits easily on a student's smartphone and is designed to provide haptic and sound feedback without requiring oversight by a trained professional. The MBraille designers considered the student as part of an entire system, including their needs and the limitations of their environment. The end result, which applied a form of systems thinking, was built to maximize usability for this population.

Another example relies on the *appropriate technology* framework. This considers the resources, culture, and economic conditions of the community in which the technology will be used (Hazeltine & Bull, 2003). Innovation in this space relies on locally sourced or readily obtainable materials. A related concept is a *frugal innovation*. Such inventions find that the Global North's design methodology is "too expensive and resource-consuming,...lacks flexibility, and...is elitist and insular" (Radjou et al., 2012). Frugal innovation meets unmet needs by simplifying the design to reduce the cost and complexity of a product. By paring down technology to its core features, broad diffusion of a solution at a low cost can be achieved (Brem & Wolfram, 2014). Such forms of creativity make the highest use of scarce resources to meet a low price constraint.

One example of these can be seen in the example of a simplified Braille printer called the Braiset, developed by two inventors working in India (Rai, 2017). The printer, which won a social innovation award, modified a traditional design by reducing the number of print solenoids to one, to reduce cost (Sharma, 2019). The United States counterpart to this application states that, among other things, the utility of the invention is "to provide an improved Braille printer which will be available in the market at a very economical cost affordable by common man, simple-structured and complete-functioned" (Rai, 2019). This device was the product of frugal innovation and appropriate technology. Its significance was its simplicity and affordability. The removal of solenoids permitted a lower price to facilitate wide distribution to those who are unable to afford a full-featured device. As their social innovation award suggests, the idea of these inventors has the possibility of opening communicative freedom consistent with the UN Sustainable Development Agenda, the CRPD, and the Marrakesh Treaty.

The Braiset printer was the subject of patent applications, including one under the Patent Cooperation Treaty, which claimed various aspects of the Braille printing head (Rai, 2017). One might guess that this claim's focus on features of the single solenoid head was targeted to obtain protection for the most tangible and machine-like aspects of the device in an attempt to satisfy the traditional patent-ability requirements. Yet the International Preliminary Report on Patentability found that this printhead technology already existed separately inUnited States, Japanese, and German patent applications (Chatal, 2019). In other words, the ingenious aspect of this award-winning Braille printer, which was modeled from existing Braille printers, was its ability to strip its parts down to those which are most essential. Unfortunately, the prevailing patentability standards led this patentee to seek protection for the feature which remained (here, the printer head), which was part of the prior art. Yet a claim for the award-winning aspects of this device—which feature simplicity and low cost—would be unlikely to fare better. Notably, *removing* technology is not something that the patent system rewards. Indeed, inventors in the Global North are urged to include "additional proprietary features into their innovations" to avoid this problem (Kotha et al., 2014).

10.4 Patents as a Creative Incentive

The five patentability requirements are used to channel claimed solutions into two categories—those that will obtain patent protection and those that will not. The five requirements are novelty, nonobviousness/inventive step, patentable subject matter, utility/industrial application and adequate disclosure. These are used throughout the world and required by the World Trade Organization's Agreement on Trade-Related Aspects of Intellectual Property Rights (TRIPS Agreement). They represent the minimum standards that nations must use to filter patent applications from those that warrant a patent and those that do not. Each country has the flexibility to more specifically define these standards. In other words, the precise contours of the applications of the five requirements are subject to each member's local variations.

10.4.1 The Patent System and Technological Creativity

The patentability standards reward technological solutions that evidence specific types of advances. One example of an assistive device that fits comfortably within these standards is a refreshable Braille device that relies on an electrorheological fluid to push various pins into place so that the text can be read by vision-impaired persons (Ahn et al., 2005). The patent's owner, the Electronics and Telecommunication Research Institute in South Korea, has been engaged in research since the 1970s and has earned over $150 million in patent royalties over the years (Electronics and Telecommunication Research Institute, 2014). The

Institute has a well-established scientific infrastructure, experience, and sustained funding.

A close examination of the prevailing patentability standards demonstrates some reasons that some innovations fail to meet them, despite being highly productive for end users in resource-constrained regions. Patent law operationalizes its assumptions by ensuring that the patentability standards are difficult to meet, including the inventive step requirement. This requirement (called "nonobviousness" in the United States) requires that the proposed claim meets a high bar that excludes solutions that are likely to arise in routine product design and development. It holds that if those within the field would have inevitably generated the invention, there is no need for the government to grant a patent. It is thought that granting patents that would arise in the normal course "might stifle, rather than promote, the progress of useful arts" (*KSR v. Teleflex*, 2007). In the Global North, the inventive step depends primarily on whether the invention's success was predictable to the person of ordinary skill in the art. The inventive step requirement is most likely to be met for claims that overcome unpredictability, add to the technological state of knowledge, or address a problem that was previously thought to be unsolvable. By imposing a high bar, the inventive step requirement seeks to push scientific and technological research forward.

The inventive step requirement places a hypothetical person of ordinary skill in the relevant art at the center of the inventive step inquiry. Under this standard, this hypothetical person is imagined to be tasked to recreate the claimed invention based on the state of the art at the time the patent application was filed. According to the leading U.S. case, *KSR v. Teleflex*, the hypothetical inventor who merely combines knowledge in the art with ordinary creativity or common sense to recreate the invention fails the standard. As *KSR* explains "[w]hen there is a design need or market pressure to solve a problem...[and i]f this leads to the anticipated success, it is likely the product not of innovation but ordinary skill and common sense" (*KSR v. Teleflex*, 2007). Solutions are not sufficient if they are the product of routine testing in a predictable art. Claims that are suggested by the published literature or driven by market considerations are less likely to be considered obvious (ibid.). For inventions that are a combination of pre-existing elements, a showing that the person had a motivation or reason to combine them will render the resulting claim invalid as obvious. The standard is objective. In other words, a claimed invention must be new to the world and not obvious in light of prior art, whether the particular inventor knew about the prior art or not.

Some assistive technology solutions from emerging countries appear to meet the Global North's patentability standards. For example, a team of researchers in India filed a patent application for a Braille tablet that relies on haptics in the form of electronic pulses (Rai, 2018). This eliminates mechanical actuators, liquids, or springs that would otherwise be needed to push up dots to form raised Braille lettering. By eliminating mechanical components and substituting haptic signals, the device uses less energy, is more compact, and has fewer moving parts. Although this application has yet to be examined, assuming a lack of close prior art, this type of technological shift from moving actuators to entirely different means represents the type of technological advance that the patent system is likely to reward.

As the Braiset example illustrates, designs that simplify, substitute, or remove features to meet cost constraints are less likely to meet the Global North's standards. Further, patent law has not resolved whether user input might be viewed as prior art, a form of common-sense inquiry, or the type of "market pressure" that weighs against a finding of patentability. The Solar Ear hearing aid provides an example. This device was developed by Harry Weinstein, who began work on this problem in a rural town in Botswana (Humphrys, 2013). There, he learned that batteries were expensive and difficult to acquire in that region. He founded Godisa Technologies, which hired South African engineers to work with hearing-impaired students (ibid.). They designed the Solar Ear, which uses a battery that can be recharged in a device that uses solar power. This low-cost device relies on the appropriate technology concept of "soft design," which uses materials that can be easily modified in the field (Hazeltine, 2003). The Solar Ear is self-fitting and uses moldable materials to conform to the user's ear to eliminate the need for expert customization and fitting (Humphrys, 2013).

Although Weinstein did not attempt to patent the Solar Ear, his ability to do so is not assured. The Solar Ear is a combination of pre-existing technology, including a solar recharging unit that is modified to fit hearing aid batteries and the use of moldable materials for the hearing aid earpiece. Weinstein's travel to Botswana to investigate is irrelevant because the patentability assessment is objective. If the prohibitive cost and inaccessibility of hearing aid batteries are known to the world, this information may constitute a reason to combine well-known solar recharging technology with hearing aid batteries, thereby rendering that aspect of the Solar Ear unpatentable. Similarly, if Botswana's shortage of trained professionals to fit hearing aids fitters is known, applying principles of soft design may require nothing more than ordinary creativity. In that case, Weinstein and his team have developed an innovation that may not fare well under traditional patent law standards.

10.5 Redefining Technological Creativity

To the extent that nations wish to encourage patenting by domestic inventors, they might expand the types of creativity rewarded by the system. The TRIPS

Agreement requires that member nations implement the inventive step require-ment. It includes flexibility that permits nations to more precisely define the standards at the domestic level. However, the Agreement does not authorize nations to grant patents that fail to claim a solution with an inventive step (TRIPS, 1994, Article 27). Nonetheless, there is sufficient flexibility to allow some crafting at a local level.

How might a modified standard look? This question is not easy to answer. As the U.S. Supreme Court writes, the system conceives of invention as an additive process that "advances, once part of our shared knowledge, define a new thresh-old from which innovation starts once more" (*KSR v. Teleflex*, 2007). The patent-ability requirements are geared to reward additions to the state of art, and only rarely simplifications. Early patent cases teach that the substitution of one mater-ial for another, without more, is not sufficient to demonstrate entitlement to a patent. In the early *Hotchkiss v. Greenwood* (1851), the Supreme Court held that a patent claim to a doorknob that used wood rather than porcelain was not a patentably distinct invention, even though the substitution resulted in a device that was both "better and cheaper." Today, substitutions of less expensive parts are not considered patentable unless they have a technologically synergistic result (*KSR v. Teleflex*, 2007). A modification made solely to lower an item's cost weighs against a finding of patentability (*DyStar v. Patrick*, 2006). Under these rules, an invention like the Braiset printer—which removed solenoids to keep its cost down—faces an uphill battle.

The patent application filed by the IIT-Delhi for the DotBook, a tactile com-puter tablet, has a similar problem under current standards. The DotBook is com-pact and displays a single line of text at a time. By displaying a single line of text, the cost of the device is kept low. Its ergonomic design comes packaged with soft-ware, including email, a calculator, web browser, and easy integration with third-party applications. According to the patentee, this design was the product of "multiple user trials with many of the smallest needs and preferences being taken care of based on user feedback" (IIT-Delhi, n.d.). It offers a ready-to-go, easily portable solution for about $800, where comparable versions designed in the Global North cost at least three times as much.

Consistent with prevalent patent law standards, IIT-Delhi filed a patent appli-cation for the DotBook as "A System for Generating Refreshable Tactile Text and Graphics," claiming the various pieces of hardware that make up the tablet (Rao et al., 2013). This has an open-ended claim, with a Braille dot configuration, elec-tronic parts that drive the system, a filtered cooling unit, and shape memory alloy wires to push the Braille dots into position, among other things (ibid.). The International Preliminary Report on Patentability, drafted by an Australian exam-iner, applied the traditional patentability standards and found that the claims were likely to fail the inventive step requirement (Odeh, 2013). Specifically, the examiner located an earlier U.S. patent that disclosed a similar invention but was missing one feature—that is, a filter for the cooling unit (ibid.). Regarding this filter, the examiner found that a filter was well known in the art and therefore incapable of justifying compliance with the inventive step requirement (ibid.). This analysis seems correct under the current patentability standards, which requires that claims provide more than the routine addition of something already known in the art to deliver predictable results—here, a filter to keep dust out of a cooling unit.

Undoubtedly, the availability of an $800 DotBook will provide opportunities for those with visual challenges within India over foreign-made solutions that cost thousands more. There are different ways the current standard might be modified to capture this product. This requires patentability standards to encompass simplification and cost reduction as indicative of inventive activity. The purpose of the invention, including affirming human rights and free expression, might be weighed in the calculus. As standards shift, the types of claims that applicants draft will shift accordingly. In this case, the DotBook patentee might submit claims directed to the overall device, including its ergonomic design and carefully selected feature set to meet the communication needs of those with visual challenges. Relevant here, numerous legal documents establish the importance of distributing assistive technologies more broadly throughout the world. Also, to meet the TRIPS Agreement's obligations, such claims must provide a sufficient level of distinction from the prior art.

10.6 Patents: Impediments to Innovation?

Expanding the range of creativity eligible for patent protection is one way to grant more patents. Yet opening the door by expanding patentability standards to capture more forms of creativity might have negative implications for developing economies. There are reasons to be cautious.

The TRIPS Agreement allows entities in wealthier regions to obtain patents worldwide, including in the Global South, under the same standards as domestic inventors. These patents can impose barriers for those who wish to make, improve, or adapt to the same invention there. Given that the Global North might generate patents under a broader creativity standard faster than resource-constrained regions, this raises the possibility that those in the Global South might be prevented from experimentation and learning-by-doing by the potential assertion of patents obtained by these foreign entities. Instead, creating an alternative reward system might be the best way to incentivize and foster new solutions.

Those in the Global North have the patent advantage. Many firms and universities have longstanding expertise and established structures for filing patent applications worldwide. In contrast, some economically vibrant Sub-Saharan nations have filed only a handful (Graff, 2020). In South Africa, fewer than 20% of all patent applications are filed by residents and only 3% derive from other developing and emerging economies (ibid.). Of the remainder, the majority are filed by owners in Europe and the U.S. (ibid.). This suggests that foreign patent holders can control a significant share of the innovation within a developing country. A nation cannot skew its patentability standards to benefit domestic inventors (TRIPS, 1994, Article 4). The same standards must be applied to all.

In the past, weak patent protection has enabled nations to build up experience and infrastructure in a technology field. Both Switzerland and India have enjoyed periods when chemical inventions had either limited or no patent protection. During those times, both countries built up know-how and manufacturing capacity. Today, Switzerland is home to some of the world's finest pharmaceutical innovators. By the early 1990s, firms in India satisfied up to 80% of its domestic drug needs (Lanjouw, 1998). This phenomenon has been discussed in economic literature, which has observed that "[o]nce the country's technological ability is above a certain threshold, the imitation effect is dominated by the innovation effect, and the optimal protection of IPRs increases with the levels of development" (Chen, 2005). Strengthening patent protection too soon forecloses a developing nation's ability to experiment in a technology field without the risk of liability for infringement.

Of course, patents are not monolithic impediments to innovation. The TRIPS Agreement allows for certain limited exceptions for patent protection (TRIPS, 1994, Article 30). In this vein, Brazil has authorized an exception to its patent law that permits non-commercial uses of another's patent (Garrison, 2006). However, this exclusion is untested and unique in the patent system as a whole. Moreover, a patent's power is limited to the geographic region in which it is granted. Further, patents expire 20 years after their application dates. Exploiting these limitations allows the use of a patented invention without liability. An example from Ethiopia illustrates how a patent's limits make room for innovation. There, a patent search for hearing aid technology located several that were rich with helpful information that taught how these devices could be recreated (Foray, 2009). These patents, which had only been granted in jurisdictions outside the region, allowed researchers to learn from the information disclosed in these patents without concern that the results of these efforts would be infringing. Because these patents were legally effective only in nations outside this region, the researchers were able to use their disclosures to create an effective hearing aid without the possibility of being sued for patent infringement. The power of a patent is limited. In assessing the efficacy of the patent system, the strengths, limitations, and potential harm must be weighed.

10.7 Alternatives to the Patent System

It is rational to conclude that the patent system may not be the optimal approach toward the distribution of an invention into the most hands at the lowest cost. One alternative is to encourage innovators to provide their solutions as open source (Chao, 2015). For example, Tesla allows its patented technology to be used by any car company that, in good faith, uses these patents to build cars designed to reduce carbon emissions (Musk, 2014). Similarly, the maker of the Solar Ear hearing aid maintains its design as open source. In a statement that echoes the principles of market shaping, Weinstein explains:

> we wanted people to copy us. In fact, if somebody ends up producing a cheaper, better version of the Solar Ear and uses their distribution channel to get their products to more children with hearing loss, we will have attained our object-ive—even if it puts us out of business. (Humphreys, 2013)

There is a critical difference between Tesla's approach and Weinstein's. Tesla owns patents and imposes limitations on the use of these rights (Musk, 2014). These conditions allow Tesla to control uses outside the scope of the specified scope of acceptable uses until Tesla's patents expire. Weinstein chose not to patent the Solar Ear, so that all are free from claims of patent infringement without restriction. Notably, Tesla's approach allows control over downstream uses and distribution.

Governments can offer innovation incentives outside the patent system. Commercialization grants have become important ways for ideas to develop into prototypes, which lead to funding for the creation of a version that can be widely distributed. Such grants can favor domestic creators and institutions. Such a system can be part of a broader system of invention prizes (Shavell & Van Ypersele, 2001). Such prizes can be given to those who decide to obtain or, alternatively, forgo patent protection. Others might reward open-source approaches. The high social value of these devices, coupled with the need to distribute this technology broadly at low cost, makes this ideal for open source. In this way, the government can do the most to increase the quality of life for its citizens while encouraging a variety of inventors to work toward solutions in their areas of expertise.

10.8 Conclusion

Through a number of mechanisms, developing countries have been responsible for a number of assistive technology innovations. Many of these are the product of creativity that operates within significant financial and other constraints. These include limited research budgets, as well as the fact that finished products must be sold at low price points. Some examples, including the Braiset Braille printer, represent an appropriate technology that meets a number of needs. Nonetheless, the current patent system—which was conceived and evolved in the Global North—was not developed to reward this type of device. In other words, it can be expected that solutions that remove features to meet price constraints may be difficult to patent.

Nations wishing to encourage these types of innovations can choose from a variety of tools. This includes modifying the patent system to capture different types of creative works that are not currently rewarded. Nonetheless, this approach should be considered carefully. In part, because the Global North enjoys an asymmetric advantage in the patent system, developing countries might wish to wait to expand patent protection until they reach a significant stage of technological development. Regardless, there are other available ways to reward innovation that can both reward inventors and assist them in bringing their products to market. These include grants and prize systems, particularly those that encourage open-source disclosure of the solution's technical details.

References

Ahn, S. D., Kang, Y., Kim, C. A., Oh, J. Y., You, I. K., Kim, G. H., Baek, K. H., & Suh, K. S. (2005). *U.S. Patent No. 8,047,849*. U.S. Patent and Trademark Office.

ATScale. (2019, February). Strategy overview. ATScale2030. Retrieved June 28, 2020, from https://atscale2030.org/strategy

Brem, A. & Wolfram, P. (2014). Research and development from the bottom up: Introduction of terminologies for new product development in emerging markets. *Journal of Innovation and Entrepreneurship*, 3(1), 9.

Chatal, C. (2019, February 5). International preliminary report on patentability. *Patent Application No. PCT/IB2017/0010099*. The International Bureau of the World Intellectual Property Organization (WIPO).

Chao, T. E. & Mody, G. N. (2015). The impact of intellectual property regulation on global medical technology innovation. *BMJ Innovations*, 1(2), 49–50.

Chen, Y. & Puttitanun, T. (2005). Intellectual property rights and innovation in developing countries. *Journal of Development Economics*, 78(2), 474–493.

DyStar Textilfarben GmbH & Co. Deutschland KG v. C.H. Patrick Co., 464 F.3d 1356, 1368–1369 (Fed. Cir. 2006).

Electronics and Telecommunication Research Institute (ETRI). (2014, April 2). *ETRI tops U.S. patent evaluation for 3rd year*. Retrieved June 28, 2020, from https://www.etri.re.kr/engcon/sub1/sub1_08.etri

Foray, D. (2009). *Technology transfer in the TRIPS age*. International Centre for Trade and Sustainable Development (ICTSD).

Garrison, C. (2006). *Exceptions to patent rights in developing countries* (No. 17). International Centre for Trade and Sustainable Development (ICTSD).

Grace, C. (2004, June). *The effect of changing intellectual property on pharmaceutical industry prospects in India and China: Considerations for access to medicines*. World Health Organization (WHO). https://www.who.int/hiv/amds/Grace2China.pdf?ua=1

Graff, G. D. & Pardey, P. G. (2020). Inventions and patenting in Africa: Empirical trends from 1970 to 2010. *The Journal of World Intellectual Property*, 23(1–2), 40–64.

Hazeltine, B. & Bull, C. (Eds.). (2003). *Field guide to appropriate technology*. Elsevier Science & Technology.

Hildago, C., Klinger, B., Barabási, A., & Hausmann, R. (2007, July 27). The product space conditions the development of nations. *Science*, 317(5837), 482–487.

Hossain, G. & Yeasin, M. (2013). *Assistive thinking: Assistive technology design in disability management*. In Proceedings of the 1st International Conference on Technology Helping People With Special Needs (ICTHP).

Hotchkiss v. Greenwood, 52 U.S. 248 (1851).

Humphreys, G. (2013). Technology transfer aids hearing. *Bulletin of the World Health Organization*, 91(7), 471. World Health Organization (WHO).

IIT-Delhi. (n.d.). *DotBook: IIT Delhi launches India's first braille laptop. Indian Institute of Technology Delhi*. Retrieved June 28, 2020, from http://old.iitd.ac.in/content/dotbook-iit-delhi-launches-india's-first-braille-laptop-1

Kaluyu, A. (2018, November 21). *Serving learners with visual impairment for 10 years*. Kilimanjaro Blind Trust Africa. https://kilimanjaroblindtrust.org/2018/11/21/serving-learners-with-visual-impairment-for-10-years-and-beyond/

Kotha, R., Kim, P., & Alexy, O. (2014, November). *Turn your science into a business*. Harvard Business Review. https://hbr.org/2014/11/turn-your-science-into-a-business

KSR Intern. Co. v. Teleflex Inc., 550 U.S. 398 (2007).

Lanjouw, J. (1998). *The Introduction of pharmaceutical product patents in India: "Heartless exploitation of the poor and suffering?"* (No. w6366). National Bureau of Economic Research.

MacLachlan, M., McVeigh, J., Cooke, M., Ferri, D., Holloway, C., Austin, V., & Javadi, D. (2018). Intersections between systems thinking and market shaping for assistive technology: The SMART (systems-market for assistive and related technologies) thinking matrix. *International Journal of Environmental Research and Public Health*, 15(12), 2627.

Musk, E. (2014, June 12). *All our patent are belong to you*. Tesla. https://www.tesla.com/blog/all-our-patent-are-belong-you

Nahar, L., Jaafar, A., Ahamed, E., & Kaish, A. (2015). Design of a Braille learning application for visually impaired students in Bangladesh. *Assistive Technology*, 27(3), 172–182.

Odeh, N. (2013, August 5). Written opinion of the International Searching Authority (ISA). *Patent Application No. PCT/IN2013/000347*. The International Bureau of the World Intellectual Property Organization (WIPO).

Radjou, N., Prabhu, J., & Ahuja, S. (2012). *Jugaad innovation: Think frugal, be flexible, generate breakthrough growth.* John Wiley & Sons.

Rai, K. & Khurana, A. (2018, September 17). *Patent Application No. IN 2018/11035031.* India Patent Office.

Rai, K. (2017, August 2). *WIPO Patent Application No. WO2018025082.* The International Bureau of the World Intellectual Property Organization (WIPO).

Rai, K. (2019). *U.S. Patent Pub. No. US2019/016083.* U.S. Patent and Trademark Office.

Rao, P.V., Jain, P., Singhal, A., Balakrishnan, M., & Gupta, K. (2013, May 31) *WIPO Patent Application No. PCT/IN2013/000347.* The International Bureau of World Intellectual Property Organization (WIPO).

Sharma, S. (2019, May 12). Ex-PEC students get platinum award from Infosys Foundation for affordable Braille printer. *Times of India.* https://timesofindia.india-times.com/city/chandigarh/ex-pec-students-get-platinum-award-from-infosys-foundation-for-affordable-braille-printer/articleshow/69296442.cms

Shavell, S. & Van Ypersele, T. (2001). Rewards versus intellectual property rights. *The Journal of Law and Economics, 44*(2), 525–547.

Solomon, N. and Bhandari, P. (2015). *Patent landscape report on assistive devices and technologies for visually and hearing impaired persons.* World Intellectual Property Organization (WIPO).

The State of Patent Eligibility in America, Part II. 116th Cong. 1–2 (testimony of Rick Brandon) (2019, June 5). https://www.judiciary.senate.gov/imo/media/doc/Brandon%20Testimony1.pdf

The Transforming Braille Group LLC. (2016, April 4). *Orbit Reader 20 details. Transforming Braille.* http://transformingbraille.org/orbit-reader-20-details/

TRIPS: Agreement on Trade-Related Aspects of Intellectual Property Rights, April 15, 1994, Marrakesh Agreement Establishing the World Trade Organization, Annex 1C, 1869 United Nations Treaty Series 299, 33 ILM 1197.

United Nations (UN). (2006, December 13). International Convention of the Rights of Persons with Disabilities and its Optional Protocol (CRPD), United Nations Doc. A/61/106.

United Nations (UN). (2015). 2030 Agenda for Sustainable Development. A/RES/70/1.

U.S. AID. (2014, Fall). *Healthy markets for global health: A market shaping primer.* U.S. Aid Center for Innovation and Impact Bureau for Global Health. https://www.usaid.gov/sites/default/files/documents/1864/healthymarkets_primer_updated_2019.pdf

World Health Organization. (2018, May 18). *Assistive technology.* https://www.who.int/news-room/fact-sheets/detail/assistive-technology

World Trade Organization Agreement: Marrakesh Agreement Establishing the World Trade Organization. April 15, 1994. 1867 United Nations Treaty Series 154, 33 I.L.M. 1144.

11

Accessible ICT as a Ray of Hope for Disability Rights in Pakistan

Marva Khan, Muhammad Atif Sheikh, and Abia Akram
Lahore University of Management Sciences & Special Talent Exchange
Program, Pakistan

11.1 Introduction

Created after the partition of the Indian Sub-Continent in 1947, Pakistan is a republic comprising 220 million people, with a parliamentary form of government. A post-colonial state with a religious identity, Pakistan's legal system is a hybrid of common law and customary law, infused with Islamic law (Constitution of Islamic Republic of Pakistan, 1973, Articles 2, 2A).

Persons with disabilities (PWDs) currently constitute Pakistan's largest minority, with an annual population growth rate of 2.65% (Ahmed, 2005; British Council Policy Brief, 2014). This increase is linked to terrorism causing significant yearly disabilities (Saeed et al., 2014), inadequate healthcare (including the failure to eradicate polio), malnutrition, and better data collection. Estimates on the number of PWDs vary wildly from 3.3 to 27 million, depending on whether one relies on government statistics using unsuitable and jaded metrics restricted to a handful of poorly categorized disabilities, namely, "blind, deaf, physically handicapped or mentally retarded" (Pakistan Bureau of Statistics, 2017; Section 2 Ordinance 1981), or whether they come from independent sources (British Council Policy Brief, 2014).

Also notable is a report produced by the Research, Evaluation and Monitoring Unit of the British Council (2014), which highlighted that "1.4 million children with disabilities do not have access to either inclusive or special schools." This exclusion is not only problematic for disabled children and their families but is also causing significant losses to the state economy. The estimated GDP loss in Pakistan due to excluding PWDs from the workspace in 2014 was around $11.9–15.4 billion, or 4.9–6.6% of GDP (British Council Economist Intelligence Unit, 2014).

Sociopolitical turmoil marks Pakistan's history, which includes repeated military coups and the abrogation of two constitutions. Consequently, the government has predominantly ignored fundamental rights (Choudhury, 1956;

Accessible Technology and the Developing World. Michael Ashley Stein and Jonathan Lazar, Oxford University Press.
© Oxford University Press 2021. DOI: 10.1093/oso/9780198846413.003.0012

Singhal, 1962). Pakistan signed and ratified the United Nations (UN) Convention on the Rights of Persons with Disabilities (CRPD) in 2008 and 2011, respectively (UN HC Indicators), but did not ratify the Optional Protocol. Recent years, however, have witnessed an increase in efforts for disability-related inclusion, mainly due to advancements in accessible information and communications technology (ICT). This progress was primarily due to innovations by the private sector, which are increasingly encouraged and disseminated by the government of Pakistan. This positive trend by the executive branch of government arising via public–private partnerships for increasing disability inclusion is counterbalanced by the overall state infrastructure, which remains mostly inaccessible (Dr. S. Shahid, personal communication, September 27, 2019). Thus, despite positive efforts to increase inclusion and accessibility by private entities, both on their own and in partnership with the state, we argue that advances in ICT must be coupled with a dramatic makeover of state infrastructure and societal attitudes relating to PWDs.

This chapter delves into the evolution of disability rights and assistive technology in Pakistan. The chapter highlights various state initiatives toward increasing inclusivity for disabled persons by all three branches of the state, namely, the executive, legislature, and the judiciary. Next, the chapter traces how contemporary advancements in technology have bolstered inclusivity, with a considerable role played by private educational and corporate institutions, and non-governmental organizations. It then highlights the various avenues of public–private partnerships that have opened up because of ICT advancements. The chapter culminates with recommendations for the state and civil society to further tap into channels such as corporate social responsibility and provide a nurturing ground to the private sector to further bolster the ongoing efforts.

11.2 Evolution of Legal Status of PWDs in Pakistan

This section details the development of the attitudes of the legislature, executive, and judiciary, and posits that the executive branch has comparatively adopted more measures.

11.2.1 Evolution of Legal Safeguards for PWDs

In 1981, the then Chief Martial Law Administrator and President General Zia-ul-Haq passed the Disabled Persons (Employment and Rehabilitation) Ordinance XL (Ordinance), which only recognized four categories of disabilities, namely, "physically impaired," "visually impaired," "hearing impaired," and "mental retardation." The Ordinance mandated a 1% quota (subsequently increased) of vacancies in governmental departments for employing persons with disabilities (PWDs). The state failed to make infrastructural changes to make offices more accessible. People resisted the quota (*Ijaz Ahmad v. GM*, 2001). When the fraction of reserved seats calculated as per the Ordinance was less than one, seats for PWDs were recurrently rounded down to zero seats, leaving no reserved seats for the disabled (*Haroon-ur-Rashid v. Balochistan*, 2013, Para. 5). There was also a push to allocate reserved seats to able candidates, which was thwarted by the Lahore High Court, the highest provincial court of Punjab province (*Muhammad Goraya v. PPSC*, 1990).

Societal biases also led to misinterpretation of the Ordinance. Sometimes qualified PWDs were not considered for open merit seats and reserved seats, even when they did not seek special accommodations (*Ijaz Ahmad v. GM; Tariq v. Punjab*, 1996; *Muhammad Shafi v. MD*, 1998). Consequently, PWDs' only recourse was to go to court. Lack of inclusion of PWDs in pivotal governmental positions also reflects these biases. The first disabled Deputy District Public Prosecutor, judge, and top-tier groups of federal bureaucracy were appointed in 2018 (*Aamir Tufail v. Punjab;* Cheema, 2018; *Muhammad v. Chairman FPSC*, 2016).

The first reported judgment that expands local jurisprudence to include the right to equality and freedom of profession (Constitution, Articles 25 and 18) came in 2013. The HC stated that PWDs fall under "reasonable classification" for differential (affirmative) treatment (2013 PLC (CS) 81 BHC, Para. 11). *Hafiz v. Punjab* (2017) was a landmark judgment that assessed the inclusion of PWDs within the purview of the CRPD for the first time. It acknowledged the fluidity of disabilities, as opposed to the four generic and rigid categories recognized by the Ordinance. It also discussed providing PWDs "reasonable accommodation…through assistive technology," and cited contemporary facilities available, including accessible keyboards and speech recognition software; it also discussed adaptive technologies (Para. 14), all while urging the usage of offensive terms for PWDs to be stopped (Para. 27; also challenged in *Khan v. Punjab*, 2008). C. J. Shah expanded this jurisprudence in *M. Shafiq-ur-Rehman v. Pakistan* (2017) by mandating the government to include information on PWDs in the next population census, in line with Article 31 of the CRPD.

11.2.2 Recent Legislative Developments and Executive Measures

11.2.2.1 Recent Inclusive Legislation in Pakistan

After constitutional changes in 2010 that increased provincial autonomy, disability rights have become a provincial subject, which led to the following legislative developments in each province:

Table 11.1

Territory	New laws
Punjab	The Ordinance was promulgated as Disabled Persons (Employment and Rehabilitation) Amendment Act, 2012.
Sindh	Sindh Empowerment of Persons with Disabilities Act, 2018.
Balochistan	Balochistan Persons with Disabilities Act, 2017.
Khyber Pakhtunkhwa	The Ordinance was promulgated as Khyber Pakhtunkhwa Disabled Persons (Employment and Rehabilitation) Amendment Act, 2012.
Islamabad Capital Territory	Islamabad Capital Territory Rights of PWDs Bill, 2018, was passed by the National Assembly (lower house of Parliament) on January 10, 2020. It will only apply in the capital city.
Federal Law	Elections Act, 2017.

11.2.2.2 The Digital Pakistan Policy, 2017

The years 2008–2018—marked by two successive democratic regimes in Pakistan, along with the advent of assistive technologies by the private sector—witnessed a boom in rights-based legislation and policy reform (see Table 11.1). People working on projects for PWDs also noticed an awareness within the government of the need to improve inclusion (Dr. S. Shahid, personal communication, September 27, 2019). Digital Pakistan Policy, 2017, issued by the Ministry of Information Technology in consultation with civil society and organizations like the Special Talent Exchange Program, also elucidates this (see Section 11.3.2). This governmental policy posits general guidelines for the promotion of ICT accessibility for PWDs by state institutions across the country, such as introducing software applications, seeking funding from donors, and making provisions for accessibility in competitive exams; and generally promotes using ICT for reducing barriers for PWDs (as seen in Policy Objectives and Policy Goal 17).

The Policy also aims to promote public–private partnerships; establish an "ICT Accessibility Unit"; develop accessible software; provide ICT accessibility in all competitive examinations; and subsidize the cost of assistive technologies, among other aims (Digital Policy, pp. 26–27). However, when the political party Pakistan Tehreek-e-Insaf won the general elections in 2018 and Imran Khan became Prime Minister of Pakistan, he tried to distinguish the party from all previous political regimes by halting or setting aside earlier government projects, thus making the implementation of these policies uncertain.

11.2.2.3 Executive Departments Working Toward Inclusivity

This section highlights various initiatives, other than the Ordinance, taken up by the executive branch of the government for improving accessibility and inclusion of PWDs.

The federal government has recently started investing in start-ups geared toward providing accessibility and eradicating socioeconomic disparity through

Ignite, which is partially sponsored by Pakistan Telecommunication Authority licensees (Ignite, 2018). Ignite has been involved in ICT projects, including funding the development of a mobile system through which users can retrieve information orally using local languages, and adding an Urdu screen reader (Ignite Projects, 2018). They also funded the Augmented Reality Learning System, which aims to make special education and therapy available to children with neurological disabilities (Ignite, 2018). In another project, they involved university students and professional developers for creating ready-to-use mobile applications for PWDs, culminating with the Mobile App Awards (Ignite, 2018). Award-winning applications pertained to helping PWDs remotely control electrical appliances: Tell Me, an android-based voice application, helps people with visual impairment to know their surroundings, and Rollout provides information about restaurants and hotels that facilitate physically disabled customers (Ignite, 2018). Ignite also set up the National Incubation Center Lahore at the Lahore University of Management Sciences, and awarded the University a contract for setting up the National Incubation Center Quetta (Lahore University of Management Sciences, 2018).

The first notable initiative toward inclusivity was by the state-run media house. Founded in 1964, Pakistan Television Corporation (2019) has continued to use sign language interpretation for various news bulletins and special transmissions, such as the Prime Minister's address to the nation. However, it is not consistent in using sign language.

Another federal department that has made some efforts toward inclusion is the Federal Public Service Commission. The Commission conducts exams for joining the federal bureaucracy and other state departments. It entertains physically impaired candidates applying for all 12 occupational groups of the federal bureaucracy upon producing a disability certificate (Federal Public Service Commission Rules, Section 9 [ii]; PLC [CS] *Muhammad Yousaf v. FPSC*, 2016). The Federal Public Service Commission provides PWDs with a helper/writer; grants an extra 15 minutes per exam to visually impaired candidates (Federal Public Service Commission Rules for Competitive Exams, 2018; Section 9 [iii]); and provides an option of sitting for either paper-based or computer-based exams, a facility available at the Federal Public Service Commission (FPSC) federal headquarters in Islamabad and its four provincial offices (Section 9 [c]).

One government department working toward improving the political participation of PWDs is the Election Commission of Pakistan (https://www.ecp.gov.pk/). Ms. Nighat Siddiqui (personal communication, December 3, 2018) and Mr. Altaf Ahmad (personal communication, November 24, 2018) at the Election Commission highlighted such measures, including employing PWDs (The News, 2017), and working closely with the previous PML-N government in passing the Elections Act, 2017, one aspect of which was to make the electoral process more inclusive for PWDs. Section 93 of the Elections Act guarantees postal ballot facilities to PWDs; Section 48 stipulates adopting special measures for PWDs on the electoral roll; and Section 84 (9), read along with 2013 Guidelines for Polling Agents (p. 25), mandates agents to assist PWDs (Rehman, 2017) and give them preferential treatment in queues.

With respect to political participation of PWDs, a 2018 study highlighted that visually impaired voters only had access to printed (non-Braille) ballots; hence, they were "particularly at risk of compromised [participation] in [the] electoral processes" (CPDI, 2018). The Election Commission of Pakistan used print, electronic, and social media applications, including WhatsApp, along with other informational content for advertising postal ballot facilities for PWDs (The News, 2019). Consequently, 222,227 PWDs registered as voters. While the visually impaired voters' ballots were still at risk of being compromised, the extension allowed them to at least participate in the electoral process. The Election Commission of Pakistan also plans to use this door-to-door citizen registration facility, which was previously used for increasing women's participation, to distribute a translated Washington Group Short Set of Questions—survey/census questions designed for identifying and disaggregating PWDs according to their disability status to assess substantive equal participation (Washington Group, 2016). This will be used to assess the number of visually impaired persons in a specific area so that Braille posters can be sent to those areas specifically (Siddiqui, 2019).

Furthermore, in addition to contributing to Ignite, the Ministry of ICT and Pakistan Telecommunication Authority have launched their websites in Urdu, English, and accessible format for visually impaired persons, in consultation with PWD organizations (PTA, 2018). The Ministry also launched a mobile app development competition—utilizing their ICT Research and Development Fund—for facilitating PWDs (APP, 2017). However, since the change in government in 2018, there has been no information about the continuation of this or similar projects.

The Higher Education Commission of Pakistan developed "Policy for disabled students at higher learning institutes in Pakistan." This draft Policy, which is still open for recommendations, mandates the Higher Education Commission to provide fiscal relief to higher education institutions for providing facilities like writing assistants to disabled students (Higher Education Commission [HEC], 2018). The draft Policy does not focus on using ICT for accessibility, ignoring the Digital Pakistan Policy 2017 (see Section 2 [b][i]).

11.2.3 Provincial Measures

Apart from the new provincial laws, this part of the chapter outlines the actions of Pakistan's four provinces—Punjab, Sindh, Khyber Pakhtunkhwa,

Balochistan—specifically reviewing the measures adopted by the first two and the lack of initiatives undertaken by the latter two.

11.2.3.1 Punjab

The Punjab government is the wealthiest province in terms of budgetary share and consequently was able to take more steps toward the inclusion of PWDs. Punjab's Special Education Department appears more active for promoting education for disabled students: they have established institutes for the four legally recognized categories of impairments, and have opened teacher training colleges (Government of Punjab, 2019). The Punjab government has also created bodies like the District Rehabilitation and Training Committees (Social Welfare Department, 2015); and initiated one of the first mega public transport projects by offering reserved seats to PWDs on the Orange Line, an intracity commuter train (Sheikh, 2018). Other provinces also have special education departments; however, they are either defunct or have been mostly inaccessible for the last several years, although recent legislations, as mentioned in Table 11.1, plan to revitalize these departments (Government of Balochistan, 2019; Government of Punjab, 2019; Government of Khyber Pakhtunkhwa, 2018; Sindh Government, 2018).

Furthermore, the government can expand or replicate its ongoing projects for increasing the inclusion of PWDs. For instance, the Punjab government and Punjab IT Board initiated an e-learning scheme about the digitization of middle school syllabi and distribution of tablets to students for accessing the curriculum. If they select accessible technology, these programs can contribute toward e-learning for PWDs (Government of Punjab, n.d.) and will be easy to replicate and apply in the remaining three provinces, particularly under the Digital Pakistan Policy, by seeking assistance from the federal government. However, Pakistan Tehreek-e-Insaf's victory in the 2018 general elections, and consequent distancing from most previous governmental initiatives, has placed these ventures on hold.

11.2.3.2 Sindh

The Sindh Empowerment of PWDs Act, 2018, illustrates the Sindh government's recent progression. In 2018, the Sindh government inaugurated the Center for Autism Rehabilitation and Training in Karachi—the first in Pakistan and allegedly the largest in South Asia (The News, 2018). The Board of Secondary Education, Karachi, has ended age restriction for examinations and does not charge any fee from disabled students for registration and exams (Express Tribune, 2016). Earlier governmental measures were comparatively more exclusionary; for instance, the Board announced results of disabled students separately, awarding cash to position-holders (Express Tribune, 2016).

11.2.3.3 Khyber Pakhtunkhwa and Balochistan

The Khyber Pakhtunkhwa government initiated the Tele Education Program to provide online education to out-of-school disabled children. The Minister stated that the government would provide laptops and a meager stipend of approximately \$4.50 per month to disabled children to enable them to lead healthy lives (Zia, 2018; Samaa Digital, 2018). However, there is no data available on the implementation of these programs. The Balochistan government has not initiated any programs as yet.

11.3 Private Educational Institutions, Non-Governmental Organizations, Corporate Social Responsibility, and Public–Private Partnerships

The current status of PWDs' inclusion is attributable primarily to the private sector, mainly non-governmental organizations and other civil society initiatives, and recently bolstered by the advent of assistive technology. ICT has made providing necessary facilities to PWDs cheaper and more widespread. This section shall provide a roadmap of different initiatives of private-sector education institutions and non-governmental organizations, and highlight some critical public–private partnerships.

11.3.1 Private Education Institutions

During recent years, the private sector has emerged as a key actor in empowering PWDs, at times in collaboration with various non-governmental organizations and governmental departments. Unlike governmental policies, general elections do not always hinder private-sector initiatives. This section deals with efforts toward inclusivity beyond offering studying and employment opportunities in educational institutions.

Private educational institutes have been more active in using and developing assistive technologies to allow equal opportunity to their students. Lahore University of Management Sciences (LUMS) is one such research-active institution which has also been a recipient of several grants and donations as part of corporate social responsibility initiatives, including being the center where the first government-funded National Incubation Center was set up (Ignite). Apart from organizing conferences on inclusive higher education, such as the one-day policy dialogue on inclusive higher education organized by Marva Khan, one of this chapter's co-authors (LUMS, 2016; Moalims, 2016), LUMS has also formed an Accessibility Committee; an Office of Accessibility; an Accessibility Lab (Knowledge and Data Entry Lab) and an Assistive Technology Lab for visually impaired students and faculty, all of which work toward making infrastructural and educational facilities on campus more accessible.

11.3.2   NGOs and Other Private-Sector Initiatives

11.3.2.1   Non-Governmental Organizations, Disabled Peoples' Organizations, and ICT

This section details non-governmental organizations (NGOs) and disabled peoples' organizations (DPOs) working toward inclusion by relying on ICT or by pushing for ICT-based reforms. For instance, ConnectHear is a start-up that uses social media to convey information on various news items and sociopolitical issues using sign language (Facebook, 2020).

Founded in 1997 by PWDs, including Atif Sheikh and Abia Akram, co-authors of this chapter, the Special Talent Exchange Program is working on multiple private and public avenues for mainstreaming PWDs. Apart from the standard objectives of inclusive education, employment, and healthcare, they were active participants in formulating the Digital Pakistan Policy 2017, pushing for new legislation, along with assisting the Electoral Commission of Pakistan in many of their ventures geared toward devising inclusive electoral processes (Special Talent Exchange Program Portfolio, 2019; Election Commission of Pakistan, 2018). The Special Talent Exchange Program (STEP) has also worked toward developing accessible career webpages and conducting accessibility audits of all offices and the sales and service centers of the telecommunications company, Telenor; it also helped Telenor set up Assistive Technology Centers in 2009–2010, which function as training labs, at the National Institute of Special Education and STEP's offices (Portfolio, 2019). They also created Pakistan's first disability job center and have been involved with various other research and policy ventures (Sightsavers, 2012; Disability-Inclusive Disaster Risk Reduction Training Manual; CRPD Urdu).

STEP is also actively involved with the Ministry of Human Rights regarding the passage of the ICT Rights of PWDs Bill (2018; see Table 11.1), which has gone through several rounds of discussion, by attending all meetings and lobbying parliamentarians. This law mandates the creation of a council for redressal of PWDs' issues and aims to promote education, political, and economic participation of PWDs. The government also created an ICT Accessibility Working Group with representation from PWDs, including STEP, the government, development organizations, and the corporate sector.

Similarly, the Pakistan Disabled Foundation organized the 2017 IT Conference & International Day of the Blind, inviting numerous IT experts including a visually impaired master trainer and researchers from across the country (Pakistan Disabled Foundation, 2017), and conducted a technology and capacity-building seminar in collaboration with Azad Jammu and Kashmir Association of the Blind of Mirpur, Kashmir in 2014. It entailed a demonstration of computerized Perkis SMART Brailler and a demonstration of a Braille note taker to the conference attendees (Pakistan Disabled Foundation, 2017). Similarly, the Pakistan Association of the Blind (2019) offers computer training "short courses through Jaws software, Braille learning short courses, [and] mobility techniques courses." Thus, the NGOs are also taking cognizance of the need to increase access and inclusion.

The Lahore Businessmen Association for Rehabilitation of the Disabled runs the National Vocational and Technical Training Commission, Lahore, which offers vocational skills courses in programs like Adobe Photoshop exclusively to PWDs. It seeks to liaise between PWDs and employers, in collaboration with the Lahore Chamber of Commerce and Industry (National Vocational and Technical Training Commission, n.d.).

11.3.2.2 Conventional Initiatives

Non-governmental organizations predating Pakistan have worked toward inclusion as well. The Ida Rieu Welfare Association, Karachi, founded in 1921 for education students with visual, hearing, and speech impairments, has several projects such as the Hatim Alavi Memorial Braille Library—Pakistan's first and most extensive Braille library—the Sarah and Mujahid Hostel for Deaf and Blind Girls; Irfan Mowjee Vocational Centre; Sultan Ali Campus; School for the Blind, Rashidabad; and Shirin Keshani School-College for the Deaf (Ida Rieu). The Dr. Panjwani School-College for the Blind caters to visually impaired students, and the College is affiliated with the University of Karachi for graduate and postgraduate programs. They primarily teach and use Braille.

The Pakistan Association of the Deaf has also established a Deaf Empowerment and Education Center, offering various educational courses. Furthermore, the Deaf Sign Language Research Group facilitates sign language education by aiming to prepare all primary- and secondary-level books in sign language and developing ICT learning tools of all sign language books (Pakistan Association of the Deaf [PADEAF], 2018a).

The Aziz Jehan Begum Trust for the Blind (https://ajbtrust.org/) launched a Braille Transcription Center in 1989. The Trust also transcribes textbooks and examination papers in Braille. Previously, they set up a Recording Studio in 1986 for making talking books, with the assistance of the Japanese Embassy. They also launched Open Mind Pakistan, a trainee program for PWDs (Aziz Jehan Begum Trust, 2017; Telenor, n.d.).

The Network of Organizations Working for People with Disabilities launched *Dastoor* to bridge the gap created by unequal access to education by capacity building (NOWPDP, 2015). The Network also launched the "Disability Services Directory" that categorizes services provided by organizations working for PWDs (NOWPDP, 2015), and also signed an agreement with Habib Bank Limited to work toward making it the first inclusive bank in Pakistan (Express Tribune, 2015).

11.4 Conclusion

An accurate assessment of the status of PWDs is not possible until the government makes concrete efforts to ascertain how many PWDs live in Pakistan, along with the various types of impairments they have. However, apart from the introduction of quotas in 1981, most steps toward inclusion and facilitation adopted by the government took place in the last decade. Recent measures for promoting the political participation of PWDs looks promising if carried out effectively.

Various DPOs and NGOs have existed throughout Pakistan's history—some even predating its creation—that work toward the empowerment and inclusion of PWDs, and advancements in ICT have effectively bolstered their efforts. Technological advancements have introduced new and cheaper assistive technologies which are more readily accessible, along with enhancing connectivity among the masses themselves and with the government.

Apart from being ill-informed, governmental initiatives are often scrapped or forgotten when a different political regime comes into power, simply to discredit the earlier administration. One way forward for DPOs and other private/NGO collaborative partners of the government is to reduce the political undertones of such initiatives so that they may continue when new regimes come to power. Furthermore, in light of the lack of research and development initiatives by the government, it is even more critical that civil society makes a greater effort in acquiring and developing assistive technology. Once familiar with it, these technologies ought to be recommended to various government departments for utilization and possible dissemination among the masses, particularly from the executive branch of the government (as that has historically been the most active concerning facilitating PWDs).

While ICT advancements are imperative for the changing landscape for PWDs, these technologies are insufficient for holistic reform. Several socioeconomic factors need redressal alongside these advancements. Raising awareness is an essential need of the hour. There is a severe deficiency of understanding of the numerous types of disabilities that exist, which leads to improper diagnosis and insensitive measures by state and non-state actors, which do more harm than good. It is essential to start by making all public spaces accessible and revising school curriculums, and generally increasing the visibility of PWDs (Human Rights Commission of Pakistan, 2019). This in turn, will condition people to set aside paternalistic attitudes and see PWDs on an equal footing. Assistive technology has a crucial role in fulfilling Sustainable Development Goal 3, "ensure healthy lives and promote well-being" (Bennet, 2017). Hence, ICT is a great platform for raising awareness by designing applications to assist in early detection of disabilities, as well as for providing general information on various impairments.

Recently, the government appears cognizant of the need to improve the inclusivity of PWDs. However, due to easier detection, the physically impaired, and
mainly persons with visual disabilities, are the primary beneficiaries of most
efforts. In contrast, persons with cognitive impairments remain more difficult to
employ due to a lack of understanding of these impairments (LUMS, 2019).
Furthermore, Pakistan's infrastructure, including public spaces, is mostly
inaccessible. When most governmental projects have focused on the visually
impaired, merely asking visually impaired people to leave their homes and reach
a venue and transverse it is a problem due to infrastructural limitations.

Repeated misinterpretation of quotas introduced by the 1981 Ordinance highlights the need for changing social attitudes toward PWDs, since reserved posts
for PWDs often remained unutilized. Thus, the need to work toward removing
social stigmas associated with disabilities is essential. Societal biases and paternalistic attitudes of the abled hinder the proper implementation of favorable laws
and policies. Hence, inclusive strategies will only be effective if people are willing
to implement them and accept the application. ICT has increased the state's access
to remote areas of the country. This avenue can be utilized constructively for
spreading awareness, providing assistive technology to schools and hospitals
around the country for better detection of disabilities, and better catering to disabled children's educational needs.

While there is a need for new initiatives, it is essential to keep utilizing avenues
that have previously proven effective, like promoting corporate social responsibility. Pakistan's GDP growth rate fell from 5.5% to 3.3% in 2019 and inflation rose
to 6.8%, with the 2020 forecast rate standing at 11.5% (Asian Development
Bank, 2020). This means it is essential for the government to explore corporate
social responsibility (CSR) to reduce its costs for mainstreaming PWDs in the
society. Granting tax benefits and other reliefs to private-sector initiatives that
further such state policies, such as the foundation of the first inclusive and accessible bank, will promote new CSR ventures.

Measures such as sporadic tax relief and some types of public–private partnerships are not enough. There is a need to assimilate the scattered efforts undertaken across the country and come up with a more comprehensive access system.
This includes various DPOs, other NGOs, and public–private partnerships, which
often venture in the same direction but each with limited resources. The state
ought to increase communication with civil society to access more technical
expertise. A comprehensive network of assistive technologies and service providers will also improve the outreach of these services across the country, which,
if left in the hands of the private sector alone, will never have access to the same
resources as the government.

While PWDs in Pakistan are still largely excluded, the increasing consciousness toward rectification of this issue in the private, and now the public, sector
does offer hope. The optimal way forward is to learn from international and

domestic best practices and build on public–private partnerships for the inclusion of Pakistan's largest minority.

References

Aamir Tufail v. Government of Punjab & two others, PLC(CS) 493 LHC (2018).

Asian Development Bank. (2020). *Pakistan: Economy*. Retrieved April 29, 2020, from www.adb.org/countries/pakistan/economy

Ahmed, T. (2005). The population of persons with disabilities in Pakistan. *Asia Pacific Population Journal, 10*(1), 39–62.

Associated Press of Pakistan. (2017, January 9). *All set to announce mobile app awards winners by January end: PTA.* https://www.app.com.pk/business/all-set-to-announce-mobile-app-awards-winners-by-january-end-pta/

Asfandyar Khan & others v. Government of Punjab and others, 2018 PLD 300 LHC.

Aziz Jehan Begum Trust. (2017). *Other Projects.* https://ajbtrust.org/?page_id=1326

Baksh, M., & Mehmood, A. (2012, December). *Web accessibility for disabled: A case study of government websites in Pakistan.* Paper presented at the 10th International Conference on the Frontiers of Information Technology, Islamabad. 10.1109/FIT.2012.68

Bennet, B., et al., (2017). Assistive technologies for people with dementia: Ethical considerations. *Bulletin of the World Health Organization, 95*, 749–755. World Health Organization. https://www.who.int/bulletin/volumes/95/11/16-187484.pdf

Centre for Peace and Development Initiatives. (2018). *Electoral and political rights of persons with disabilities in Pakistan: Situational analysis & way forward.* Retrieved from http://www.cpdi-pakistan.org/wp-content/uploads/2019/03/Electoral-and-Political-Rights-of-Persons-with-Disabilities-in-Pakistan-Situation-Analysis-Way-Forward.pdf

Cheema, U. (2018, 13 May). *Yousaf set to become first blind judge in Pakistan.* Retrieved from https://www.thenews.com.pk/print/316123-yousaf-set-to-become-first-blind-judge-in-pakistan

Choudhury, G. (1956). The constitution of Pakistan. *Pacific Affairs, 29*(3), 243–252. 10.2307/2753474

Constitution of Islamic Republic of Pakistan [Statute] (1973).

Constitutional Amendment Act [Statute]. (2010).

Custom Research Report Produced for the British Council. (2014). *Moving from the margins: Mainstreaming persons with disabilities in Pakistan.* https://www.british-council.pk/sites/default/files/moving_from_the_margins_final.pdf

Dawn. (2017, March 4). *Mobile apps to help autistic children.* Retrieved from https://www.epaper.dawn.com/

Dawn News. (2017, April 3). *Mobile apps to help autistic children.* Retrieved from https://www.dawn.com/news/1324471

Digital Pakistan Policy (2017).

Disabled Persons (Employment and Rehabilitation) Ordinance XL, 1981.

Express Tribune. (2015, 5 February). *Special needs.* Retrieved from https://tribune.com.pk/story/832971/special-needs-hbl-and-nowpdp-sign-agreement/

Facebook. *ConnectHear Official.* Retrieved from web.facebook.com/pg/connect-hearofficial/about/?ref=page_internal

Federal Public Service Commission (FPSC) Rules for Competitive Exams (2018). Retrieved from http://www.fpsc.gov.pk/sites/default/files/CE-2018-Rules_.pdf

Government of Balochistan. (2019). *Special education wing.* Retrieved from http://www.swd.balochistan.gov.pk/special-education.html

Government of Khyber Pakhtunkhwa. (2018). *Social welfare.* Retrieved from http://social_welfare.kp.gov.pk/

Government of Punjab. (2019, May 23). *Welcome to Special Education Department.* Retrieved from https://sed.punjab.gov.pk/

Government of Punjab. (n.d.). *ELearn.Punjab.* Retrieved from https://www.pitb.gov.pk/elearn_punjab

Habib Bank Limited. (2018). *Diversity & Inclusion.* https://www.hblpeople.com/diversity/

Hafiz Junaid Mehmood v. Government of Punjab, 2017 PLD 1 LHC.

Hammad, T., & Singal, N. *Education of women with disabilities in Pakistan: Enhanced agency, unfulfilled aspirations* [PDF]. https://pdfs.semanticscholar.org/d740/2cfce0681306a381ecefe666173bf9d419f9.pdf

Haroon-ur-Rashid and others v. Registrar Balochistan High Court, Quetta and others, 2013 PLC (CS).

Higher Education Commission (HEC), Pakistan. (n.d.). *Policy for students with disabilities for higher learning institutes in Pakistan.* Retrieved from https://hec.gov.pk/english/services/universities/Pages/Policy.aspx

Human Rights Commission of Pakistan (HRCP). (2019, March). *State of human rights in 2018.* hrcp-web.org/hrcpweb/wp-content/uploads/2019/04/State-of-Human-Rights-in-2018-English-1.pdf

ICT Rights of Persons with Disabilities Bill (2018). National Assembly of Pakistan. Retrieved from: http://www.na.gov.pk/uploads/documents/1545385013_141.pdf

Ida Rieu Welfare Association. (n.d.). *The Hatim Alavi Memorial Braille Library.* Retrieved from https://idarieu.org/hatim-alivi-library/

Ida Rieu Welfare Association. (n.d.). *About us.* Retrieved from https://idarieu.org/about-us-3/

Ida Rieu Welfare Association. (n.d.). *Dr. Panjwani School-College.* Retrieved from https://idarieu.org/dr-panjwani-school/

Irshad, H., Mumtaz, Z., & Levay, A. (2011). Long-term gendered consequences of permanent disabilities caused by the 2005 Pakistan earthquake. *Disasters, 36,* 452–464.

Ignite (n.d.). Retrieved from https://www.ignite.org.pk/projects/project_details. php?id=14

Ignite (n.d.). Retrieved from https://www.ignite.org.pk/component/tprojects/ project/104.html

Ignite (n.d.). Retrieved from https://ignite.org.pk/component/publication/?view=pub lication&layout=newsdetail&id=75&Itemid=182#.WzDxxqczZPY

Ignite (n.d.). *Projects*. Retrieved from https://www.ignite.org.pk/component/ tprojects/project/16.html; http://cle.org.pk/dial

Ijaz Ahmad v. General Manager Sui Northern Gas Limited, 2001 CLC 28 Lahore.

Lahore University of Management Sciences (LUMS). (2017, April 10). *LUMS launches three mobile apps for autistic children at World Autism Awareness Day*. Retrieved from https://hr.lums.edu.pk/news/general-news/lums-launches-three-mobile-apps-autistic-children-world-autism-awareness-day

Lahore University of Management Sciences (LUMS). (2016). *Annual report* (Vol. 2015–2016, Rep.). http://fliphtml5.com/zvdq/vzsr/basic

Lahore University of Management Sciences (LUMS). (2018, February 25). *LUMS awarded contract by Ignite to set-up National Incubation Center Quetta*. https:// niclahore.lums.edu.pk/lums-awarded-contract-by-ignite-to-set-up-national-incubation -center-quetta/

Lahore University of Management Sciences (LUMS). (n.d.). *Overview*. Retrieved from https://alap.lums.edu.pk/

Lari, Z. S. (2000). *Self-empowerment of women with disabilities in Pakistan* [Conference session]. International Special Education Conference. University of Manchester: UK.

Mental Health Ordinance (2001).

Muhammad Iqbal Goraya v. Punjab Public Service Commission, 1990 PLC (CS) 634 LHC.

Muhammad Rashid Shafi v. Managing directors & others, 1998 PLC (CS) 848 LHC.

Muahammad Shafiq-ur-Rehman v. Federation of Pakistan and others, 2017 PLD 558 LHC.

Muhammad Yousaf v. Chairman Federal Public Service Commission, W.P. No.7572/2016 LHC.

Miles, M. (1990). Special education in Pakistan. *International Journal of Disability, Development and Education, 37*(2), 159–168. 10.1080/0156655900370208

Moalims. (2016, February 2). *Policy dialogue on inclusive higher education conference at LUMS*. Retrieved from http://www.moalims.com/news/75453/16/policy-dialogue-on-inclusive-higher-education-conference-at-lums

National Assembly of Pakistan.(2018, 21 December). *Orders of the day*. Retrieved from http://www.na.gov.pk/uploads/documents/21-12-2018%20(Friday).pdf

National Database & Registration Authority (NADRA) Official Website. (2019). *National database & registration authority*. Retrieved from https://www. nadra.gov.pk/

National Incubation Center (NIC) Lahore. *NIC Lahore*. https://niclahore.lums.edu.pk/

National Vocational and Technical Training Commission (NAVTTC). (n.d.). *Success stories*. Retrieved from http://www.navttc.org/SuccessStories.aspx?cat=1

Network of Organizations Working for People with Disabilities (NOWPDP). (2015–2019). *Disability services directory*. Retrieved from http://directory. nowpdp.org

Network of Organizations Working for People with Disabilities (NOWPDP). (2015–2019). *Dastoor*. Retrieved from https://www.nowpdp.org/dastoor/

Network of Organizations Working for People with Disabilities (NOWPDP). (2015–2019). *Training opportunity*. Retrieved from http://www.nowpdp.org/ trainingopportunity/

Pakistan Association of the Blind. (2019). *Training*. Retrieved from http://pabnpk. com/train.htm

Pakistan Association of the Deaf (PADEAF). (2018a). *Deaf empowerment education center*. Retrieved from http://www.padeaf.org/deaf-empowerment-education-center-2/

Pakistan Association of the Deaf (PADEAF). (2018b). *Deaf sign language*. Retrieved from http://www.padeaf.org/deaf-sign-language-research-group/

Pakistan Bureau of Statistics. (2017). *Disabled population by nature of disability*. Retrieved from http://www.pbs.gov.pk/content/disabled-population-nature-disability

Pakistan Disabled Foundation. (2017). *Capacity building seminar*. Retrieved from http://pdf.org.pk/technology-and-capacity-building-seminar-organized-by-pdf-in-collaboration-with-ajk-association-of-the-blind-of-mirpur-kashmir/

Pakistan Disabled Foundation. (2017). *Two-day IT conference*. Retrieved from http:// pdf.org.pk/two-days-it-conference-international-day-of-the-blind-2017/.

Pakistan Population Census. (1998). *Population census organization*. Statistics Division, Government of Pakistan.

Pakistan Today. (2018, June 27). *People with disabilities can avail postal ballot facilities: ECP*. Retrieved from https://www.pakistantoday.com.pk/2018/06/27/people-with-disabilities-can-avail-postal-ballot-facilities-ecp/

Pakistan Telecommunication Authority. (2018). *PTA launches new website accessible for persons with disabilities*. Retrieved from

www.pta.gov.pk/en/media-center/single-media/pta-launches-new-website-accessible -for-persons-with-disabilities

Pakistan Television. (2019). *About*. Retrieved from www.ptv.com.pk/About

Punjab Information Technology Board (PITB). (n.d.). *Registration of visually impaired persons* (VIPs). Retrieved from https://swd.punjab.gov.pk/news_vips

Rehman, S. U. (2017, December 15). *CEC for full participation of disabled persons in election process*. Retrieved from https://www.brecorder.com/2017/12/15/387175/ cec-for-full-participation-of-disabled-persons-in-election-process/

Research, Evaluation and Monitoring Unit (REMU). (2014). *Moving from the margins: Mainstreaming persons with disabilities in Pakistan*. Retrieved from the British Council, :https://www.britishcouncil.pk/sites/default/files/mainstreaming_persons_with_disabilities_0.pdf

Robert, M. (2000). *Disability issues, trends, and recommendations for the World Bank*. World Bank.

Saeed, L., Syed, S., & Martin, R. (2014). Historical patterns of terrorism in Pakistan. *Defense and Security Analysis*, *30*. 10.1080/14751798.2014.921450

Samaa Digital. (2019, March 29). *KP govt starts tele-education program for special children in Mardan*. Retrieved from https://www.samaa.tv/news/2018/03/kp-govt-starts-tele-education-program-for-special-children-in-mardan/

Sheikh, A. (2018, May 16). *Lahore's Orange Line Metro Train conducts trial run*. Retrieved from https://www.dawn.com/news/1408039/lahores-orange-line-metro-train-conducts-trial-run

Sightsavers. (2018, June). *How partnership working helps us reach marginalized communities*. Sightsavers. www.sightsavers.org/stories/2019/01/how-working-in-partnership-helps-us-reach-the-most-vulnerable-communities/

Sindh Empowerment Act (2018).

Sindh Government. *Rozgar*. Retrieved from http://istd.sindh.gov.pk/initiatives/381 / https://www.sindhrozgar.gos.pk/

Singhal, D. (1962). The New Constitution of Pakistan. *Asian Survey*, *2*(6), 15–23. doi:10.2307/3023614

Social Welfare Department. (2015). *Disabled assessment & rehabilitation*. Retrieved from https://swd.punjab.gov.pk/dar

STEP Portfolio, 2019.

Suresh, T. (2014, August). *Moving from the margins: Mainstreaming persons with disabilities in Pakistan*. British Council. Retrieved from: https://www.britishcouncil.pk/sites/default/files/moving_from_the_margins_final.pdf

Syed Babar Ali School of Science and Engineering LUMS. (n.d.). *Knowledge and data entry lab*. Retrieved from https://sbasse.lums.edu.pk/labs/knowledge-and-data-engineering-lab

Tariq Aziz v. Government of Punjab through Secretary Education & Others, 1996 PLC (CS) 189 LHC.

Telenor. (n.d.). *Open mind Pakistan*. https://www.telenor.com.pk/about/corporate-social-responsibility/disability

The Express Tribune. (2016, May 4). *Catering to special needs students*. Retrieved from https://tribune.com.pk/story/1096744/catering-to-special-needs-students-with-aural-oral-impairments-take-first-final-exam-under-bsek/

The Express Tribune. (2017, April 3). *Three apps unveiled to support autistic children*. Retrieved from https://tribune.com.pk/

The News. (2017a, February 13). *AIOU to provide free education to the disabled.* Retrieved from https://www.thenews.com.pk/print/186077-AIOU-to-provide-free-education-to-the-disabled

The News. (2017b, December 15). *ECP enabling disabled persons the right to vote: CEC.* Retrieved from https://www.thenews.com.pk/print/256339-ecp-enabling-disabled-persons-to-exercise-right-of-vote-cec

The News. (2018, September 8). *Sindh CM Inaugurates Autism Rehabilitation Center.* www.thenews.com.pk/latest/365703-sindh-cm-inaugurates-autism-rehabilitation -center

United Nations Human Rights Office of the High Commissioner. (1996–2014). *OHCHR dashboard.* Retrieved from http://indicators.ohchr.org/

United Nations Treaty Collection. (2006, December 13). *Optional Protocol to the Convention on the Rights of Persons with Disabilities.* Retrieved from https://treaties. un.org/Pages/ViewDetails.aspx?src=TREATY&mtdsg_no=IV-15-a&chapter =4&lang=_en&clang=_en

Washington Group on Disability Statistics. (2016, January 18). *Short set of disability questions.* Retrieved from http://www.washingtongroup-disability.com/washing-ton-group-question-sets/short-set-of-disability-questions/

Zia, A. (2018, April 4). *Online learning: Education opportunity for disabled out of school youth.* Retrieved from https://tribune.com.pk/story/1676782/1-online-learning-education-opportunity-disabled-school-youth/

12

Design Approaches for Creating Person-Centered, Context Sensitive, and Sustainable Assistive Technology with the Global South

Mario Andrés Chavarria and Klaus Schönenberger
EssentialTech Centre, École Polytechnique Fédérale de Lausanne, Switzerland

Anthony Mugeere
Department of Sociology and Anthropology, Makerere University, Kampala-Uganda

Samia Hurst and Minerva Rivas Velarde
IEH2, Université de Genève, Switzerland

12.1 Introduction

Approximately one billion people around the world live with disabilities, 80% of whom are in low- and middle-income countries (LMICs; World Health Organization, 2011). Yet only 5–15% of those requiring assistive devices and technologies have access to them (World Health Organization, 2019). The World Health Organization, in its effort to close the assistive technology gap, created the Global Cooperation on Assistive Technology, whose main purpose is to maintain or improve a disabled person's functioning and to enhance their overall wellbeing (World Health Organization, 2019). In this chapter, of the various types of assistive technology, we are specifically focusing on assistive technology to improve personal travel and mobility. The Global Cooperation on Assistive Technology initiative at World Health Organization focuses on five interlinked areas (5Ps): (1) *People*: user involvement; (2) *Policy*: development of tools to support countries in developing national policy; (3) *Products*: encouraging countries to develop a list of national priority products; (4) *Provision*: integration of assistive product service provision into the health system; and (5) *Personnel*: building the capacity of their community-level workforce (World Health Organization, 2019). However, involving users is not enough if devices continue to be based on a "bodily deficit,"

Accessible Technology and the Developing World. Michael Ashley Stein and Jonathan Lazar, Oxford University Press.

focusing purely on addressing impairment, which reflects an outdated understanding of disability reflected in the medical model of disability. Design of assistive technology would be better served by adopting a more comprehensive view on the experience of disability, such as the capabilities approach (Sen, 2008). This will enable assistive technology (AT) to focus on a person's wellbeing and provide tools that offer users the freedom to live lives they have a reason to value.

Our work is dedicated to developing a critical understanding of the underlying assumptions, motivations, and values that inform disability and technology research design. The chapter is divided into two parts: In the first, we outline relevant legal frameworks, then we review the methodological shortcomings of well-intentioned technology "solutions" that fail when they are implemented among the intended population due to the lack of a proper understanding of the final user and their local context. In the second part, we present lessons learned from the design and implementation of assistive devices that have overcome common barriers such as high cost and maintenance needs, dependence on access to electricity or internet, and inaccessible features. We also argue that it is necessary to gain a better understanding of disability to enhance context-sensitive and person-centered technology design. We hope to contribute to shifting the focus on the development of assistive devices to a new design approach that is person centered and context sensitive, and renders the product sustainable.

12.2 International Frameworks for AT for Mobility and Independence

It has been at least 15 years since the Convention on the Rights of Persons with Disabilities (CRPD) was enacted. Article 20 set the obligation for signatory countries to ensure personal mobility with the greatest possible independence for persons with disabilities (PWDs), and required states to facilitate access to assistive devices. More recently, the Dubai Declaration (2018) was adopted by the International Telecommunication Union, which promotes the adoption of user-centered technology (ICT) as a means to ensure equitable, affordable, inclusive, and sustainable development of telecommunication and information and communications technology networks, applications, and services. The synergies between these two frameworks, as well as regional and domestic legislation, show positive progress, but there is still much to be done. It is important that those who produce the technology take into account all aspects of accessibility, mobility, and wellbeing for PWDs. Furthermore, we believe that domestic governments and the international community are moving toward a deeper reflection of the underlying philosophy of terms such as inclusive design, usability, and accessibility (Persson et al, 2015). Engineering design can produce social trans-formation for people with disabilities (Williams, 2019), and approaches such as human systems engineering and user-centered design based on community needs can help make progress toward the goals of the CRPD and Dubai Declaration.

In Uganda, for instance, the government enacted the National Policy on Disability (2006) which—among other things—established a rehabilitation and resettlement scheme that includes vocational rehabilitation services and sheltered workshops that focus on employable skills training and orthopedic workshops for the provision of AT devices such as calipers, wheelchairs, and white canes to facilitate the mobility and independence of persons living with physical impairments (MOGLSD, 2006). Other African countries such as Burkina Faso and Niger have also provided a legal framework that promotes the mobility and independence of persons living with disabilities by focusing on access to public buildings and transport, such as the provision of tax breaks designed to encourage accessibility measures (Handicap International, 2010). Other aspects of the same legislative framework in Uganda include the Urban Code or Building Code, which takes into account persons living with disabilities in the construction of new buildings, while holders of the disability card are also entitled to a reduction in travel expenses on public transport.

12.3 Available Assistive Technology and Innovation for Personal Mobility

Over the last decade, we have seen enormous technological progress on assist-ive devices for navigation and mobility. While the legal framework is encour-aging and comprehensive, the design of such devices often overlooks the experience of disability and the everyday needs and constraints of disabled people from the Global South. To elaborate on this issue, let us take as an example a recent technology development for blind persons. We will present an overview of, first, available hardware-based navigation systems and, second, software-based navigation systems; lastly, we will discuss the shortcomings of existing solutions with regard to the needs of LMICs.

12.3.1 Hardware-Based Navigation Systems for the Visually Impaired

The visually impaired population encompasses those with moderate or severe vision impairment or blindness, which refers to persons with visual acuity worse than 3/60 (we acknowledge that terminology and identity labels differ from country to country). Usually, people with visual impairment depend on a probing cane to help them move around a city safely. This method has been used for decades and it is the most common and affordable solution for the visually impaired. However, the use of a cane has major limitations, such as people only being able to detect trials and obstacles right in front of them by swinging the cane, while trying to feel what the tip of the cane is touching on the ground. This leaves the users with no information about their surroundings, outside what is at ground level and right in front of them. This poses an additional threat to the user as well as to the "mobility ecosystem" of the city. Over the last few years, we have seen the launch of highly innovative technologies that aim to overcome the above-mentioned barriers. These technologies, mainly electronic travel aid systems, will be now described and discussed:

12.3.1.1 Electronic Travel Aid Systems

These systems aim to increase the mobility of users and provide additional information about their surroundings. These systems use different types of sensors to acquire data from the surroundings of the user and processing algorithms to convert these into audio or tactile signals. The most basic, and affordable, versions of the electronic travel aid systems are the ultrasound canes/glasses. These systems use ultrasound sensors to detect obstacles and warn the user by a sound signal. More complex electronic travel aid systems provide the user with more detailed information about their surroundings; for example, one of the earliest and most known models of electronic travel aid (ETA) is the vOICe. The vOICe navigation system uses a camera (mounted on a pair of glasses) to monitor the wearer's surroundings and an algorithm to process the acquired data and convert it into a time-multiplexed auditory representation (Milotta et al., 2015). Not all the ETA systems use audio signals; other devices use tactile interfaces, such as the forehead display of the AuxDeco system (www.eyeplus2.com), or electronic signals, such as the vision-tongue display of the BrainPort Vision Pro (www.wicab.com).

12.3.2 Software-Based Navigation Systems for the Visually Impaired (VI)

Alongside ETA systems, there are various software applications for visual users. With the proliferation in the use of smartphones among the VI population, the navigation apps (e.g., RightHear [https://right-hear.com/], Lazarillo [https://lazarillo.cl/en/], etc.) have become popular, affordable solutions for VI users navigating urban environments. However, these apps only provide map- or global positioning system-based information and cannot provide real-time information about the surroundings such as collision warnings, changes in the path, etc. In general, the software-based solutions can be divided into two main groups: the ones based on artificial intelligence and the ones requiring the assistance of sighted employees/volunteers. The former type requires high processing power and, therefore, expensive hardware components. An example of this kind of app is *Seeing* from Microsoft (https://www.microsoft.com/en-us/ai/seeing-ai). The *Seeing* app, running on a smartphone with camera and internet access, can help the user to read communication and correspondence on paper, identify bank notes and colors, and describe objects in the vicinity of the user. The second type of app is based on video-call communication between the user and a sighted assistant such as *Be my eyes* (https://www.bemyeyes.com/), a free app that connects blind and low-vision people with sighted volunteers and company representatives for visual assistance. The mobile apps that include visual recognition of the environment, both artificial intelligence and video-call-based, require high-speed and high-quality internet connection.

12.6 Mismatch Between Existing Solutions and the Needs of LMICs

The tools we have mentioned are excellent, technologically speaking, and can be very useful. However, they fail to reach the large majority of the VI population due to major design flaws. Their key features may be focused on the needs and contexts of high-income countries, or perhaps they are made for a very selective pool of users that ultimately exclude the average user from the Global South. For potential users in most LMICs, the first barrier to accessing these solutions is the cost (WHO, 2018). Most ETA systems are expensive, with prices in the thousands or even tens of thousands of U.S. dollars, which is several times higher than the average monthly (or even annual) salary of a worker in a low- or middle-income country. In the African context, this can be illustrated by considering the cost of hearing aid technologies. In Northern Nigeria, for instance, the least expensive hearing aid costs almost a month's average salary (McPherson & Clark, 2017). Market research in Kenya also found that the main reason people with disabilities did not use splints was the cost (Cassit Orthopedics Limited, 2016). These costs are aggravated by high import duties and informal charges levied on medical appliances such as hearing aids in some developing countries, e.g., Ghana, where a 15% tax is imposed on imported hearing aids (McPherson & Clark, 2017; Borgl, 2015). In Sub-Saharan African countries and other LMICs, there are minimal to non-existent accurate statistical estimates on the availability of AT (e.g., wheelchairs, white canes, prostheses, and hearing aids) for PWDs (Mji, 2019). The poor access to such technologies is largely a result of a lack of knowledge, resources, services, and products (Visagie, 2019). Research among 400–500 households in South Africa, Namibia, Malawi, and Sudan found that the most common sources of AT were government health services (37.8%), "other" (international humanitarian aid, development charities, and religious organizations [29.8%]), and private health services (22.9%; Visagie, 2016).

Shifting AT design would require full and meaningful participation of the variety of voices and realities existing among PWDs. International and domestic disabled peoples' organizations (DPOs) have a crucial role in using international standards as emancipatory tools. The articles of CRPD and other international standards related to disability need to be reconstructed to make sense of the priorities, needs, and struggles of PWDs in LMICs.

PWDs are overrepresented among those living in poverty, and acquiring a smartphone or sustaining the cost associated with high-quality internet is often unattainable (Hanass-Hancock et al., 2017; WHO, 2011). There are significant knowledge gaps regarding the level of penetration of access to the internet, mobile phones, and ICT in LMICs. The Digital Accessibility Rights Evaluation Index (Global Initiative for Inclusive Information and Communication Technologies [G3ict], 2020) shows that, despite worldwide progress, LMICs are still lagging behind in achieving equitable access to ICT for PWDs. Trends of poverty and lack of progress on accessibility overall call into question the relevance of apps and other technology that relies on maximum internet speed and smartphones with high-performance processing needs.

In addition to the lack of affordability and risk of increasing vulnerability, other factors that prevent the adoption of the mentioned technologies (Golledge et al., 2004; Elli et al., 2014; Tapu et al., 2018) are:

Complexity: Common ETA systems and their interfaces are difficult to use and require special training to be understood and used properly. In addition, they can interfere with other senses; for example, audio interfaces can obstruct the users' hearing, hindering their ability to receive auditory feedback from the surroundings (this can be avoided by delivering auditory stimuli via bone conduction without obstructing the external ear; MacDonald et al., 2006).

Discomfort: All wearable components should be small, unobtrusive, and light so that the devices can be carried for long periods. However, the assistive hardware components are usually heavy and voluminous, such as the tactile interface of the AuxDeco, which puts a big weight on the user's forehead, or the vision-tongue display, which has to be carried inside the mouth.

Aesthetics: Users of wearable assistive aids often consider visual appeal and acceptability as more important than the technology's potential benefit (Golledge et al., 2004; Elli et al., 2014); that is, anything that makes them look different, stand out, and attract unwanted attention, such as big electronic devices on their faces, will be rejected.

Information saturation: The amount of information provided by the assistive device must be evaluated carefully to avoid sensory and cognitive saturation of the user. This is a common case among lower-cost devices such as ultrasound canes, which do not have the processing capabilities to discern between useful and unnecessary information (Saha et al., 2019). These devices might perform well in laboratory settings, but they are useless in complex urban environments (Elli et al., 2014; Collins, 1985).

Socioeconomic context: Failing to take into account the local context can render an excellent technology useless. For example, users in countries with high levels of criminality will not use any aid system that requires carrying high-tech devices in visible places. Examples include ETA systems or navigation apps requiring them to take out their smartphones in public, which could put them in danger of being mugged. This issue is not uncommon: PWDs in LMICs are particularly vulnerable to abuse, violence, and crime (Murray et al., 2013; Vilalta, 2016).

Disability is both a cause and consequence of poverty. Historically oppressed social groups such as indigenous persons, ethnic minorities, refugees, and migrants tend to have higher rates of disability than non-minority groups (Anderson, 2016; Rivas Velarde, 2018). Disability scholars from the Global South have heavily criticized disability theory and the disability rights movement, as most of the writing in dominant discourse about disability came from the Global North (Rivas Velarde, 2018; Hickey, 2018; Meekosha, 2011). Trends in AT suggest that historical neglect is coming to the fore once again, as the voices of those from the Global South are absent. Current trends are enforcing double layers of discrimination, where neither the needs nor realities or constrains face by disabled people in the Global South are considered. Technology and innovation serve only a privileged few. Disabled peoples' organizations from the Global South have a crucial role in addressing these patterns of neglect end ensuring meaningful representation of all voices, particularly of those facing double layers of discrimination.

12.7 Shifting the Focus on the Development of Assistive Devices

In order to deepen our understanding regarding the priorities and needs of PWDs in the Global South, we undertook a preliminary consultation that aimed to inform the development of our research questions and set up a research project partnership between research institutions in Switzerland and Colombia. This exploratory work took place between October and November 2019. We set up informal meetings and talks with DPOs, academic institutions, rehabilitation centers, and government officials to learn more about the everyday priorities of organizations of persons who are blind or have low vision regarding assistive devices, as well as mapping the available technology in the market.

We learned that there is a significant mismatch between the needs of the users and the available technology in the Colombian market; there, innovative navigation systems such as AuxDeco systems that were donated to DPOs tend to be in disuse. We enquired as to why people were not interested in using them and were provided with a range of reasons, including that they attract unwanted attention and are uncomfortable to wear. Equally important was that the systems were of no use to them since, for instance, they did not allow them to move around safely in the streets of Cali, Colombia. A key reason for this was that the artificial intelligence (AI) features did not perform efficiently enough in average middle-income settings; their navigation technologies are based mostly on street maps which, in the city that we investigated, do not contain enough information for safe navigation. As an example, sometimes the streets are not perfectly perpendicular or do not have a uniform width, i.e., opposite corners are not aligned; such small details are not present in standard maps, making it difficult for the VI user to find the next walking path when crossing streets and leaving them dangerously wandering in the middle of the street.

We also observed, as shown in Figure 12.1, other barriers to using the infrastructure, such as a damaged sidewalk, obstacles in the path such as tree branches, holes, electric cables, and the absence of a pedestrian crosswalk; these were constant through their daily navigation even in the more privileged parts of town. This not only confirmed the previously discussed literature (Hanass-Hancock et al., 2017; WHO, 2011; Murray et al., 2013; Vilalta, 2016) but also raises serious concerns that go beyond the performance of a specific tool or app. It also shows that perhaps new AT may be reinforcing the exclusion of PWDs in the Global South. Assistive technology design should examine the fact that the large majority of PWDs cannot even afford basic assistive tools, such as the walking cane.

During this exploratory phase, we listened to the research priorities and concerns of DPOs and collected some examples of assistive devices that have focused on the priorities, needs, and concerns of PWDs in LMICs, but we also looked at relevant examples of low-cost technologies in high-income countries that will now be discussed individually. For instance, the Perception and Intelligent Systems (PSI) group (UNIVALLE-Colombia) developed a wearable navigation system for the blind adapted to the urban environment in Latin America, based on AI algorithms (Díaz et al., 2020). The developed system is capable of detecting

Figure 12.1 Illustrations of the Barriers Faced by A Blind Person and His Guide in Cali, Colombia **Left:** Absence of aid infrastructure to navigate through the city. **Center:** Obstacles and damages on the few available aid infrastructure. **Right:** Obstacles in the path, such as tree branches, holes, electric cables, etc.

walkable spaces, obstacles, and objects of interest, such as doors, chairs, staircases, and computers, among others, and planning a path that allows the users to reach their target locations in a safe way (purposeful navigation). The navigation system was successfully tested with blind users in Cali, Colombia, who gave positive comments about its portability, navigation in indoor and outdoor environments, and its ability to run locally (not in the cloud/external servers), which allowed it to work in places without high-quality internet access or GPS-denied environments. This system reflects that co-design with users from the Global South with local scientists seems to be more likely to be accepted; however, further testing and formal evaluation are required. In this preliminary research, we learned more about the link between innovation and public policies. The Colombian government has implemented different public policy initiatives to close the technology access gap in relation to the population with disabilities, such as:

ConVerTic (www.convertic.gov.co): Provides licenses (free of charge) of different aid software for the VI.

Cine para Todos (https://cineparatodos.gov.co): Ensures access to people with sensory or physical disabilities to cinema (free of charge) by incorporating accessible elements into the films.

Centro de Relevo (https://centroderelevo.gov.co): A public platform for the communication between hearing and non-hearing persons based on video calls with sign language interpreters.

These policies were developed in collaboration with DPOs, and are examples of good practices in the use of ICT to enable accessibility and social inclusion.

12.8 Design Methods for Creating AT That Is Person Centered, Context Sensitive, and Sustainable

Context-sensitive, sustainable, and person-centered AT design requires gaining an understanding of how the assistive devices enhance each person's capabilities. The previously mentioned examples show higher involvement of users, which leads to devices and software that will likely match with what individuals value in their lives. Current dominant definitions of functioning and disability, such as the one contained in the International Classification of Functioning (WHO, 2001), are heavily focused on bodily deficit, which tends to be the view present in design approaches. However, personal agency, social determinants of health, and variations on the lived experience of disability are less present in many existing design approaches (Mitra, 2003, 2014, 2017). We suggest implementing the capabilities approach to address the shortcoming of the International Classification of Functioning (Mitra, 2003, 2014, 2017; Bickenbach, 2013).

The capabilities approach has been predominately used in the disability literature, as presented by Sen (1993, 2008) as a theoretical framework which focuses on the significance of individuals' capability to live the kind of lives they value. The capabilities approach stands upon two normative claims that the primary importance that individuals shall enjoy the freedom to achieve wellbeing as a core moral imperative, and that this freedom shall be understood in terms of a person's capabilities to do and be what they have reason to value (Sen 1993, 2008). This means focusing on peoples' real opportunities to do what they value in their terms and in their context. The capabilities approach (CA) is a useful approach to understanding the social participation, agency, and culture in the experience of impairments and disability which are not sufficiently present in the ICF (Mitra, 2003, 2014, 2017; Bickenbach, 2013). The CA offers a promising framework to improve AT design, as it provides a framework to study and understand the interfaces between agency, environment, bodily impairment, and AT. People may live with the same or similar impairments; however, the experience of disability is highly dependent on the context. Persons with different impairments face different barriers and, as such, would benefit from different interventions and technology features. The CA would allow AT design to incorporate social constraints, such as susceptibility to violence; economic hindrance, such as the inability to buy a smartphone and pay for reliable broadband on an ongoing basis; as well as culture and users' views regarding what is important for them and the capabilities they would value the most and how AT can enable them. CA shifts the focus of AT from its outdated approach of overcoming impairment by using technology to equalize or mimic the functioning of non-impaired persons, to instead focus on enhancing freedom and wellbeing. Some of the examples outlined in this chapter, including the work of Perception and Intelligent Systems group UNIVALLE and the upcoming example of AGILIS from EPFL, borrow some elements from the CA: they are seen to be driven by user priorities and a sound understanding of the users' contextual enablers and constraints, but there is as yet no evidence of the impact of these tools on wellbeing.

12.8.1   Understanding the Users

To create AT that is person centered and context sensitive, starting from a deeper understanding of functioning as described above, one has to take into account several factors.

The existing literature (Hanass-Hancock et al., 2017; WHO, 2011; ICT, 2020; Murray et al., 2013; Vilalta, 2016) and preliminary conclusions from our exploratory research indicate that it is very important to ensure high involvement of a significant variation of potential users during the concept design phase—primarily people with disabilities but also, when appropriate, their care givers, clinicians, therapists, and educators. Early and frequent prototyping can help with this, to ensure that what is being built will provide a measurable impact on the user's quality of life. Testing involving daily routines in a user's life is especially important to understand both positive and negative impacts. Also, the documentation and interfaces should be designed so the intended users can not only use it easily, but also *learn* to use it in an easy way, without having to rely on other people. In the future, more needs to be done to examine the scalability and the long-term impact that the revised tool assistive might have on the quality of life and wellbeing of its users. Interdisciplinary cooperation will be essential in putting in place appropriate methodologies to document and evaluate the outcomes of these initiatives and pave the way for the development of efficient assistant technology, ultimately closing the gaps in access. Furthermore, what is learned in design may also benefit other populations (e.g., ETA systems may also benefit people without disabilities in other contexts, such as navigating in poor light conditions or autonomous driving).

12.8.2   Designing Hardware Devices

All devices should take into account international standards in order to guarantee their safety and effectiveness. One example of a standard that should be considered is the ISO/DIS 21856 standard focusing on assistive products (ISO, 2020). This standard refers to general requirements and test methods, providing guidance and specifications for the general design of assistive products and technologies. Depending on the application and intended impact of the solution under design, there are different specialized standards that should be consulted; for example, the ISO 9241 series covers both the hardware and software ergonomics aspects of human–system interactions:

Ergonomics is the scientific discipline and systematic study concerned with the understanding of the interactions among human and other elements of a system, and the profession that applies theory, principles, data and methods to design in order to optimize human well-being and overall system performance. (ISO, 2010)

The ISO 9241 series is divided into several "Parts" addressing different sub-topics. Some of the most relevant for the subject of this chapter are (see also Table 12.1):

Part 100 (ISO/TR 9241-100:2010): Introduces the designer to—and provides the designer with—standards related to software ergonomics, with a special focus on understanding and specifying user requirements as well as designing and evaluating user interfaces. This is a crucial point in order to develop universally accessible technology that provides appropriate software ergonomics considering bodily diversity.

Part 171 (ISO 9241-171:2008): Promotes increasing the usability of systems for a wider range of users by providing ergonomics guidance and specifications for the design of accessible software. Additionally, it addresses issues associated with designing accessible software for people with diverse physical, sensory, and cognitive abilities, including those who are temporarily disabled, and the elderly.

Part 210 (ISO 9241-210:2010): Provides requirements and recommendations for human-centered design principles and activities for both hardware and software components of interactive systems.

Part 910 (ISO 9241-910:2011): Provides a structure for the analysis and understanding of different features of the tactile/haptic interaction.

Table 12.1 Summary of the Discussed ISO Standards

ISO standard/code	Topic
ISO/DIS 21856	Assistive products—General requirements and test methods
ISO 9241 Series	Ergonomics of human–system interaction *Part 100: Introduction to standards related to software ergonomics* *Part 171: Guidance on software accessibility* *Part 210: Human-centered design for interactive systems* *Part 910: Framework for tactile and haptic interaction* *Part 920: Guidance on tactile and haptic interactions*
ISO 23599:2019	Assistive products for blind and vision-impaired persons—Tactile walking surface indicators
ISO 14971:2019	Medical devices—Application of risk management to medical devices
ISO 9999:2016	Assistive products for persons with disability—Classification and terminology

Part 920 (ISO 9241-920:2009): Provides guidance on the design and evaluation of tactile and haptic hardware and software (and combinations of hardware and software) interactions, including: the design/use of tactile/haptic inputs, outputs, and/or combinations of inputs and outputs, the tactile/haptic encoding of information (text, graphical data, and controls), the layout of tactile/haptic space, etc.

The ISO standards also provide guidelines for large infrastructure developments, such as the ISO 23599:2019, which provides product specifications for tactile walking surface indicators and recommendations for their installation in order to assist in the safe and independent mobility of VI persons.

When designing assistive and medical devices, it is crucial to identify the hazards associated with the device, to estimate and evaluate the associated risks, to control these risks, and to monitor the effectiveness of the controls. The ISO 14971:2019 provides guidelines for the application of risk management to medical devices (including software) and specifies the terminology, principles, and processes for risk management. This standard is applicable to all phases of the device's life cycle. It also applies to risks associated with a medical device, such as bio-compatibility, data and systems security, electricity, moving parts, radiation, usability, etc.

In technology development, providing precise and reliable documentation is essential. Technical documents can be very complex; therefore, it is important to use the appropriate, clearly defined terminology to improve the understanding of the document; for example, in the case of AT, the ISO 9999:2016 establishes a classification and terminology of assistive products for PWDs.

12.8.3 Ethical and Social Considerations

It is important to address ethical and social considerations; for example, is the device reliable enough for the user to depend on it in unknown locations? Will the characteristics of the device put the user at risk if used in neighborhoods where crime is high? Do the benefits of the device justify its cost? Is the device financially accessible to most patients with visual or hearing impairments? Is the information captured by the device managed in a way that does not violate anyone's privacy? Is the look and feel of the device such that the user does not feel embarrassed or uncomfortable to wear it?

12.8.4 Social Entrepreneurship

It is crucial to ensure that the developed technology remains sustainably available. This means the creation of a complete value chain in parallel to the technology development and a business model which will allow the creation of a social business to industrialize the technology or the transfer of the intellectual property to an existing company interested in deploying and scaling up the technology (Makohliso et al., 2020). These models often promote cooperation between the North and South, enabling collaboration between communities and worldwide experts. Social entrepreneurship is often overlooked in research and literature regarding innovation and AT in the Global South. It is important to address this knowledge gap and generate evidence of how a good idea goes beyond a prototype or an academic paper to become a sustainable long-term solution.

12.8.5 Cooperation Between the North and South

It is important to address that in AT there are points where the needs in the South and North will coincide, points where it will half match, and points where it will not match at all; for example, making low-cost systems is useful in both hemispheres, since there are also poor communities in high-income countries. An example of a half-matched requirement is concealed devices; in low-income countries it is necessary for security reasons and aesthetics, while in high-income countries (with a safer environment) it is only necessary for aesthetics. Finally, something that does not match at all is the barrier in the use of navigation apps in low-income countries for security reasons (this does not exist in safer countries). In sum, directly transferring technology from one context to another (South to North or North to South) without re-contextualization, can be problematic and render it useless.

The AGILIS project from the International Committee of the Red Cross, Switzerland, is a successful example of context-sensitive, person-centered AT development that has taken measures to ensure sustainability by creating a social business: AGILIS (https://essentialtech.center/project/agilis/) developed a prosthetic foot, considering the context of LMIC in order to create a solution that can be universally accessible. After development of the technology, the International Committee of the Red Cross decided to launch the Rehab'Impulse initiative (www.rehabimpulse.org) to deploy the developed devices among PWDs in all regions of the world. The products are provided at affordable prices, meeting the needs of both service users and providers in LMICs, enabling the first step toward social inclusion. AGILIS was first tested in Vietnam; today it is in the process of being deployed to a variety of humanitarian settings where the International Committee of the Red Cross is present (Figure 12.2).

12.8.6 Broader Impact

The development of innovative new technologies for people with disabilities can generate positive publicity, raising awareness about their current conditions such

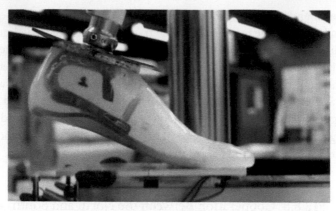

Figure 12.2 Low-Cost, Innovative Prosthetic Foot Developed by the AGILIS Project (©M.Janier/EPFL)

as their needs and the barriers they face in their daily lives. This might lead to the creation or improvement of technology procurement, as well as social programs and policies focused on aiding people with disabilities. For local governments, the implementation of high-tech solutions, especially to help poor constituents in the country, can generate reputational benefits and pride. This effect is quite notably visible, for example, in Rwanda, where the government has gone out of its way to be seen as at the edge of research on "drones for good" and other high-tech solutions (www.africandroneforum.org/). Our experience is that technology (via the media, who are eager to report on innovation) can raise awareness about the problems faced by vulnerable people in a positive way. In turn, this can inspire aid programs and/or legislation changes to improve it further (Thackeray & Hunter, 2010).

12.9 Conclusion

This chapter is a call to action for DPOs, engineers, disability scholars, rehabilitation practitioners, policymakers to shift the way in which technology is designed, tested, deployed, and sustained. It aims to set up the groundwork to be further developed and tested. Our recommendations are not exhaustive and intend to stimulate discussion and analysis regarding the underlying values in the design of AT.

Our preliminary research suggests upcoming cutting-edge technology is often designed for high-income countries and is expected to work in LMICs. This common and erroneous assumption has led to products that do not work properly, nor do they have the intended impact. Poor or superficial understanding of the lived experience of disability and everyday constraints faced by potential users in the Global South, as well as limited knowledge about infrastructure and impact of high maintenance costs reinforce the exclusion of PWDs in LMICs. This hinders the development of sustainable AT solutions that could allow PWDs in LMICs to achieve the kinds of lives they have reason to value.

In Africa, research has so far demonstrated that lack of policy and poor policy implementation, lack of AT awareness, supply chain challenges, lack of trained service providers, and lack of money greatly hamper access to these technologies (Visagie, 2019). There is therefore need for more support for governments and non-governmental organizations to prioritize meeting the unmet demand and expand the range of these technologies, especially in the rural areas (Matter & Eide, 2018; Marino et al., 2015). Uganda, Rwanda, and Ethiopia are some of the African countries that promoted the use of AT devices through policy and program interventions, but there is little to suggest that PWDs are fully reaping their benefits. For Ethiopia, local governments have been mandated to provide continuous training of professionals and staff working in rehabilitation services on the use of AT devices for persons living with disabilities (Ministry of Labour and Social Affairs, 2012). Under this intervention, the decentralization system has been used as a basis for setting up orthopedic centers to provide prosthetic appliances such as crutches, artificial limbs, wheelchairs, and other technical aids and devices to PWDs. In Rwanda, the Rwanda Revenue Authority and the Rwanda Utilities Regulatory Authority exempt taxes on all AT devices to promote the mobility and independence of persons living with disabilities (Rwanda Government, 2015). These exemplify some concrete actions and steps that governments and DPOs can follow to close the AT gap.

Academics, including bio-engineers, disability scholars, anthropologists, and economists, among others, shall work alongside PWDs toward developing rigorous methodologies to gain a better understanding of how AT could enable PWDs to live the lives they would like to live. Design of AT shall be focusing on enhancing the user's agency, bodily integrity, and capabilities, and not trying to "fix disabled bodies." To address the AT gaps, it is necessary to design technology with and for its potential users. The potential users and their community (local healthcare personnel, volunteers, family, employers, etc.) are aware of the issues, needs, and available resources of the area in which they live, and so their contribution shall be at the heart of AT design (Norman, 2019).

It is also crucial to ensure adaptability to the global context. Adding customizing tools or features to the developed technologies empowers the user to configure the device to change depending on their needs or environment, without the need of specialized personnel. This can only happen by putting in place a synergic multi-disciplinary approach. The design must focus upon the entire issue under consideration, engaging all the involved disciplines as a whole and not as isolated components. In other words, all the project stakeholders (technology experts, designers, bio-ethicists, social scientists, potential users, healthcare personnel, industry and policymakers) should work together as a big team, rather than smaller sub-sections, encouraging constant interaction and communication. Additionally, no activity should be developed in isolation but as a synergic part of a complex sociotechnical system.

To ensure that a prototype reaches the market, it is necessary to incorporate a viable business model. In order to assure that the technology will remain sustainably available, it is vital that developers include, in parallel to the technology development, the development of a complete value chain and business model. By the end of the development project, the technology and business plan can be used to create a new social start-up that will industrialize the technology, or be transferred to an existing company which is interested in deploying and scaling the technology. Entrepreneurship is at the heart of the strategy to create a long-lasting impact on the shortage of efficient AT.

Designing with and for those in LMICs will not exclude the fact that the developed technology can also be useful for those in high-income countries. This is because, firstly, PWDs are also at higher risk of poverty in high-income countries; therefore, affordable assistive devices adapted to low-resource environments will enhance access for this part of the population. Secondly, the advantages of usability and processing capacity are perhaps appreciated and beneficial to those well-off consumers. It is crucial to achieve a better understanding of how AT can enable PWDs to live the lives that they have a reason to value. Disability is very diverse, and there is a lack of evidence as to what type of technology works for different people in different contexts. Consequently, we need to gain insight into bodily integrity, and how functioning is perceived by persons living with impairments, in order to develop better technologies that can be a tool for self-determination.

References

Anderson, I., Robson, B., Connolly, M., Al-Yaman, F., Bjertness, E., King, A., ... & Yap, L. (2016). Indigenous and tribal peoples' health (The Lancet–Lowitja Institute Global Collaboration): A population study. *The Lancet*, *388*(10040), 131–157.

Bickenbach, J. (2013). The international classification of functioning, disability and health and its relationship to disability studies. In N. Watson, A. Roulstone, C. Thomas (Eds.), *Routledge handbook of disability studies* (pp. 65–80). Routledge.

Borg, J. & Östergren, P-O. (2015). Users' perspectives on the provision of assistive technologies in Bangladesh: awareness, providers, costs and barriers. *Disability and Rehabilitation: Assistive Technology*, *10*(4), 301–308. http://dx.doi.org/10.310 9/17483107.2014.974221

Cassit Orthopedics Ltd. (2016). *Medical personnel questionnaire*. Cassit Orthopedics Ltd.

Caycho, R. (2020). The flag of imagination: Peru's new reform on legal capacity for persons with intellectual and psychosocial disabilities and the need for new understandings in private law. *The Age of Human Rights Journal, 14*, 155–180.

Collins, C. C. (1985). On mobility aids for the blind. In D. Warren and E. Strelow (Eds.), *Electronic spatial sensing for the blind* (pp. 35–64). Springer.

Díaz Toro, A., Campaña Bastidas, S., & Caicedo Bravo, E. (2020). Methodology to build a wearable system for assisting blind people in purposeful navigation. *3rd International Conference on Information and Computer Technologies (ICICT)*, San Jose, CA, pp. 205–212. 10.1109/ICICT50521.2020.00039

Elli, G., Benetti, S., & Collignon, O. (2014). Is there a future for sensory substitution outside academic laboratories? *Multisensory Research, 27*(5–6): 271–291

Frediani, G., Busfield, J., & Carpi, F. (2018). Enabling portable multiple-line refreshable Braille displays with electroactive elastomers. *Medical Engineering and Physics, 60*, 86–93

Global Initiative for Inclusive Information and Communication Technologies. Retrieved May 2020 from https://g3ict.org/digital-accessibility-rights-evaluation-index/country_classification

Golledge, R., Marston, J., and Loomis, J. (2004). Stated preference for components of a personal guidance system for nonvisual navigation. *Journal of Visual Impairment and Blindness, 98*(3), 135–147.

Hanass-Hancock, J., Nene, S., Deghaye, N., & Pillay, S. (2017). "These are not luxuries, it is essential for access to life": Disability related out-of-pocket costs as a driver of economic vulnerability in South Africa. *African Journal of Disability, 6*, 1–10.

Handicap International. (2010). *Legal framework governing disability rights: Burkina Faso, Mali, Niger, Senegal, Sierra Leone and Togo, West Africa*. Report May 2020 retrieved from https://afri-can.org/wp-ontent/uploads/2019/08/legal_frame_ork_summary_gb1.pdf

Hickey, H., & Wilson, D. L. (2017). Whānau hauā: Reframing disability from an Indigenous perspective. *Mai Journal, 6*(1), 82–94.

International Organization for Standardization (ISO). (2010). *Ergonomics of human-system interaction—Part 100: Introduction to standards related to software ergonomics*. Retrieved from https://www.iso.org/obp/ui/#iso:std:iso:tr:9241:-100:ed-1:v1:en

International Organization for Standardization (ISO). (2020). Retrieved from www.iso.org

Leo, F., Ferrari, E., Baccelliere, C., Zarate, J., Shea, H., Cocchi, E., Waszkielewicz, A., & Brayda, L. (2019). Enhancing general spatial skills of young visually impaired people with a programmable distance discrimination training: a case control study. *Journal of NeuroEngineering and Rehabilitation, 16*, 108.

MacDonald, J., Henry, P., & Letowski, T. (2006). Spatial audio through a bone conduction interface: audición espacial a través de una interfase de conducción ósea. *International Journal of Audiology, 45*(10), 595–599.

Marino, M., Pattni, S., Greenberg, M., Miller, A., Hocker, E., Ritter, S., & Mehta, K. (2015). *Access to prosthetic devices in developing countries: Pathways and challenges.* IEEE 2015 Global Humanitarian Technology Conference. Retrieved from http://ieeexplore.ieee.org/stamp/stamp.jsp?tp=&arnumber=7343953

Makohliso, S., Klaiber, B., Sahli, R., Tapouh, J. R. M., Amvene, S. N., Stoll, B., & Schönenberger, K. (2020, July 4). Medical technology innovation for a sustainable impact in low- and middle-income countries: A holistic approach. https://doi.org/10.31224/osf.io/2dytg

Matter, R., & Eide, A. (2018). Access to assistive technology in two southern African countries. *BMC Health Services Research, 18*, 792–802.

McPherson, B., & Clark, J. (2017). *Preferred profile for hearing-aid technology suitable for low- and middle-income countries.* WHO. http://apps.who.int/iris/bitstream/10665/258721/1/9789241512961-eng.pdf

Meekosha, H. (2011). Decolonizing disability: Thinking and acting globally. *Disability & Society, 26*(6), 667–682.

Meekosha, H. & Shuttleworth, R. (2009). What's so "critical" about critical disability studies? *Australian Journal of Human Rights, 15*(1), 47–75.

Milotta, F. L. M., Allegra, D., Stanco, F., & Farinella, G. M. (2015). An electronic travel aid to assist blind and visually impaired people to avoid obstacles. In G. Azzopardi & N. Petkov (Eds.) *Computer Analysis of Images and Patterns.* CAIP 2015. Lecture Notes in Computer Science, vol 9257. Springer.

Ministry of Labour and Social Affairs. (2012–2021). *Ethiopia national plan of action (ENPA) of persons living with disabilities.* Addis Ababa.

Mitra, S. (2003, September 8–10). *The capabilities approach of disability* [Paper presentation]. Third Conference on the Capabilities Approach From Sustainable Development To Sustainable Freedom, University of Pavia, Italy.

Mitra, S. (2014). Reconciling the capability approach and the ICF: A response. *ALTER: European Journal of Disability Research, Forthcoming,* Available at SSRN: https://ssrn.com/abstract=2377333

Mitra, S. (2017). *Disability, health and human development.* Springer Nature.

Ministry of Gender, Labour and Social Development (MOGLSD, 2006). National Policy on Disability in Uganda, Kampala.

Mji, G. & Edusei, A. (2019). An introduction to a special issue on the role of assistive technology in social inclusion of persons with disabilities in Africa: Outcome of the fifth African Network for Evidence-to-Action in Disability conference. *African Journal of Disability, 8*(0), a681. https://doi.org/10.4102/ajod.v8i0.681

Murray, J., de Castro Cerqueira, D. R., & Kahn, T. (2013). Crime and violence in Brazil: Systematic review of time trends, prevalence rates and risk factors. *Aggression and Violent Behavior, 18*(5), 471–483.

Norman, D. & Spencer, E. (1 January 2019). *Community-based, human-centered design.* Retrieved from https://jnd.org/community-based-human-centered-design/

Persson, H., Ahman, H., Yngling, A. A., & Gulliksen, J. (2015). Universal design, inclusive design, accessible design, design for all: different concepts: one goal? On the concept of accessibility: historical, methodological and philosophical aspects. *Universal Access in the Information Society, 14*(4), 505–526.

Rivas Velarde, M. (2018a). The Convention on the Rights of Persons with Disabilities and its implications for the health and wellbeing of indigenous peoples with disabilities: A comparison across Australia, Mexico and New Zealand. *Disability and the Global South, 5*(2), 1430–1449.

Rivas Velarde, M. (2018b) Indigenous perspectives of disability. *Disability Studies Quarterly, 38*(4). http://dx.doi.org/10.18061/dsq.v38i4.6114

Rwanda Government (2015). *ICT and Disabilities in Rwanda*. Kigali.

Saha, M., Saugstad, M., Maddali, H. T., Zeng, A., Holland, R., Bower, S., Dash, A., Chen, S., Li., A, Hara, K, & Froehlich, J. (2019, May). Project sidewalk: A web-based crowdsourcing tool for collecting sidewalk accessibility data at scale. In *Proceedings of the 2019 CHI Conference on Human Factors in Computing Systems* (pp. 1–14).

Sen, A. (1993). Capability and well-being. In M. Nussbaum & A. Sen (Eds.), *The quality of life*. Clarendon Press. 10.1093/0198287976.001.0001

Sen, A. (2008). The idea of justice. *Journal of Human Development, 9*(3), 331–342.

Sonar, H. Gerratt, A., Lacour, S. P., & Paik, J.(2020). Closed-loop haptic feedback control using a self-sensing soft pneumatic actuator skin. *Soft Robotics, 7*(1): 22–29.

Tapu, R., Mocanu, B., & Zaharia, T. (2018). Wearable assistive devices for visually impaired: A state of the art survey. *Pattern Recognition Letters.* https://doi.org/10.1016/j.patrec.2018.10.031

Thackeray, R. & Hunter, M. (2010). Empowering youth: Use of technology in advocacy to affect social change. *Journal of Computer-Mediated Communication, 15,* 575–591.

Varao Sousa, T., Carriere, J., & Smilek, D. (2013). The way we encounter reading material influences how frequently we mind wander. *Frontiers in Psychology, 4,* 892.

Vilalta Perdomo, C., Castillo, J., Torres, J. (2016). *Violent crime in Latin American cities. IDB-DP-474.* Institutions for Development Sector. Inter-American Development Bank.

Visagie, S., Eide, A. H., Mannan, H., Schneider, M., Swartz, L., Mji, G., Munthali, A., Khogali, M., van Rooy, G., Hem, K.-G., & MacLachlan, M. (2016). A description of assistive technology sources, services and outcomes of use in a number of African settings. *Disability and Rehabilitation: Assistive Technology, 12,* 705–712. http://dx.doi.org/10.1080/17483107.2016.1244293

Visagie, S., Matter, R., Kayange, G., Chiwaula, M., Harniss, M., & Kahonde, C. (2019). Perspectives on a mobile application that maps assistive technology resources in Africa. *African Journal of Disability, 8*(0), a567. https://doi.org/10.4102/ajod.v8i0.567

Williams, R. M. & Gilbert, J. E. (2019). "Nothing about us without us": Transforming participatory research and ethics in human systems engineering. In R. D. Roscoe,

E. K. Chiou, & A. R. Wooldridge (Eds.), *Advancing Diversity, Inclusion, and Social Justice Through Human Systems Engineering* (pp. 113–134).

World Health Organization (WHO). (2001). *International classification of functioning, disability and health*. World Health Organization.

World Health Organization (WHO). (2011). *World report on disability 2011*. World Health Organization.

World Health Organization (WHO). (2018). Assistive technology. Retrieved from: https://www.who.int/news-room/fact-sheets/detail/assistive-technology

World Health Organization (WHO). (2019, August 22–23). *Global perspectives on assistive technology: proceedings of the GReAT Consultation 2019, Volume 1*. World Health Organization.

13

Ludic Design for Accessibility in the Global South

Manohar Swaminathan and Joyojeet Pal
Microsoft Research India

Man only plays when in the full meaning of the word he is a man, and he is only completely a man when he plays.[1]

(Friedrich Schiller)

13.1 Introduction

One rule that was clearly written out for the tech world as it watched Steve Jobs present new devices at one of Apple's mega events was that technology adoption was about the experience it provided (Grant & Sharma, 2011). Without undermining the instrumental efficiency of achieving the ends that a product sets out to enable, Apple's products decisively reframed the culture of usability—the value of the experience itself was brought to the center of design in technological devices (Burgess, 2012). Play, as a conceptual framework in thinking about product experiences, has been central to the design of technical artifacts for several years, and while several frameworks have evolved, including participatory design, value-centric design, reflective design, and ludic design, the fundamental concept of creating enjoyable experiences has remained constant (Sengers et al., 2005).

Research and product development around accessibility, on the other hand, have traditionally been driven by a utilitarian ethic—approaching it from a "greatest good" perspective that enables instrumental functions for individuals in their access to society around them. In this approach, which is often inherently ableist, technology is an offset for impairment. Through this lens, "improving quality of life" has typically been equated to moving toward functional parity with the non-disabled in the architectural or virtual worlds. This approach is pronounced in low-resource settings, such as in parts of the Global South, which have economic, infrastructural, and policy constraints. These constraints undermine the users' experiences in two important ways: First, the mainstream market products that make it to users may be best set up for the infrastructural realities

Accessible Technology and the Developing World. Michael Ashley Stein and Jonathan Lazar, Oxford University Press.
© Oxford University Press 2021. DOI: 10.1093/oso/9780198846413.003.0014

of where these products are designed—for instance, design artifacts may assume access to paved and safe sidewalks. Second, the lack of research and product development capacity of nation states in the Global South may shortchange their ability to invest in the development of product experiences that are designed for the cultural and social preferences of their populations. For instance, screen readers, developed and optimized largely for Roman scripts and environments with high computing power, are critical tools for making the digital world accessible to the blind community. A majority of the world's blind population is in the Global South, and for the vast majority of this population, English is not their primary language. However, there is very little development in text-to-speech technologies for most of the world's languages. For instance, there are about 20 Indian languages that have native speakers ranging from 20 million to 190 million. However, Hindi, with 190 million speakers, is the only language that has support for text-to-speech services offered by Amazon and Google.

Since the living public environment around us is largely inaccessible, accessible technologies are often the essential bridge that makes most forms of architectural and communication access possible for people with disabilities. Thus, people with disabilities often find themselves in near ubiquitous proximity to such devices or environments. Consequently, the enjoyability of accessible experiences has important consequences for peoples' willingness to use them—and indeed a significant body of work (Riemer-Reiss & Wacker, 2000; Shinohara & Wobbrock, 2011) has shown that awkward or non-pleasurable technology experiences have historically presented adoption challenges among populations of people with disabilities, even when the users clearly need and want to adopt the new technology.

We propose that the design of new technologies for people with disabilities should be driven neither incrementally, simply by the incorporation of accessibility features into mainstream technology, nor through a function-driven approach that ignores the experience of technology use. We argue for re-thinking both the design process and the usage scenarios of accessibility technology around the notion of playfulness and exploration rather than the received course of utility alone. We propose the notion of ludic design as a framework to re-think accessibility, specifically building on past work on play and enjoyment in the process of interaction with the world around oneself. We propose this as a framework to approach the usability of technology, but also the commercial logic of technology adoption, which is deeply tied to the former. We argue that, irrespective of structural impediments to the end-to-end implementation, there are steps to thinking about playfulness in design that can be incorporated both at the start of and midway into the design process.

Central to this approach to thinking about design artifacts is the notion of "homo ludens" or a "man of play," defined by Huizinga (1950) and used by Gaver (2002) in proposing ways of thinking about ludic design.

In order to truly leave work behind, we need to embrace an open-ended, self-motivated form of play. This is an engagement that has no fixed path or end, but instead involves a wide-ranging conversation with the circumstances and situations that give it rise. (Gaver, 2002; Pal & Chirumamilla, 2013)

The notion of initially learning to use a technology through exploration, or continuing to get better at it over time in the same way, is commonplace in the predominate, mainstream experience of technology use, and builds on a body of work that shows that exploration is a superior means of learning for using new technology (Lazar & Norcio, 2003). In contrast, while getting used to an accessible device also requires a significant amount of exploration, there is an important difference in the engineering of most accessible technologies. Typically, assistive technologies either have been custom designed for people with disabilities with very specific affordances (the abilities enabled by or suggested by the design of a technological artifact) in mind or end up being adapted for people with disabilities from mainstream commercial products. There is often a further level of adaptation in moving these assistive technologies to the Global South, which, too, is pegged to the functions they seek to enable. Gaze-controlled interfacing, which has recently become a standard part of Windows 10, is a potential boon to people with severe muscular impairments or those who have lost the use of their hands, enabling them to have full access to the digital world. However, in the Global South, such technology is invariably introduced with very specific educational goals for children (Jeevithashree, 2019; Wong & Lam, 2017).

13.1.1 The Role of Play

Playfulness within design does not simply relate to more usable or interactive functionality in a design artifact. It is a broader concept that relates to the healthy development of the learner or user of artifacts, starting from early childhood learning and continuing into adulthood (Gordon, 2014). There is also a significant body of work that links play to effective learning among both children and adults (Van Vleet & Feeney, 2015; Proyer, 2011), and there are strong links between play and playfulness and creativity (Russ, 2003) and innovation (Bateson, 2014).

The value of play in learning extends to other domains of accessible technology such as rehabilitative learning: for instance, virtual reality play that results in positive learning in children with cerebral palsy is described by Reid (2002), while the use of a robots for therapy is illustrated by Howard (2017) and Lathan (2001).

The role of play in development and wellbeing synthesized from different fields is presented in Gordon (2014). The example quote below discusses the role of play in the first year of a child's development:

Freedom to play without inhibition or constriction is a key ingredient for joy, interest, passion, and vitality later in life (Marks-Tarlow, 2012; Fredrickson, 2001; Panksepp, 2004). As psychologist Alan Schore put it, play creates a "positively-charged curiosity that fuels the burgeoning self's exploration of novel socio-emotional and physical environments" (Schore, 1994).

Our research is based on the premise that play and playfulness are central to what makes us human, and that by separating playfulness and exploration from the design experience, we fail the intended end users of our products. Given the separation of a large number of marginalized identities, particularly individuals with disabilities, from the design process, we propose that a re-think of the design of accessible technologies is needed. We propose that a ludic thinking approach can serve to more broadly understand the role of play across various learning experiences of people with disabilities, which can in turn be useful in approaching the design process.

The focus of our work is in regions where scalable deployment of technologies for accessibility faces major challenges. Disability and poverty are tightly coupled—people with disabilities are likelier to experience poverty, and those who are poor and have disabilities are likely to face further marginalization and exclusion from access to social and economic resources (Pellegrini, 1990; Yeo & Moore, 2003; Grech, 2016). In addition, disabilities are also likely to be misdiagnosed or detected later than in the Global North, resulting in development delays, preventable disabilities, delayed interventions, and further exclusion from peers and from the mainstream.

Cultural attitudes toward disability, including stigma that leads to the invisibility of people with disabilities from the public, are an additional factor that not only undermines peoples' ability to participate in society, but also is a barrier against the state being proactive in ensuring accessible spaces and infrastructure. The formal education system, which provides a space for socializing and learning about the world around us, and the healthcare system that provides for quality of life are both inaccessible for the vast majority of people with various disabilities.

The poverty or exclusion of people with disabilities from the mainstream has major implications for the deployment of technologies for accessibility. Adults with disabilities in these settings have a compounded deficit: lack of a specialized education, and a lack of access to accessible technologies in one's extended social circles. These can come together to make the adaptation of assistive technologies additionally challenging. Disability rights and capacity building have been championed by strong disabled peoples' organizations (DPOs) in various parts of the Global South. However, the corresponding skill-development work and introduction of assistive technologies for people with disabilities has been excessively focused on channeling people toward livelihoods and employability rather than enabling an individual's aptitude-based long-term self-realization. This, in development contexts, is often justified by pointing to the reality of constrained resources and infrastructure. This has the potential to direct the learning experiences of people with disabilities toward strict instrumental goals and channel them toward sustenance-level assistance for subsistence-level livelihoods, perpetuating this cycle. Thus, the utilitarian lens of accessible technology deployment can have long-term negative consequences.

Ludic design as a framework for thinking about accessible technologies is particularly relevant to technology use by children. For children, this includes introducing technology that is both fun to use and cognizant of their play practices. Indeed, play practices of children have long been considered central to their long-term development, and are even enshrined in international law. Accessibility professionals have argued that earlier introduction to technology has better longer-term impacts and increases peoples' ability to adapt to new technology.

13.2 Ludic Design for Accessibility

Researchers and developers working on accessibility seek positive impact on technology adoption by people with disabilities. While customization for the needs of the few who may not be served by off-the-shelf solutions is a central tenet of quality accessibility work (Kintsch & DePaula, 2002; Hurst & Tobias, 2011), the overall goal is always to maximize the number of people whose interactions with society can be enabled by better technologies (Plos et al., 2012). Our proposition of ludic design for accessibility contends that play and an enjoyable experience need to be considered as part of a technological interaction irrespective of whether the use case is of a single or small number of users, or of a mass-manufactured scenario.

An outcome of an artifact or solution from a ludic design approach will have two mutually synergizing aspects:

- At a minimum, the end users will have an enjoyable experience with the technology, making the artifact or system desirable to use.
- The goal of design should be to enable various forms of skill acquisition and long-term learning as a side effect of extended play by the users.

13.2.1 Framework for Ludic Design

The challenge for researchers using the ludic design approach is to ensure that the use-case goals they have in mind are not undercut by the intent to enable play, and vice versa. To this end, we propose ludic design for accessibility as a five-point framework that helps us to think about and to create artifacts for accessibility. We build on the characterization of play as described by Huizinga (1950) and propose these as underpinning attributes of any ludic design for accessibility (LDA).

(1) *Free activity*: An activity standing quite consciously outside "ordinary" life as being "not serious," but at the same time absorbing the player intensely and utterly; the player has the complete freedom to engage with the activity or not.

(2) *Interest agnostic*: An activity not necessarily connected with material interest; no profit is necessarily to be gained by it.

(3) *Bounded*: An activity that proceeds within its own proper boundaries of space and time.

(4) *Social*: An activity that promotes the formation of social groupings.

(5) *Desired side effect*: An activity that delivers some benefit or skill development to the end users of the artifact or activity, without detracting from the above elements.

The first four steps are Huizinga's formal description of pure play, while the fifth is the aspect that enables the delivery of the intended benefit to the target audience.

The notion of pure play is illustrated by the game called Calvinball in the comic strip Calvin and Hobbes by Bill Watterson. In the game, the players make up rules as they go, and however bizarre the rules may be, both the players abide by them. And, as Calvin says, "the only permanent rule in Calvinball is that you can't play it the same way twice" (Watterson, 1990). Calvinball presents an analogy for a kind of play that is central to most peoples' lived experiences—things that are purely play, without the structure of sport—whether rolling tires down a street, or blowing balloons, where the inherent pleasure is in the process.

13.2.2   Disability, Play, and the Global South

Play is recognized by scholarship and practice as highly valuable for children's cognitive and social development (Tarantino, 2018). Children have a high capacity for gaining competence in novel endeavors and creative expansion on those (National Scientific Council on the Developing Child, 2007). However, there has been less discussion on play as enabling a creative and overall positive state of being for adults, despite the case being made (Milteer et al., 2012). We argue that the cases of playful participation in successful design artifacts derive from a deeper reality of the human experience, in that play is a natural state of being. There are many examples of this in the technological artifacts we use as adults. Casually engaged browsing of video content, such as Instagram, with no clear goal other than momentary pleasure, or non-competitive gaming, are areas where this is evident.

The notion of LDA is of particular importance, as well as of unique challenge, because of the role of play in the lives of people with disabilities. Growing up, children with disabilities may be left out of both structured and unstructured play (Bateson & Martin, 2013) for a complicated mix of reasons related to the ways in which formal learning occurs. This can have significant consequences for cognitive development (Holt, 2007), including the long-term dissuasion from the idea of learning through play.

> Play forms an important foundation for the development of skills in all children. Unfortunately, for infants and children with disabilities, real play may be absent or diminished, replaced by therapies and/or special instruction. Infants and young children with disabilities experience barriers to play that are created by the nature of their disability. Parents of these children may feel they do not have time to play, given the demands of intervention and education. Alternatively, they may not know how to facilitate play with a child with a disability.
>
> (Lane and Shelly 1996)

People with disabilities are disadvantaged in accessing play-based learning also partly because of the nature of care, where needed. There is a high correlation of disability with poverty, and there is typically care involved which may not be available through the state. A child with disability is primarily cared for by the parents and immediate family, who fall further into poverty because of the disability in the family (Dalal & Pande, 1999). Resources are consequently focused on survival and sustenance from childhood, often turning into a vicious cycle of exclusion in school due to their "being different" (which is compounded by their lack of the play experiences that other children have), eventually resulting in an adult life that is likewise focused on livelihood and employability skills rather than leisure or play.

Thus, we need to be cognizant that the process and outcomes of ludic design for people with disabilities in a Global South setting is likely to present a vastly different set of challenges and opportunities than parts of the West, where play-based, exploratory use of technology is already the norm. Ludic design has the potential to enable people with disabilities in the Global South to indulge in play, and while that in and of itself is valuable, the play may result in the eventual acquisition of important skills. The nature of play, that of inducing formation of social groupings, is also appropriate in many parts of the Global South that still rely on a community-based living experience where the units of support are neighbors or villages, rather than families arranged into nuclear units (Ghai, 2015).

There are two challenges facing the ludic designer. The first is to create accessible play that satisfies four elements—it is a free activity, it is interest agnostic, it has boundaries, and is social. These in and of themselves pose a significant challenge. However, this brings us to the second challenge, since there is more to be accomplished. The inherent goal of LDA is to have a desired side effect of some "benefit," as understood through some curricular learnings that users are expected to achieve. This turns the largely understood design process on its head—instead of starting from the intended benefit and arriving at an assistive technology or solution to deliver the benefit, the designer has to conceptualize an enjoyable play experience in which the curricular benefits (say, addition, Boolean logic, etc.) are a likely second-order effect. This second challenge requires exploration of a larger design space and the need to re-design play to incorporate the benefits without losing the four elements of play.

13.2.3 Ludic Design, Gaming, and Gamification

Ludic design has similarities to gamification but, as will be clear, it is distinctly different: Gamification has two key components: a task to be completed or worked on; and elements of game mechanics creatively added around the task to incentivize people to engage with it, with the goal of maximizing completion of the task or improving its quality.

By intent, gamification seeks to achieve a material end goal, with incentives for players who participate in the form of money, badges, or some other form of recognition. These incentives, along with a strong element of competition, drive peoples' participation in such efforts. We subscribe to the distinction between play and competitive games as articulated by Gray (2009) and keep play central to our efforts. Another big difference is that while play is often characterized as competitive or agnostic, this view is also inherently ableist. To quote from Gray (2009):

> In my theory, contest is a morphing of play with something that is close to the opposite of play—a drive to beat and dominate others. When we combine these two opposites, play becomes more serious (and thereby more acceptable to contemporary adults) and domination becomes more playful—not entirely a bad thing, but not the same as pure play.

The difference between gamification and the ludic design approach can be captured with an example. Modern health apps encourage people to exercise and often do so with some comparative metric in mind (against other individuals, against a median for their age, or against their own past performance). These are play, in a sense, since they appeal to an individual's sense of enjoying a certain activity. One could interpret some of the key attributes of LDA as being present—it is a free activity, it has boundaries, and it is social. But it is arguably not interest agnostic, and the benefit is not the side effect.

Instead, the example of a child rolling a tire with a stick is an example of a playful activity that ends up contributing to dexterity, athleticism, and control. However, arguably unlike "gaming," the activity is fulfilling in and of itself, requiring no scores, no audience, no comparison. This is play in its purest form, incorporating all the critical elements of play. Thus, LDA is not gamification.

LDA is also not accessible gaming but, as seen in the illustrative example, could be an important component in situations where an existing game or play is being appropriated by the ludic designer. There are significant efforts around the world to bring the joy of games and sports to people with disabilities, recognizing the benefits as well as the negative impacts of exclusion from games (Chakraborty, 2017). Many of the common competitive sports in the world now have digital game versions that are accessible for people with disabilities.

While gaming, as defined by digital games played for recreational purposes, has important elements of non-utilitarian benefit for its consumers, it is generally defined as being the end in and of itself. There have been many efforts to make digital gaming accessible (Microsoft, 2018; Swaminathan et al., 2018; Zhao et al., 2018). These have indeed extended the possibility of playful appreciation of technological artifacts, and fulfilled most of the key attributes of LDA. However, there needs to be a more conscious effort to enable the side effect in a more constructive way rather than as expertise related to the skills one can acquire through a technological artifact.

13.2.4 LDA, an Illustration

We use an illustrative example to clarify the framework of LDA. Suppose the intended side effect is to inculcate numeracy into young children who are blind or low vision. Sighted children pick up concepts of big and small, one, many, or absence of things, ordering of things by size, etc., as part of growing up from infancy to childhood by visual experiences around them. Children who are born blind may have a harder time picking up these concepts, which are typically visually communicated, and may be disadvantaged when they start primary schooling. Hence enhancing numeracy is a desirable end goal of a ludic designer.

Figure 13.1 shows the LDA process with enhancing numeracy of blind children as the intended side effect. The figure shows the flow of activities around the use of playing cards, chosen as the instrument by the ludic designer (from a set of possible others, including, for instance, Ludo, Snakes and Ladders, etc.). The designer and the players, including both children with vision impairments and sighted, enter the play and go through three iterative processes: collaborative redesign of the cards to make them accessible with the designer playing a lead role, evolving and agreeing to the rules of play, again seeded with some rules by the designer, and repeated play. At every stage, all the aspects of pure play are

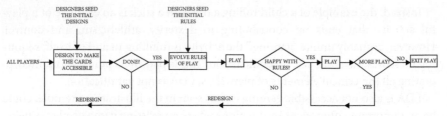

Figure 13.1 Ludic Design Process for the Numeracy Example

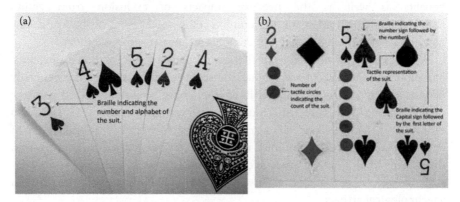

Figure 13.2 (a) Off-the-Shelf Braille Playing Cards (b) An Accessible Design for Braille Learners

maintained. As the children enjoy the play and play it often, an evaluator can join the play and in the process of playing evaluate the numeracy skills attained by the players, without breaking the fourth wall of play.

There are Braille playing cards available off the shelf (Figure 13.2a), and these use a combination of two Braille characters to identify each of the 52 cards. But it requires the players to be literate in Braille. One can make other tactile markings to make the cards accessible to both sighted and Braille non-literate children who are blind. Figure 13.2b shows an example of an inclusive design evolved during play with children.

13.3 LDA in Practice

Though we articulate LDA here for the first time, there have been some projects that have independently, though not explicitly, followed the LDA framework presented in Section 13.2. The use of Minecraft for autism is one such illustrative example.

Minecraft is the second-best-selling video game of all time, behind Tetris, and has, as of mid-2018, 91 million active monthly users across all platforms. One can

Figure 13.3 A Calming Quiet Garden in Minecraft (https://katerinegland.com)

surmise that the huge popularity of Minecraft could be due to the fact that it closely exhibits the four attributes of pure play: Minecraft is a free activity, with kids and adults indulging in it for countless hours of their own volition, with no material gain as a goal (see Figure 13.3). It has well-defined boundaries, being the archetypal sandbox world with well-defined rules and a virtual day–night cycle. One of the key reasons for Minecraft's popularity is the fact that diverse communities spontaneously form around activities in the Minecraft world.

The fifth element of the ludic design approach is illustrated by how Minecraft has been appropriated by the community to support young adults with autism spectrum disorder to engage better with each other inside the clearly artificial world of Minecraft (built out of unit-sized blocks made of different materials mined by the players) and eventually in the real world. Individuals can create "mods" for Minecraft and this feature has been utilized to create Autcraft, a Minecraft server that is a safe environment for children and adults with autism and their families to play Minecraft (Ringland et. al, 2016).

Autcraft is an excellent example of appropriation of a mainstream artifact into an assistive technology, with many documented benefits to the user community. The use of Minecraft to help socialize individuals with autism spectrum disorder is described in (Riordan & Scarf, 2017), which also describes how prosocial behavior can be encouraged even outside the game world.

13.3.1 Seven Steps to Ludic Design for Accessibility

We propose a seven-step process to facilitate a more systematic incorporation of an LDA approach to thinking about design. Each of these steps involves conscious thinking by the design team to incorporate ways of weaving play more closely into the end output of their work. These steps are the same, irrespective of whether the LDA is for an existing design artifact being re-conceptualized or for a new product altogether. Likewise, the steps and processes remain the same irrespective of whether the design thinking is for a mainstream use scenario or for something to be used very specifically by a single or small number of users. Along the way, we highlight some of the exciting research questions that need to be addressed.

(1) INVOLVE members of the community of users as the first step in the design process. Perhaps the greatest disservice a designer can do for the design process is imagine that they have answers for how a population of users may appropriate a technology, without the active participation of the community in the design process. This is especially true in operationalizing LDA, since it involves being able to understand what is joyful and pleasurable to the likely end user, not just what is usable for a defined end goal.

(2) EXAMINE play in the lives of the intended users of a technology, not just as part of the instrumental use-case scenario—i.e., "how may one enjoy the live process of engaging in this technology?"—but rather put an effort into examining what people understand as play in interactions similar to the ones that the technology aims to enable. In this, designers are asked not just to be contextual examiners, but to use play as a central concept of examination during their work with users.

(3) IDEATE on the pitfalls of design decisions, and ways in which play may be undermined by the design artifact. What are the assumptions with each element of the design? How can these impact the use experience? What are comparable products or technological settings that can be used to examine the potential pitfalls of the use cases?

(4) LIBERATE the use case from a strict orientation of rules. Designers should allow for flexibility for people to adapt the technology to the ways of use that will best suit them; to adapt specific elements of a product that may be playful and valuable to one user, but not to another. This does not mean that the design needs to be entirely open-ended; designers should have a vision for how the technology may be used, but also be open to the idea that people may use their own imagination to appropriate the technology as they see fit.

(5) ENABLE collaborative use of the technological device, even where it is intended primarily for use by individual users. The ability to collaborate is central to the social elements of play that are part of LDA; thus, using a technological milieu as a means of interpersonal connection is a central tenet of the LDA process.

(6) EVALUATE how, by intense engagement in the play, the intended goals of the design are being met or not. This requires devising the means to measure a trait or a skill that is the goal before and after extended engagement with the play and being able to establish the benefit of such play. The challenge is to design these measurements to be part of the play and not break the fourth wall of the play.

(7) ITERATE on the process with different play scenarios. Designers should assume that LDA is not a goal unto itself but that, like accessibility, it is only something to aspire to, not something that is achievable in its entirety. Designers must constantly re-think the design as time goes by, imagining new ways in which users can play with or enjoy elements of an experience.

Like in all human-centered design activities, these seven steps are not meant to be followed strictly linearly, but will have forward and reverse loops among all the steps, and hence the iteration is a continuous process of going forward and backward.

13.3.2 LDA and the Global South

We believe that the ludic design framework and process is universally applicable. However, we also believe that the Global South and ludic design are ideally matched for many reasons discussed throughout the paper. We highlight this with reference to the seven-step process and also indicate how lessons learned here can be applied to the Global North.

13.3.2.1 Involve and Examine

The voices of people with disabilities in the Global South are rarely heard in any national or international fora (Grech, 2012), and critical studies and understanding of the lived reality of people with disabilities is the essential first step in any efforts to create technology solutions—as discussed in Rembis (2016): "Disability and poverty, and the interactions between these are constantly renegotiated across space, time and people, including by disabled people themselves" (p. 61). It is crucial for Grech that theorists and other researchers and scholars not only recognize, but vigorously engage with the notion that disabled people actively make their own meanings and affect their own lived experiences, even in the most constraining environments.

Second, designers and technologists working to build assistive technologies need to work closely with the community that they are seeking to help. This broadens and deepens the nature of solutions that can be arrived at (Bennett et al., 2018).

Our personal experience has been that every person with a disability that we have asked about their play-life—what they play, whom they play with, etc.—has uniformly been delighted and surprised, said that it is the first time anyone has asked them about their play, and proceeded to share their personal stories with great warmth: a great start for collaboration.

13.3.2.2 Ideate and Liberate

The diversity of individuals, their peers and family, the chronic shortcomings of the infrastructure, the constraints on every kind of resource needed, all make it impossible for any artifact with a fixed set of rules and requirements to be deployed in the Global South. However, the tenets of pure play, that of the players deciding on the rules of time and space on the fly, the flexibility of the rules, give us the best chance of arriving at any practical solutions. Ludic design aims to transfer agency to the end users, where it rightly belongs.

13.3.2.3 Enable

Culturally, in the Global North, the quest is for solutions that cater to independence. As Holmes and Maeda (2018) describe it,

> In the United States, there is a deep attachment to the idea of a rugged, lone pioneer...venturing out into the great unknown to make their way in the world....These stories of independence rarely reflect the truth of our lives, which are full of dependencies....Interdependence is about matching complementary skills and matching contributions.

This is the lived reality of people with disabilities across the world. Thus, designing for interdependence is key to successful utilization of the solutions. Interdependence as a frame for assistive technology (AT) design has recently been introduced by Bennett et al., (2018) to complement the traditional emphasis on independence. By emphasizing the importance of social groups, ludic design intends to mesh with the reality of the lives of people with disability and leverage existing interdependence by co-opting the participation of the immediate family and support groups into the design process and in the use of the end artifact.

13.3.2.4 Evaluate and Iterate

This is the toughest challenge to be faced in the Global South. Without clear demonstration of the benefits of ludic design, it is difficult to direct the scarce resources of the community, including the time and efforts of the persons with disabilities (PWDs) and their immediate community, as well as the material resources needed to build out the solutions. This is a major research challenge as well.

We suggest that the designer be open to measures of success that may be evolved and articulated by the community during the course of the ludic design process and not be tied to any preconceived metric that the project could have started out with. A good measure of the effectiveness of the ludic designed artifact is the willingness and enthusiasm of the community to continue to engage with the artifact for reasons and benefits that they believe are important to them. For example, a solution intended to enable a child with low vision to become proficient in screen-reader usage may transform the child from a socially isolated child into the gregarious and sought-after storyteller of the neighborhood. The designer has to apply the principle of pure play even in the evaluation steps and be willing to change the rules of the game in consultation with others in play.

There are populations in countries like the United States (people with disabilities in inner-city neighborhoods, people experiencing homelessness, undocumented migrants, etc.) who can be clearly classified as the Global South in terms of their lived realities, and lessons learned in the Global South can be applied with suitable modifications. In the Global North, the rights-based model for supporting people with disabilities has taken strong roots and is supported by the resources and infrastructure in most communities. We might learn that designing for interdependence and with the involvement of social groups might result in better outcomes than the current approach to ATs.

13.4 Conclusion

We have argued for re-thinking both the design process and the usage scenarios of accessibility around the notion of playfulness and exploration rather than on utility alone. We have proposed a ludic design framework for accessibility with five key attributes, building on top of past work on play. To enable designers to put these attributes into practice, we have provided a seven-step design process.

While our goal is to present LDA as a means to re-think the design process, it is equally valuable to use LDA or play as a critical toolset to examine accessibility as it relates to peoples' life experiences. For this, we need to deeply examine play in the lives of people with disabilities. How does one understand play or leisure as part of one's daily lives? What are examples of playful activities in one's day? How do these relate to learning? How does one remember play growing up? How does play influence interpersonal relationships with people both with and without similar disabilities? These questions can in turn be used to examine ways in which play is part of or excluded from the daily technology use of people.

References

Bateson, P. (2014). Play, playfulness, creativity and innovation. *Animal Behavior and Cognition*, 1(2), 99–112. 10.12966/abc.05.02.2014

Bateson, P., & Martin, P. (2013). *Play, playfulness, creativity and innovation*. Cambridge University Press. 10.1017/CBO9781139057691

Bennett, C. L., Brady, E., & Branham, S. M. (2018, October). Interdependence as a frame for assistive technology research and design. In *Proceedings of the 20th international ACM SIGACCESS conference on computers and accessibility* (pp. 161–173). ACM.

Burgess, J. (2012). The iPhone moment, the Apple brand, and the creative consumer: From "hackability and usability" to cultural generativity. In L. Hjorth, J. Burgess, & I. Richardson (Eds.), *Studying mobile media: Cultural technologies, mobile communication, and the iPhone* (pp. 28–42). Routledge.

Chakraborty, J. (2017). How does inaccessible gaming lead to social exclusion? In J. Lazar & M. A. Stein (Eds.), *Disability, human rights and information technology* (pp. 212–223). University of Pennsylvania Press.

Dalal, A. K., & Pande, N. (1999). Cultural beliefs and family care of the children with disability. *Psychology and Developing Societies, 11*(1), 55–75. https://doi.org/10.1177/097133369901100103

Fredrickson, B. L. (2001). The role of positive emotions in positive psychology: The broaden-and-build theory of positive emotions. *American Psychologist, 56*, 218–226.

Gaver, B. (2002). Designing for homo ludens. *i3 Magazine, 12*, 2–6.

Ghai, A. (2015). *Rethinking disability in India*. Routledge.

Gordon, G. (2014). Well played: The origins and future of playfulness. *The American Journal of Play, 6*(2), 234–266.

Grant, D., & Sharma, A. (2011). Narrative, drama and charismatic leadership: The case of Apple's Steve Jobs. *Leadership, 7*(1), 3–26.

Gray, P. (2009). *Play makes us human I: A ludic theory of human nature*. Retrieved from https://www.psychologytoday.com/sg/blog/freedom-learn/200906/play-makes-us-human-i-ludic-theory-human-nature

Grech, S. (2012). Disability and the majority world: A neocolonial approach. In D. Goodley, B. Hughes, & L. Davis (Eds.), *Disability and social theory* (pp. 52–69). Palgrave Macmillan.

Grech, S. (2016). Disability and poverty: Complex interactions and critical reframings. In S. Grech & K. Soldatic (Eds.), *Disability in the Global South: The critical handbook* (pp. 217–235). Springer.

Howard, A., Chen, Y.-P., & Park, C.-H. (2017). From autism spectrum disorder to cerebral palsy: State-of-the-art in pediatric therapy robots. *Encyclopedia of Medical Robotics, 4*, 241–261. https://doi.org/10.1142/9789813232327_0010

Holmes, K., & Maeda, J. (2018). *Mismatch: How Inclusion Shapes Design*. MIT Press. https://mitpress.mit.edu/books/mismatch

Holt, L. (2007). Children's sociospatial (re)production of disability within primary school playgrounds. *Society and Space, 25*(5), 783–802. https://doi.org/10.1068/d73j

Huizinga, J. (1950). *Homo Ludens: A study of the play element in culture*. The Beacon Press.

Hurst, A., & Tobias, J. (2011, October). Empowering individuals with do-it-yourself assistive technology. In *Proceedings of the 13th international ACM SIGACCESS conference on computers and accessibility* (pp. 11–18). ACM.

Jeevithashree, D., Saluja, K. S., & Biswas, P. (2018, May). Gaze controlled interface for limited mobility environment. In *Proceedings of the 2018 ACM conference companion publication on designing interactive systems* (pp. 319–322). ACM.

Kintsch, A., & DePaula, R. (2002). A framework for the adoption of assistive technology. *SWAAAC 2002: Supporting learning through assistive technology*, 1–10.

Lathan, C., Vice, J. M., Tracey, M., Plaisant, C., Druin, A., Edward, K., & Montemayor, J. (2001). Therapeutic play with a storytelling robot. In *CHI'01 Extended Abstracts on Human Factors in Computing Systems* (pp. 27–28). ACM.

Mistrett, S. G., & Lane, S. J. (1996). Play and assistive technology issues for infants and young children with disabilities: A preliminary examination. *Focus on Autism and Other Disability Disabilities*, *11*(2), 96–104.

Lazar, J., & Norcio, A. (2003). Training novice users in developing strategies for responding to errors when browsing the web. *International Journal of Human-Computer Interaction*, *15*(3), 361–377.

Marks-Tarlow, T. (2012). The play of psychotherapy. *American Journal of Play*, *4*, 352–377.

Milteer, R., Ginsburg, K., & Mulligan, D. (2012). The importance of play in promoting healthy child development and maintaining strong parent-child bond: Focus on children in poverty. *Pediatrics*, *129*(1), e204–e213.

National Scientific Council on the Developing Child (NSCDC). (2007). *The timing and quality of early experiences combine to shape brain architecture: Working paper no. 5*. NSCDC.

Pal, J. & Chirumamilla, P. (2013). Play and Power: A Ludic Design proposal for ICTD. In *Proceedings of the Sixth ICTD: Full Papers – Vol. 1* (pp. 25–33).

Palmer, M. (2011). Disability and poverty: A conceptual review. *Journal of Disability Policy Studies*, *21*(4), 210–218.

Panksepp, J. (2004). *Affective neuroscience: The foundations of human and animal emotions*. Oxford University Press.

Pellegrini, A. (1990). Elementary school children's playground behavior: Implications for children's social-cognitive development. *Children's Environments Quarterly*, *7*(2), 8–16.

Plos, O., Buisine, S., Aoussat, A., Mantelet, F., & Dumas, C. (2012). A universalist strategy for the design of assistive technology. *International Journal of Industrial Ergonomics*, *42*, 533–541.

Proyer, R. (2011, August). Being playful and smart? The relations of adult playfulness with psychometric and self-estimated intelligence and academic performance. *Learning and Individual Differences*, *21*(4), 463–467.

Reid, D. T. (2002). Benefits of a virtual play rehabilitation environment for children with cerebral palsy on perceptions of self-efficacy: A pilot study. *Pediatric Rehabilitation, 5*(3), 141–148. 10.1080/1363849021000039344

Rembis, M. (2016). Disability studies. *The Year's Work in Critical and Cultural Theory, 24*(1), 174–197.

Riemer-Reiss, M. L., & Wacker, R. R. (2000). Factors associated with assistive technology discontinuance among individuals with disabilities. *Journal of Rehabilitation, 66,* 44–50.

Ringland, K. E., Wolf, C. T., Boyd, L. E., Baldwin, M. S., & Hayes G. R. (2016). Would you be mine: Appropriating Minecraft as an assistive technology for autism. In *Proceedings of the 18th international ACM SIGACCESS conference on computers and accessibility* (pp. 33–41). ACM.

Riordan, B., & Scarf, D. (2017). Crafting minds and communities with Minecraft. *F1000Research, 5,* 2339. 10.12688/f1000research.9625.2

Russ, W. S. (2003). Play and creativity: Developmental issues. *Scandinavian Journal of Educational Research, 47*(3), 291–303.

Schore, A. N. (1994). *Affect Regulation and the Origin of the Self: The Neurobiology of Emotional Development.* Lawrence Elbaum Associates.

Sengers, P., Boehner, K., David, S., & Kaye, J. (2005). Reflective design. In *Proceedings of the 4th decennial conference on critical computing: Between sense and sensibility* (pp. 49–58).

Shinohara, K., & Wobbrock, J. (2011). In the shadow of misperception: assistive technology use and social interactions. In *Proceedings of the SIGCHI conference on human factors in computing systems* (pp. 705–714). ACM.

Swaminathan, M., Pareddy, S., Sawant, T. S., & Agarwal, S. (2018). Video gaming for the vision impaired. In *Proceedings of the 20th international ACM SIGACCESS conference on computers and accessibility* (pp. 465–467). ACM.

Tarantino, B. (2018). Calvinball: Users' rights, public choice theory and rules mutable games. *Windsor Yearbook of Access to Justice, 35,* 40–68.

Van Vleet, M., & Feeney, B. C. (2015). Play behavior and playfulness in adulthood. *Social and Personality Psychology Compass, 9,* 630–643. 10.1111/spc3.12205

Watterson, B. (1990). *Calvin and Hobbes.* https://www.gocomics.com/calvinandhobbes/1990/05/27

Microsoft. (2018). *XBox AC, Xbox Adaptive Controller.* https://www.microsoft.com/en-us/p/xbox-adaptive-controller/8nsdbhz1n3d8

Yeo, R., & Moore, K. (2003). Including disabled people in poverty reduction work: "Nothing about us, without us." *World Development, 31,* 571–590.

Wong, L. M. S., & Lam, C. S. Y. (2017). Enhancement of reading ability using self-controlled gaze interface in a digital medium for children with neuro-oculomotor challenges. In L. Hall, T. Flint, S. O'Hara, & P. Turner (Eds.), *Proceedings of the 31st British Computer Society Human Computer Interaction Conference* (HCI '17). BCS

Learning & Development Ltd. Article 76, 7 pages. https://doi.org/10.14236/ewic/HCI2017.76

Zhao, Y., Bennett, C. L., Benko, H., Cutrell, E., Holz, C., Morris, M. R., & Sinclair, M. (2018). Enabling people with visual impairments to navigate virtual reality with a haptic and auditory cane simulation. In *Proceedings of the 2018 CHI Conference on Human Factors in Computing Systems*. ACM.

14

Multi-Country Comparison of ICT and Educational Accessibility for Blind Students

Sachin Pavithran
U.S. Access Board

Maria Hernandez Legorreta
Educational Consultant

14.1 Introduction

Access to information and communications technology (ICT) for all students across the globe is life-changing and critical to consider. ICT and assistive technology are linked to education, employment, independent living, and societal inclusion for students who are blind and have low vision (Telecentre Foundation, 2019; Scherer & Glueckauf, 2005). Blind and low-vision students rely on technology to access information that is available to the sighted public. As such, ICT serves as a venue to access the world for people with disabilities.

This chapter describes the availability of accessible education and ICT for blind and low-vision students in primary schools (K-12) within three different countries: India, the United Arab Emirates, and Mexico. These countries provide examples of policies, barriers, successes, commonalities, differences, and outcomes when implementing ICT in public education. These common challenges are ubiquitous, as evidenced by a report on the implementation of ICT under the Convention on the Rights of Persons with Disabilities (CRPD) in 137 countries (G3ict, 2017), and demonstrates the imperative need for more access to education and ICT for students with disabilities.

14.2 Mexico, India, and the United Arab Emirates

Mexico is a template and policy leader for Latin America. Although it has a growing economy, Mexico is considered a developing country. India has one of the largest expanding economies in the world, but not without huge equity disparities. The United Arab Emirates boasts great economic prosperity in major cities.

Accessible Technology and the Developing World. Michael Ashley Stein and Jonathan Lazar, Oxford University Press.
© Oxford University Press 2021. DOI: 10.1093/oso/9780198846413.003.0015

Education is a priority and receives significant funding, but there are substantial challenges, especially in rural areas.

Each of these countries has unique ICT experiences and successes, yet they share implementation and access challenges by blind and low-vision students. All struggle with delivering ICT and education to rural communities. Their educational policies for inclusive education are modeled after the U.S. Individuals with Disabilities Education Act, but implementation is lacking. All three countries receive international pressure to make inclusive education a reality through the CRPD.

While the United Arab Emirates may not be considered a developing country due to its flow of wealth in Dubai and Abu Dhabi, underdevelopment and disparities exist everywhere else. Expensive luxury development and automobiles in Dubai are in stark contrast to the camels and dirt roads in the rest of the kingdom. Rural areas in the United Arab Emirates face some of the same challenges as developing nations in terms of infrastructure and education.

India, Mexico, and the United Arab Emirates (UAE) share a history of European colonization, which tends to have a lasting impact, including a struggling economy (Grier, 1999). India was under British rule from 1757 to 1947 (Elphinstone, 2014). The UAE was under British rule from 1820 to 1971 (Morton, 2016) and remained small, mostly autonomous kingdoms within a larger country when it gained independence. Mexico was under Spanish reign from 1521 until 1810 (Escalante et al., 2013).

While colonization often diminishes the number of languages spoken, India has at least 22 languages (English and Hindi being the most common), with 2,000 ethnic groups (Gulf News India, 2018). Mexico has 62 recognized indigenous languages, although Spanish is the official spoken language (World Atlas, 2019a). Arabic is the UAE's official language, but English is taught in schools and used for commerce (World Atlas, 2019b).

India, Mexico, and the UAE all experience growing economies, diverse populations, wealth disparities, warm climates, inclusive education laws, and misperceptions about disability. The most important aspects for people with disabilities are the availability of employment opportunities, accessible education, and ICT.

14.3 Definitions, Policies, and Laws

ICT provides access to information through complex systems and networks, with a focus on communications technologies. This includes internet, cell phones, wireless networks, and other communication modes (TechTerms, 2018). ICT can facilitate universal educational access, bridge learning divides, support teacher development, enhance learning quality and relevance, strengthen inclusion, and improve education governance (United Nations Educational, Scientific, and Cultural Organization, 2019). ICT advances have a profound impact on whether or not people with disabilities are able to participate in education, employment, and community settings. Without ICT access, blind and low-vision students will not fully experience inclusive education and be prepared for the workforce.

Laws and policies regarding education, disability rights, accessibility, and inclusion in the UAE, Mexico, and India are strongly influenced by United Nations (UN) instruments and goals. Yet they all face their own challenges in making these policies come to fruition.

The CRPD recognizes the importance of accessibility to the physical, social, economic, and cultural environment; health; education; information; and communication in enabling PWDs to fully enjoy all rights and freedoms (UN) General Assembly, 2006a). Article 24 of the CRPD recognizes the rights of PWDs to receive equal access to free, compulsory, and inclusive primary and secondary education (UN General Assembly, 2006b). All three countries ratified the CRPD over a decade ago (International Center for Not-for-Profit Law, 2018). Ratification entails commitment to bring legal frameworks and policies into compliance. States parties must take legislative and administrative measures to implement the CRPD and abolish discriminatory laws, regulations, and practices against PWDs (UN General Assembly, 2017). Unfortunately, implementation often presents challenges for countries struggling with their general and special education frameworks.

CRPD Article 9 addresses equal access to the physical environment and ICT and other services available to the general population in urban and rural areas (UN General Assembly, 2006a). The Global Initiative for Inclusive Information and Communication Technologies (G3ict) was established to support ICT accessibility in accordance with the CRPD. Public procurement has increased as an important policy tool for governments to promote accessible ICT equipment, software, applications, and services necessary for digital inclusion and CRPD implementation (G3ict, 2015).

The UN 2030 Agenda for Sustainable Development outlines 17 sustainable development goals designed to address critical world problems and transformational change (UN General Assembly, 2015). Worldwide government officials gathered in 2015 to create educational reform, including an implementation framework. The delegates examined education initiatives and policies, human rights treaties, the Muscat Agreement, conference outcomes, and the Global Education Monitoring Report. Although significant educational progress had been made, countries were far from reaching Education for All goals (UN Educational, Scientific, and Cultural Organization, 2016). Solutions that may work in the developed world may not be practical or feasible in developing countries. Even developing countries with similar economies may face different obstacles. Most countries, especially those with fewer resources, have experienced problems in implementation, as demonstrated below.

14.4 Evaluation of ICT Accessibility in 137 Countries

An evaluation report of CRPD ICT accessibility compliance in 137 countries is covered in much more detail in Chapter 3: Global Trends for Accessible Technologies in the Developing World, and provides further illustration of the lack of ICT and educational access. Notably, the average ICT country compliance rate was 41%. Only 21% of countries felt they had the capacity for ICT implementation G3ict, 2017). This report demonstrates that ICT accessibility work is critically needed. If a country is not accessible, its primary schools likely are not either. In regard to accessibility policies for blind individuals, 33% of countries had no implementation, 41% minimal, 16% partial, 9% substantial; no countries had full implementation levels (G3ict, 2017). ICT and assistive technology accessible in primary and secondary schools is not implemented in 51% of countries, while 13% had minimal, 30% partial, and 1% full implementation. Only 18% of countries had accessible websites. There were 56% of countries with screen readers. Only 34% had libraries for the blind, and 30% had assistive technologies available to university students (G3ict, 2017).

The availability of general and accessible education varies greatly worldwide based on many factors, including infrastructure, poverty, literacy, policy, economy, and conflicts (G3ict, 2017). Countries struggling with infrastructure and providing quality general education will also struggle with quality education and ICT for students with disabilities. These numbers confirm that implementation is a barrier for many CRPD signatory countries, and that ICT is not getting into the hands of people who are blind and have low vision. The first step is developing policies that dictate what should be done, including implementation plans and appropriate funding. If countries do not have any policies in place, implementation will not occur.

Mexico, India, and the UAE are all working toward inclusion of students with disabilities, including accessible ICT and materials. Each country faces unique challenges. These countries' history, economy, education laws, policies, ICT, and accessibility to blind and low-vision students in the K-12 setting will be discussed below.

14.5 India

14.5.1 Background and History

India is home to 1.3 billion people living in one of the oldest inhabited regions in the planet (Stein & Arnold, 2010; Ancient Encyclopedia, 2012). After gaining independence from British rule in 1950, India became the largest democracy (Sorenson, 2018), and is expected to become the world's most populated country by 2028 (BBC News, 2018a).

There are an estimated 12 million blind and low-vision residents, one-third of the world's blind population (India News, 2019). Nearly 20% are school-aged children (Census Commissioner, India, 2018). However, this staggering growing number is considered conservative, given the social stigma attached to blindness, unwillingness to disclose disability, and inability to locate blind individuals.

14.5.2 Social Stigma and Unemployment

Social stigma around disability and blindness can be an obstacle to accessing education. Misunderstanding, shame, and stigma are reflective of attitudes around the world, whether countries have large or small economies. Many Indian families keep their children home because of shame and stigma (A. K. Mittal, personal communication, February 23, 2018). Families have been the care givers, breadwinners, and service providers of people with disabilities in traditional societies. Expecting families to forego their role and give control to the state can be difficult (Byrd, 2010). Developing and implementing policies will not be as beneficial and straightforward without considering the impact on all societal factors.

India has the fourth largest GDP worldwide and generates more than $250 billion in annual exports (Sorenson, 2018). Although it boasts the second largest world labor force with an increasing middle class, 20% live in abject poverty (Sorenson, 2018). Only 37.6% of people with disabilities are employed, compared to 62.5% of the population (WHO, 2011).

This employment disparity is seen across both developed and developing countries (Gudlavalleti et al., 2014). This inequality would lessen with more accessible inclusive education as educational attainment leads to employment outcomes (Howe et al, 2013). India has successfully implemented quality education in some public schools, but experienced difficulty doing so in others, including ICT availability.

14.5.3 Education

The Constitution of India mandates free and compulsory education for all children aged six to 14, regardless of religion, race, caste, or language (Government of

India, 1950). However, 40% of school-aged children do not attend school (Byrd, 2010). This is a reoccurring issue in countries where the general educational system is not effectively serving all students. It is difficult to implement special education policy in an already struggling system.

The PWDs Act of 1995 authorized free integrated education with equal access for children with disabilities through 18 years of age to all levels, and mandated teacher training in disabilities (Extraordinary Gazette of India, 1996). However, this Act was before students with disabilities were integrated, and a "separate but equal" educational system started. Teacher training was a good initiative, but general education teachers had minimal contact with students with disabilities mainly taught in segregated settings.

The purpose of India's 2000 flagship education policy, Sarva Shiksha Abhiyan, was to provide universal access to primary education and make learning resources available, including textbooks and computers. The policy mandates that children with disabilities be identified and provided with supports, including aids, appliances, monitoring, and supervision (Department of School Education and Literacy, 2019).

There are two schooling options for students with disabilities: integrated typical schools and segregated special schools. Standard integrated schools have seen a minimal number of students with disabilities enroll, as supports and services are not provided. Blind, low-vision, and deaf students have benefited most from these schools (Byra, 2010), and are usually integrated with little support. Teachers may feel unprepared to accommodate students with other disabilities. There are about 2,500 schools serving many students with disabilities.

This segregated approach may be one reason why employment opportunities are so low. Both types of schools are mainly located in urban areas, making it difficult for rural students to attend. Not all students with disabilities are being served in either the integrated or special schools. The special education law has made inroads, but has not been carried out to the extent intended. A great discrepancy exists between rural and urban settings, and each state has varying success with implementation (Byrd, 2010). While the law exists, the funding is not always available, and public awareness of resources are not widely known.

14.5.4 ICT Coverage and Rural Accessibility

India has one of the largest and fastest-growing internet markets worldwide, with over 460 million users, expected to be 635.8 million by 2021 (Internet World Stats, 2017; The Statistics Portal, 2018). Due to its large communications infrastructure, India is predicted to "leapfrog" their development to catch up with industrial countries (Miller, 2001). However, these gains are primarily expected in the business sector. The reality of access for individuals and education is much bleaker. Only 26% of Indians have access to the internet (https://www.statista.com/statistics/255135/internet-penetration-in-india/). Accessibility varies by region, with the majority using mobile devices. Men are twice as likely as women to be internet users. City populations are twice as likely to have internet access as those living in rural areas (The Statistics Portal, 2018).

Both rural and urban access to education and ICT is vital. Internet options help students with disabilities access the technology they need to complete schoolwork. Schools that serve students with disabilities with technology are mainly exclusive to urban settings, making the disparities widen and impeding educational attainment. Reaching students in rural settings needs to be a priority, as 66% of India's population lives in rural communities with digital connectivity and infrastructure challenges (Trading Economics, 2019).

14.5.5 Actuality

The enrollment of students with disabilities has been extremely slow, even with new educational policy (Shanmugan, 2011). This is often due to access issues, misconceptions about disability, family shame, lack of support, and unavailability of rural schools. Most children with disabilities (90%) are not attending school, which is five times the general rate (Gabel & Chander, 2008). This problem is exacerbated by the lack of: accessible facilities and lesson content; well-educated, sensitive, and competent teachers; and inclusive assessment methodologies. These factors contribute to the huge drop-out rate for the tiny percentage of children with disabilities attending school (Shanmugan, 2011). Not surprisingly, illiteracy rates among people with disabilities are at 52%, compared to 35% for the general population. The overlap of having a disability and living in a rural setting makes the attainment of education a hard thing to come by. It becomes difficult to access educational ICT when access to education is an issue.

14.5.6 Policy and Implementation Gaps

Policy implementation gaps are exacerbated or caused by a lack of financial commitment to ensure inclusive education. A 2008 World Bank evaluation reported that spending on inclusive education was only 1%. While India is a hub of technology and innovation, the disability focus has been on finding cures instead of accessible technology to participate in education and society. Many countries focus on finding cures with technology instead of enabling ICT access to education.

As in other developing countries, supplementary services are provided sparingly through local organizations, not through government assistance or public schools. The National Association for the Blind of India has schools for the blind and provides public school support. The Department of Education and National Association for the Blind distribute fee-based technology, limiting access based on income. They provide Braille writers to students based on availability, which can be unreliable. The National Association for the Blind offers a talking book program, but it is unclear how many students use it. It also trains teachers and students with Braille instruction.

The Secondary Education Board required schools to have technology in a minimum number of classrooms by 2010, but rural schools are struggling to implement this. There are policies, programs, and mechanisms for blind and low-vision students to get ICT, but the barriers make it difficult to access and connect devices in a meaningful way.

One of India's strengths in ICT implementation is their inclusive education policy. Unfortunately financing this policy and implementing technology in classrooms has not yet come to fruition. Significant barriers to access include shame and misconceptions about disability and access to schools in rural areas, where most of the population lives. These barriers also occur across the UAE and Mexico, as described below.

14.6 United Arab Emirates

14.6.1 Background and History

The UAE was founded in 1971 as a federation of seven kingdoms. The economy is based on oil and gas, tourism, trade, aviation, commerce, finance, renewable energy, and telecommunications. Although a conservative Islamic country, the UAE has become more open and tolerant of other cultures and religions (Alborno, 2017). This is primarily because the many immigrants coming for employment with their families make UAE one of the more diverse places in the world.

Due to its large GDP of $414 billion (Trading Economics, 2019), the general employment rate is 77% (Najm, 2019). However, the employment rate of people with disabilities is extremely low at 7% (Haza, 2019), often due to discrimination and lack of job accommodations. People with disabilities receive a small monthly allowance that forces many to live with their extended families. The government hopes to substantially improve public, private, and higher education by investing 20% of the budget: $4.4 billion in 2017, expected to be $7.1 billion by 2023 (Najm, 2019). This expenditure has a dramatic impact on ICT availability and education in general. The large affluent international population pressure the government to provide high-quality education so their children can attend prestigious colleges (Najm, 2019).

While the UAE is considered a country, each "city" is a kingdom, with autonomous policies and practices. There is little sharing of money between kingdoms. Meaningful change is difficult with the wealth concentrated in Dubai and Abu Dhabi, and not spread to other rural kingdoms.

14.6.2 ICT Coverage

The UAE has an estimated 8.8 million internet users, one of the highest in the Arab world. Although widespread, the internet is heavily regulated and controlled to target political or religious opposition (BBC News, 2018c). Students do not have open access to social media and information, which substantially limits communication.

14.6.3 Education Law and Policy

The first disability rights law was in 2006 (Federal Law 29), establishing equal opportunity in education, healthcare, training, and rehabilitation, and integrating people with disabilities into public and private schools. However, negative attitudes and wording ("suffering from," "deficiency," "infirmity," "normal"), impede their participation. As with India, misconceptions about disability can hinder access to education.

Federal Law 29, based on the U.S. Individuals with Disabilities Education Act, is seen internationally as a good special education framework. The practice of taking and implementing policies without considering the regional context, history, and relevance makes implementation difficult to accomplish. The law states that people with disabilities can enroll in any school without exception (UAE Ministry of Education, 2019). Support centers monitor the success of students with disabilities. Five of the eight schools serving students with disabilities are in Dubai.

The goal of inclusion has a long way to go, but UAE is making progress. In the 2014–2015 school year, 156 schools across the country were "inclusive"; however, this may mean they educate one student with a disability or many, as those numbers are not reported.

The progressive and ambitious "School for All" initiative, based on Law 29 and the CRPD, directs how UAE schools administer special education (Alborno, 2017). Similar to the United States, students are assessed, and those eligible receive special education services. An Individual Education Plan is developed, including accommodations, which can be computers, text-to-speech software, and more (UAE Ministry of Education Special Education Department, n.d.). Listing ICT in policy with examples of what a student can receive is a unique aspect that Mexico and India lack. This may be due to the fact that the law is based on the Individuals with Disabilities Education Act, but this does not necessarily mean adequate implementation.

Dubai has an ambitious goal of making the city physically accessible and educationally inclusive by 2020. The 2017 Dubai Inclusive Education Policy Framework states that access to quality education for all students should be provided by effectively meeting their needs in responsive, accepting, respectful, and supportive ways (Ponce de Leon, 2017). In theory, ICT specificity and inclusive education policy would mean that students with disabilities were receiving support and ICT to fully access the curriculum with their peers.

14.6.4 Actuality

The UN Economic and Social Commission for Western Asia (2017) states that the UAE's framework conforms to the CRPD, and guarantees equal access of PWDs to the general education system. A recent study (Alborno, 2017) found ICT to be underutilized in primary schools due to a lack of teacher training. Teachers did not know how to use devices or were concerned about breaking them. They attempted to improvise, but with limited success. While funding was available, teacher training on specific devices was not, which proved to be a full implementation barrier. Funding for training and support is vital to access technology.

Other inclusion barriers include the lack of: qualified special education professionals, proper training for teachers in mainstream classrooms, knowledge about inclusion among senior administrators, financing for resources and services in private schools, and awareness of the issues that students face in inclusive settings (Anati & Ain, 2012). These specific barriers can serve as a blueprint for how the UAE can improve its inclusive practices that ultimately affect ICT access and use. A comprehensive system that ensures everyone understands inclusion on a broader level and is able to implement it can be the way to move forward.

Another barrier is the cost to families, as students with disabilities must pay more to attend schools. A special education rule states that students with disabilities may pay regular school fees plus an additional 50%. The UAE is working to lower costs, but adds a "reasonable" fee for students with disabilities, which can be interpreted broadly (Williams, 2019). This initial cost to enroll is a huge barrier to educational achievement. While many people in Dubai and Abu Dhabi are wealthy, this is not the case in rural areas. A family's socioeconomic status determines whether children with disabilities can enter school, notwithstanding the other barriers that arise after enrollment. Charging a higher price to attend school puts students with disabilities and their families at a huge disadvantage to sighted peers from the start.

The UAE is succeeding in some areas, while improvement is needed in others. The UAE has a good educational framework and funding to provide high-quality ICT devices and training. However, in practice, UAE has some similar barriers to India. While resources are allocated to purchase ICT, it appears that barriers include costs associated with special education, teacher preparation (both general and special education), lack of special educators, access in rural areas, attitudes and stigma around disability, and implementing policy.

14.7 Mexico

14.7.1 Background and History

Approximately 116 million people call Mexico home (BBC, 2018). Mexico boasts the second largest economy in Latin America. In 1976, immense offshore oil reserves were discovered, making Mexico a major oil exporter. However, much of rural Mexico remains impoverished. Large shanty towns are on the peripheries of major cities (BBC News, 2018b). More than 50% of children live in poverty and 5% of children between 6 and 14 years old do not attend school (Alvarez, 2016). Mexico has been in crisis from a drug war which has monopolized needed resources over the last 12 years (CNN, 2018).

14.7.2 ICT Coverage and Accessibility

Mexico's internet usage has steadily increased, with over 70 million active users, projected to be 91.6 million by 2021 (The Statistics Portal, 2017). Approximately 51% of students have access to the internet, mostly in the central or northern regions. Most users access the internet through mobile devices. There is considerably less coverage in rural areas lacking the needed infrastructure (Alvarez, 2016). As with India and the UAE, rural communities in Mexico have significantly less access to education, ICT, and resources.

14.7.3 Education Law and Policy

Mexico has fairly progressive educational laws, but implementation is lacking. General Education Law Article 39 mandates that special education students be placed in inclusive classrooms with accommodations. The Mexican National Public Education Secretariat established two educational systems to support students with disabilities. One includes service units designed to provide curricula modifications and supports to eliminate or minimize learning barriers within regular schools. The second comprises specialized resource centers for students with visual, auditory, physical, and mental disabilities integrated into classrooms. These segregated centers do not include interaction with peers (Russo & Lozano, 2015) and generally provide lower-quality educational rigor than inclusive settings (OECD, 2000). Students with disabilities can be included in mainstream schools with proper supports and services (OECD, 2000).

14.7.4 Actuality

The system is not providing inclusive education to blind and low-vision students. More infrastructure, accommodations, tools, and trained teachers are needed for accessibility. It is estimated that 80% of blind children do not attend public school (Alvarez, 2016). Most of these excluded children seek education from private institutions and civil society (Martinez, 2017; Alvarez, 2016). Accessible and general technology is not considered a right, is limited, and is not widely provided by the government. The mainly overarching laws do not specify how to provide services for students with disabilities.

The Mexican Supreme Court recently ruled in case 714/2017 (Campana Latinoamerican por el Derecho a la Educacion, 2018) that excluding students with disabilities from public education is unconstitutional (Expansion, 2018) and that two systems (general and special education) is also unconstitutional. They cited the CRPD with no exception for disability or environment. This ruling has the potential to significantly affect students and their use of educational ICT. New guidelines, regulations, and funding are needed to execute this enormous decision. Thus far, there has been no educator training or additional funding and resources allocated for implementation.

Teachers state that students are not denied entry, and are able to enroll, but that there are no resources, teachers, or services to help them. There has been no change thus far, as students are not enrolling in their public schools and educators feel unprepared to teach students with sensory disabilities. In 2018, President Obrador of Mexico has made promises about educational reforms, but what changes will actually occur has yet to be seen.

14.8 Conclusion

ICT and accessible education availability vary greatly worldwide based on infrastructure, poverty, literacy, policy, economy, conflicts, and other factors (G3ict, 2017). Countries struggling with infrastructure and providing quality general education will also struggle with quality education for students with disabilities and ICT.

Rural communities generally have significantly less access to education, ICT, and resources. Illiteracy and lack of education are detrimental to the lives of blind and low-vision people in all countries, and opportunities for inclusive education are critical for these groups to be able to live independent and productive lives, free from poverty, abuse, and exploitation.

ICT is vital to facilitate universal educational access, bridge learning divides, support teacher development, enhance learning quality, and strengthen inclusion (UN Educational, Scientific, and Cultural Organization, 2019). Access to ICT in primary education could reduce discrimination and provide the necessary tools for equal opportunity, accessibility, and full inclusion and participation in society.

The education policy objectives of India, the UAE, and Mexico are to provide equitable and inclusive education for all students. The policies in all three countries aim to make inclusive education a reality; however, implementation seems to be a common problem. Accessible educational ICT is not generally addressed in primary school settings, especially for blind and low-vision students. A comprehensive system that ensures everyone understands inclusion on a broader level and is able to implement it can be the way to move forward.

Although the benefits of ICT access are essential, obstacles include large populations, prejudice, inadequate teacher training, lack of infrastructure, and poverty in rural areas. Another difficulty is policy without implementation plans and adequate financial commitment. Education is generally impacted by similar issues as ICT implementation.

Mexico, India, and the UAE provide a broad representation of issues that other countries around the world may be facing. The evaluation of CRPD implementation in 104 countries exemplifies common barriers and the crucial need for more access to education and ICT. The unavailability of education, and the lack of curricula and ICT access, is most likely the main contributing factor for the disproportionately low employment rate of blind and low-vision students compared to their sighted peers. ICT access leads to post-secondary employment for blind and low-vision students (Howe, 2013).

Countries have to consider innovative ways to bring ICT and educational access for all students. If countries do not have any policies in place, implementation will not occur. The impact of policies on all societal factions must also be considered. Although policies are an appropriate beginning, they will not magically solve problems, including the lack of educational opportunities and infrastructure in rural areas, teacher training, and stigma around disability. This will take a much more concentrated and coordinated effort as well as adequate financial commitment by governments, local organizations, international organizations, and other stakeholders in the education system.

References

Alborno, N. E. (2017). The 'yes...but' dilemma: Implementing inclusive education in Emirati primary schools. *British Journal of Special Education, 44*(1), 26–35.

Alvarez, C. G. P. (2016). Education technology in Mexico: Developing new tools for students in Latin America. *User Experience Magazine, 16*(2). Retrieved from http://uxpamagazine.org/educational-technology-in-mexico/

Anati A. & Ain, A. (2012). Including students with disabilities in UAE schools: A descriptive study. *International Journal of Special Education, 27*(2), 75–85.

Ancient History Encyclopedia. (2012). *Ancient India.* Retrieved from https://www.ancient.eu/india/

British Broadcasting Corporation (BBC) News. (2018a). *India country profile.* Retrieved from https://www.bbc.com/news/world-south-asia-12557384

British Broadcasting Corporation (BBC) News. (2018b). *Mexico country profile.* Retrieved from https://www.bbc.com/news/world-latin-america-18095241

British Broadcasting Corporation (BBC) News. (2018c). *United Arab Emirates county profile.* Retrieved from http://www.bbc.co.uk/news/world-middle-east-14703998

Byrd, E. (2010). India, families, and a special school. *TEACHING Exceptional Children Plus, 6*(3), 11.

Campana Latinoamerican por el Derecho a la Educacion. (2018). *Mexico: La Exclusion de estudiantes con discapacidad del Sistema educativo general es discriminatoria y, por tanto, inconstitucional.* Retrieved from https://redclade.org/noticias/mexico-la-exclusion-de-estudiantes-con-discapacidad-del-sistema-educativo-general-es-dis-criminatoria-y-por-tanto-inconstitucional/

Census Commissioner, India. (2018). *Disabled population by type of disability, age and sex.* Retrieved from http://www.censusindia.gov.in/2011census/population_enu-meration.html

Cable News Network (CNN). (2018). *Mexican drug war fast facts.* Retrieved from https://www.cnn.com/2013/09/02/world/americas/mexico-drug-war-fast-facts/index.html

Department of School Education & Literacy. (2019). *SSA Shagun.* Retrieved from http://www.ssa.nic.in/index.html

Escalante-Gonzalbo, P., García-Martinez, B., Jáuregui, L., Vazquez, J.Z., Speckman-Guerra, E., Garciadiego, D. J., & Aboites-Aquilar, L. (2013). *A new compact history of Mexico* (1st edition). El Colegio de México. Retrieved from https://catalyst.library.jhu.edu/catalog/bib_5873871

Expansión (2018). *Excluir a Estudiantes con discapacidad de clases regulares es anti-constitucional.* Retrieved from https://expansion.mx/nacional/2018/10/05/excluir-a-estudiantes-con-discapacidad-de-clases-regulares-es-anticonstitucional

Extraordinary Gazette of India. (1996). *The Persons with Disabilities (Equal Opportunities, Protection of Rights and Full Participation) Act.* Retrieved from ind51207_(2)disableact1995.pdf (indiacode.nic.in)

Global Initiative for Inclusive Information and Communication Technologies (G3ict). (2015). *CRPD implementation: Promoting global digital inclusion through ICT procurement policies & accessibility standards.* G3ict Publication & Reports.

Global Initiative for Inclusive Information and Communication Technologies (G3ict). (2017). *Convention on the Rights of Persons with Disabilities (CRDP) 2016 ICT Accessibility Progress Report.* Retrieved from https://g3ict.org/publication/2016-crpd-ict-accessibility-progress-report

Gabel, S. & Chander, J. (2008). *Inclusion in India education.* Retrieved from https://www.researchgate.net/publication/236221765

Government of India. (1950). *The Constitution of India.* Retrieved from https://www.india.gov.in/sites/upload_files/npi/files/coi_part_full.pdf

Grier, R. (1999). Colonial legacies and economic growth. *Public Choice, 98*(3/4), 317–335. Retrieved from http://www.jstor.org/stable/30024490.

Gudlavalleti, M., John, N., Allagh, K., Sagar, J., Kamalakannan, S., & Ramachandra, S. (2014). Access to health care and employment status of people with disabilities in South India, the SIDE (South India Disability Evidence) study. *BMC Public Health, 14*(1), 1125. https://doi.org/10.1186/1471-2458-14-1125

Gulf News India. (2018). *Census: More than 19,500 languages spoken in India as mother tongues.* Retrieved from https://gulfnews.com/world/asia/india/census-more-than-19500-languages-spoken-in-india-as-mother-tongues-1.2244791

Haza, R. (2019). *93 per cent of disabled emirates are unemployed, says minister. Special Olympics World Games.* Retrieved from The National, https://www.thenational.ae/uae/93-per-cent-of-disabled-emiratis-are-unemployed-says-minister-1.837508

Howe, J. (2013). *Analysis of the disparity in post-secondary educational attainment and employment between individuals with visual impairment and the general population.* The University of Arizona. ProQuest Dissertations Publishing. Retrieved from http://search.proquest.com/docview/1436970349/

India News. (2019). *Number of blind to come down by 4m as India set to change blindness definition.* https://www.hindustantimes.com/india-news/india-to-change-definition-of-blindness-reduce-number-of-blind-by-4-million/story-HxHKeH3XpfPBEtSr2moerO.html

International Center for Not-for-Profit Law (ICNL). (2018). *Civic freedom monitor: India.* Retrieved from http://www.icnl.org/research/monitor/india.html

Internet World Stats. (2017). *India: Internet usage stats and telecommunication market report.* Retrieved from https://www.internetworldstats.com/asia/in.htm

Martinez, V. (2017, June 3). *Educational system excludes blind.* Retrieved from http://www.milenio.com/estados/sistema-educativo-excluye-a-invidentes

Miller, R. (2001). *Leapfrogging?: India's information technology industry and the internet.* World Bank. https://openknowledge.worldbank.org/handle/10986/13954

Morton, M. (2016). *Keepers of the golden shore: A history of the United Arab Emirates.* Reaktion Books.

Najm, M. (2019). *United Arab Emirates: Education*. Retrieved from https://www.export.gov/article?id=United-Arab-Emirates-Education

Organisation for Economic Co-operation and Development (OECD). (2000). *Inclusive education at work: Students with disabilities in mainstream schools*. OECD Publishing. https://doi.org/10.1787/9789264180383-en

Ponce de Leon, J. (2017, November 23). *Private schools, colleges in Dubai should be fully inclusive by 2020*. Gulf News. Retrieved from https://gulfnews.com/news/uae/education/private-schools-colleges-in-dubai-should-be-fully-inclusive-by-2020-1.2129581

Russo, C. J. & Lozano, R. (2015, Feb). Special-education law in Mexico and the United States. *Educational Leadership Faculty Publications, 177*, 33–36. Retrieved from https://ecommons.udayton.edu/cgi/viewcontent.cgi?article=1176&context=eda_fac_pub

Scherer, M. & Glueckauf, R. (2005). Assessing the benefits of assistive technologies for activities and participation. *Rehabilitation Psychology, 50*(2), 132–141. Retrieved from https://www.researchgate.net/profile/Marcia_Scherer2/publication/232570391_Assessing_the_Benefits_of_Assistive_Technologies_for_Activities_and_Participation/links/5419d4d50cf25ebee98881b1/Assessing-the-Benefits-of-Assistive-Technologies-for-Activities-and-Participation.pdf

Shanmugam, R. (2011). Access to higher education and workplaces for persons with disabilities. In *Enabling access for persons disabilities to higher education and workplace: Role of ICT and assistive technology*. Never-the-less. http://includ-ed.eu/sites/default/files/documents/journal-_enabling_access_for_persons_with_disabilities_to_hi.pdf

Sorenson, T. (2018). A way out. *Utah State Magazine, Summer 2018*, 20–25.

Stein, B. & Arnold, D. (2010). *A history of India*. (2nd ed.). John Wiley & Sons.

TechTerms. (2018). *ICT*. Retrieved from http://techterms.com/definition/ict

Telecentre Foundation. (2019). *ICTs and assistive technology in education: Paving the way for the integration and inclusion of people with disabilities*. Retrieved from https://www.telecentre.org/icts-and-assistive-technology-in-education-paving-the-way-for-the-integration-and-inclusion-of-people-with-disabilities/

The Statistics Portal. (2018). *Internet usage in India: Statistics & facts*. Retrieved from https://www.statista.com/topics/2157/internet-usage-in-india/

The Statistics Portal. (2017). *Internet usage in Mexico: Statistics & facts*. Retrieved from https://www.statista.com/topics/3477/internet-usage-in-mexico/

Trading Economics (2019). *India rural population*. Retrieved from https://trading-economics.com/india/rural-population-percent-of-total-population-wb-data.html

United Arab Emirates Ministry of Education (2019). *Education for people of determination*. Retrieved from https://www.government.ae/en/information-and-services/education/education-for-people-with-special-needs

United Arab Emirates Ministry of Education Special Education Department. (n.d.). *School for all: General rules for the provision of special education programs and services (public & private schools)*. Retrieved from https://www.moe.gov.ae/English/SiteDocuments/Rules/SNrulesEn.pdf

United Nations Educational, Scientific, and Cultural Organization (UNESCO). (2016). *Education 2030: Incheon declaration and framework for action for implementation of sustainable development goal 4*. Retrieved from http://unesdoc.unesco.org/images/0024/002456/245656e.pdf

United Nations Educational, Scientific and Cultural Organization (UNESCO). (2019). *ICT in education*. Retrieved from https://en.unesco.org/themes/ict-education

United Nations Economic and Social Commission for Western Asia. (2017). *Monitoring the compliance of national legal frameworks in Arab countries with the convention on the rights of persons with disabilities December 2017: Executive summary*. Retrieved from https://www.unescwa.org/sites/www.unescwa.org/files/page_attachments/monitoring-compliance-national-legal-frameworks-crpd-advance-copy-en.pdf

United Nations General Assembly. (2006a). Convention on the Rights of Persons with Disabilities. Retrieved from https://treaties.un.org/Pages/ViewDetails.aspx?src=TREATY&mtdsg_no=IV-15&chapter=4&clang=_en

United Nations General Assembly. (2006b). Convention on the Rights of Persons with Disabilities. Retrieved from www.usicd.org/index.cfm/convention#article24 https://treaties.un.org/Pages/ViewDetails.aspx?src=TREATY&mtdsg_no=IV-15&chapter=4&clang=_en

United Nations General Assembly. (2015). *Transforming our world: 2030 agenda for sustain development*. Retrieved from http://www.un.org/ga/search/view_doc.asp?symbol=A/RES/70/1&Lang=E

United Nations General Assembly. (2017). *Monitoring the compliance of national legal frameworks in Arab countries with the convention on the rights of persons with disabilities*. Retrieved from https://www.unescwa.org/sites/www.unescwa.org/files/page_attachments/monitoring-compliance-national-legal-frameworks-crpd-advance-copy-en.pdf

Williams, C. (2019). *Special needs education: Trends and future outlooks in the UAE*. The Knowledge Review. Retrieved from https://theknowledgereview.com/special-needs-education-trends-future-outlook-uae/

World Atlas. (2019a). *What languages are spoken in Mexico?* Retrieved from https://www.worldatlas.com/articles/what-languages-are-spoken-in-mexico.html

World Atlas. (2019b). *What languages are spoken in the United Arab Emirates?* Retrieved from https://www.worldatlas.com/articles/what-languages-are-spoken-in-the-united-arab-emirates.html

World Health Organization. (2011). *World report on disability*. Retrieved from www.who.int/publications/i/item/9789241564182

15

Ride-Hailing as Accessible Transit

A Case Study of Blind Users in India

Vaishnav Kameswaran
University of Michigan

Joyojeet Pal
Microsoft Research India

15.1 Introduction

Accessible transportation services are central to the quality of life for people with visual impairments, enabling participation in public spaces by allowing access to health, education, and job opportunities (Marston & Golledge, 2003; Pal et al., 2015; Accessible Mass Transit—American Foundation for the Blind, 2017). Their importance is further underscored in Article 9 of the United Nations (UN) Convention on the Rights of Persons with Disabilities (CRPD), which cites accessible transportation services as a key requirement for independent living (UN CRPD, 2006). Despite this, in many parts of the Global South, public transportation services, which are overcrowded, chaotic, and inaccessible, are difficult to use for people with visual impairments. However, the recent entry of ride-hailing services such as Uber, and its Indian counterpart, Ola, has attempted to change this landscape by providing its users with convenient access to cabs via mobile apps. Although in academic literature ride-sharing and ride-hailing are used interchangeably, with both referring to services like Uber and Lyft, we draw a distinction between these terms. We use ride-hailing to refer to when a rider hires a personal driver to take them to a destination, as opposed to ride-sharing, where a rider shares a vehicle with others (e.g., UberPool).

In this chapter, we examine the impact of these services on people with visual impairments in India. Overall, we found that our participants benefited from the increased independence, accessibility, flexibility, and safety that resulted from their use of Uber and Ola cabs. However, using ride-hailing services was not without its challenges, and this included finding the exact location of the cab on arrival, choosing and negotiating destination addresses, and dealing with drivers. Finally, we discuss how addressing some of these challenges through means like

Accessible Technology and the Developing World. Michael Ashley Stein and Jonathan Lazar, Oxford University Press.
© Oxford University Press 2021. DOI: 10.1093/oso/9780198846413.003.0016

driver incentives and training will improve the experiences of these services for people with visual impairments in metropolitan India.

15.2 Transportation in India

India has a vast and varied transportation network, with most metropolitan cities in the country providing access to urban mass-transit services as well as door-to-door services. Urban mass-transit services include buses and metro trains, while door-to-door services include auto-rickshaws and cabs. Auto-rickshaws are three-wheelers which operate in many parts of India, and both auto-rickshaws and cabs are hailed from the side of the street, often by waving to catch the driver's attention. However, urban mass-transit services are mostly overcrowded and chaotic, making them dangerous and hard to use. For instance, in Mumbai and Delhi, mass-transit services run at twice their capacity and are consequently difficult and unsafe (Pucher et al., 2004). Furthermore, the lack of accessible infrastructure like public announcement systems makes it additionally difficult to determine when the bus or train will arrive.

Streets in many urban areas in the country lack sidewalks, resulting in pedestrians needing to navigate in the path of vehicular traffic to access transit. Door-to-door services like auto-rickshaws have to be hailed from the side of the street, whereupon the auto-rickshaw driver and the rider typically negotiate over the price before embarking on the journey—a process that sometimes needs to be repeated multiple times prior to single-trip commencement. Using both urban mass-transit services and traditional door-to-door services is challenging for sighted people, and these challenges are magnified for people with visual impairments.

Ride-hailing services have attempted to change this transportation landscape by providing access to private fleets of cars via mobile apps. There are two prominent ride-hailing services in the country—Uber, active since 2010 and operating in 40+ cities, and Ola cabs, operational since 2010 and prevalent in over 110 cities. The ride-hailing infrastructure in India is different to that in the United States. For instance, while most drivers in the U.S. are part-time, a majority of the drivers in India work full-time. Uber and Ola cab drivers in India are often employed by transportation contractors, as opposed to the United States, where drivers own the vehicles they drive. They also need to be driving cars that have permits to operate as taxis and have yellow license boards, making them easy to identify. Thus, in contrast to drivers in the Global North, who are often described by riders as being "just like me" (Dillahunt et al., 2017; Kameswaran et al., 2018a), drivers in India are typically from the working class, while riders tend to be middle- and upper-class Indians. The Uber interface in India, like in the case of the United States, allows its users to set pick-up and destination locations, estimate how long their rides will take, choose between different cab types, and rate drivers. The dominant mode of payment in India is cash, and while there is no systematic published empirical evidence on this, there have been plenty of news reports suggesting that cabs tend to cancel more frequently in India based on the type of payment or drop location (Dixit, 2018; Griswold, 2018).

15.3 Motivation and Methods

In early 2016, we discovered multiple posts online describing the positive impact of ride-hailing services on people with visual impairments. For instance, a post from a user of Uber and Ola cabs on a popular youth media platform in India, YouthKiAwaaz, read:

> App-based cabs significantly contributed in doing away with nightmarish experiences which visually [impaired] persons used to face while going outside and looking for public transportation which are still largely inaccessible to PWDs…these modes of transport [public] are made for able-bodied persons who's expected to run and catch the bus…but for someone like me who can't manage these things, day to day travel becomes challenging and sometimes frustrating. (UI, 2016)

Posts like the one above motivated this study, and consequently we conducted a qualitative study consisting of interviews and observations to better understand the impact of the services on blind people in India. The interviews focused on eliciting details about our participant's ride-hailing practices, including their experiences with ride-hailing services, challenges they encountered while using them, subsequent workarounds, and the impact of these services on their perceptions of dependence and independence. The interviews were semi-structured and elicited both narrative accounts of peoples' experiences with ride-hailing (via questions such as "Tell us about your last bad experience with Uber and Ola cabs?") and conceptual responses regarding notions of independence resulting from their use of Uber and Ola cabs (through questions like "How would you describe your sense of dependence/independence when you used Uber and Ola cabs?").

Also, we separately conducted observations of people with visual impairments using ride-hailing services to complement the interview data. In these observations, we accompanied people with visual impairments on their regular journeys, which started when the participants used the mobile app to book a cab and concluded when they reached their final destination. These observations allowed us to capture some of the finer details of peoples' experiences with ride-hailing, which would have been hard to capture in the interviews, including their use of the mobile app and interactions and conversations with the driver.

In all, 30 people with visual impairments participated in the interviews, at which point we reached theoretical saturation and found no new emergent themes. A sub-set of these interviewees (n = 8) participated in the observations. We recruited participants through AccessIndia (n = 15)—an online listserv for people with disabilities in India, personal contacts (n = 6), and subsequent snowballing (n = 9). The interviews, conducted in English, included a combination of face-to-face and Skype/phone calls. The first author conducted observations in Bengaluru, a city in South India and the primary field site for the study. Most trips were from our participants' homes to their workplaces, and here we accompanied them as travel companions—assisting/helping them when asked. Bengaluru is one of the most populous cities in India, and one of 53 urban agglomerations in the country (Wikipedia Contributors, 2020). Transit services in Bengaluru are similar to other metropolitan cities in the country, and include buses, metro trains, auto-rickshaws, and ride-hailing services.

15.4 Demographics

Participants (n = 30) came from eight metropolitan cities in the country and included 24 men and six women. The youngest participant in the study was 24 years old, while the oldest was 53 at the time of the study. All participants belonged to relatively affluent classes—all had formal higher education and those that were not students held full-time jobs. This is not representative of the large population of people with visual impairments in India and the Global South—a majority of whom are low-income and lack formal education (The World Bank, 2007). In fact, past work has highlighted the challenges with digital accessibility research that inordinately captures a relatively wealthy class of people with disabilities (Pal et al., 2013, 2017). Nonetheless, our data capture an important perspective of a rider group whose ride-hailing experiences are under-studied and for whom the limited access to accessible transportation is a significant barrier to inclusion and participation in everyday activities.

15.5 Findings

We found that specific features of ride-hailing services offered several benefits over existing transportation services—including an increased sense of independence, safety, door-to-door transit accessibility, flexibility, and reduced emotional work (Kameswaran et al., 2018b). However, using these services was not without its challenges, and we observed that participants struggled with picking and choosing destinations and finding the precise location of the cab on arrival. Further, the lack of disability awareness among drivers, who were central to the accessibility of the services, also posed occasional difficulties for the participants.

15.6 Benefits of Ride-Hailing Services

15.6.1 Increased Independence

Ride-hailing had a positive impact on the lives of many of our participants: they were able to get out and about a lot more with the likes of Uber and Ola cabs. One of the major advantages of these services was that they no longer needed to rely on sighted people to get around, and this made them feel more independent (Kameswaran et al., 2018b).

> It's [Uber] liberating—it's fun. You are not dependent on a [personal car] driver to come. As a blind person, you are dependent. I can pick up the phone—book a cab get out…it has removed me from the tyranny of depending on a human for transport. Now, I have a mainstream solution which works. (P14)

This is in contrast with other forms of transportation, where people with visual impairments sought help from sighted people at multiple junctures during the same journey. For instance, bus users sought help in getting to the bus stop (because streets lacked sidewalks and were inaccessible), and while on the bus they also had to ask about approaching stops to determine if they had arrived at their destination.

Although many of our participants had participated in Orientation and Mobility training, which acquaints them with public transportation systems, they still noted how overcoming the inaccessibility of public transportation proved to be challenging and necessitated finding the right kind of help at the right time. On the other hand, ride-hailing users noted the increased autonomy in the case of Uber and Ola cabs, where they were no longer contingent on the availability of others to get around. Moreover, they had the freedom to go where they wanted, at a time of their choosing. Participants also noted how, in addition to giving them the confidence to get around on their own, ride-hailing also instilled confidence in others, i.e., their friends, family, and peers, that they could travel safely. As P12 said,

> [I]t has increased our pride, it has increased our prestige before outside world, these people can also travel here and there without anybody's help. Other people including my family and my colleagues and others on the road, they feel that he

cannot go alone...People get anxious about myself—whether I will be reaching the correct location or not—that is one problem, that confidence I am able to give to my people around. (P12)

In addition to the increased sense of independence, Uber and Ola cabs offer practical benefits such as increased accessibility, flexibility, safety, and reduced emotional work, which we will now highlight.

15.6.2 Increased Accessibility and Flexibility

First, the fact that participants could book a cab "on their own" from their mobile phone and receive notifications when their cab arrived was in contrast to mass-transit services and auto-rickshaws. In the cases of buses and trains, people with visual impairments often need assistance in getting to the bus stop or train station and in identifying the bus or train when it has arrived, because many bus stops or train stations in India lack public announcement systems.

Moreover, asking for help to get a bus does not always have predictable results, and people with visual impairments often find themselves at the mercy of strangers to assist them. Using an auto-rickshaw has a different set of challenges— it requires the traveler to venture out to main roads, listen intently to the unique sound of an auto, and signal intent (often by waving) to board. As with buses, respondents reported frequent reliance on others to board autos. P27 clearly stated the relative advantages of the hailing process in the case of Uber and Ola cabs in comparison to auto-rickshaws:

> While we have to find the auto on the road it's very difficult, like we have to take help from the sighted person, and we have to wait for the auto on the road....And definitely, it is more comfortable or more accessible to book a cab. (P27)

There are other implications to people being able to book cabs by themselves—in addition to enabling them to travel, they can also use these services to book cabs for others, thus allowing them to contribute to their families and friends, who people with visual impairments were otherwise dependent on to access transportation, which further resulted in an increased sense of self-esteem.

> I live with a set of parents who are nervous even after I have been blind for 25 years and being a 48-year-old woman, my parents are still nervous about me, being out on the road on my own...it [Uber] has now given them the confidence that I will book cab for them sitting over here. (P37)

There is uncertainty associated with hailing cabs and autos from the side of the street, where riders need to wait until an unoccupied cab/auto passes by and the driver expresses willingness to take them to their destination. Uber and Ola cabs did away with this uncertainty to an extent, by offering an estimate of how long a car would take to reach the rider (assuming that the driver does not cancel). Moreover, they offered increased flexibility, which meant that our participants could now book a cab from the comfort of their homes and workplaces at a time of their choosing.

Some of our participants, like P6, had regular arrangements with auto-rickshaw drivers or others, whom they employed to drive a car they owned. All of them recalled times when they were stood up without notice and had to resort to Uber or Ola cabs to get around. Further, others noted the increased flexibility of these services at times when other forms of transportation are unavailable or hard to access; for instance, late at night.

> And flexibility, it has offered me more flexibility—like suppose I wish to leave work early, I don't need to call—I can call the auto person who offers me pick up and drop but in case he is not available, you know I don't have to bother much, I can just book an Uber or Ola or taxi and I can go home early if I want to. (P6)

Planning for trips is also made easier by the availability of information like an estimated time of arrival, which enables more accurate planning than public transportation. Although finding, hailing, and boarding public transportation is challenging for sighted travelers as well, for people with visual impairments these challenges are magnified, and they are likely to take longer to board a bus or find an auto-rickshaw. Furthermore, using buses, trains, and auto-rickshaws requires a certain amount of local knowledge (for instance, locations of stops and the local language to seek help when required, which can be difficult in a diverse country like India), but the consistent way in which Uber and Ola cabs operate across geographical contexts makes them a convenient option for our participants when traveling.

15.6.3 Increased Safety and Reduced Emotional Work

Participants also noted the increased sense of safety they felt while using ride-hailing services. This stemmed in part from their ability to track their journey using maps and share their trips with other people (in Uber). Further, in the case of auto-rickshaws, the auto-rickshaw drivers decide the route to take to a destination, which made it hard to determine where one was going, which had safety implications.

Okay before this I used to travel using auto on my own and now I think I feel much more safer... One more reason could be the fact that I am using Google Maps now... The routes that are shown on their [drivers] thing [phones] is the same that is shown on my phone (P2)

Finally, the automatic fare calculation and access to integrated digital payment systems in Uber and Ola cabs resulted in a reduction of emotional work. This is in contrast to auto-rickshaws, where a necessary step is bargaining with the auto-rickshaw driver about the price of the journey. Although auto-rickshaws often carry digital meters, auto-rickshaw drivers do not necessarily use them, and auto-rickshaw drivers and riders negotiate on the price of the trip at the start of the journey. Often, one has to repeat this process with multiple auto-rickshaw drivers before actually boarding an auto. As P9 elucidates,

I have to tell my security in the flat to get me an auto... then I have to bargain with the guy, boss "I have to go here, there"—"How much?" I have to bargain— see now that is not there. I take an Uber, put the location here, put the other location it shows me the price... So that's the comfort I am talking about. (P9)

Consequently, the automatic fare calculation in Uber and Ola cabs liberated our participants from troubles associated with bargaining and reduced the chances of cheating. Although some Uber and Ola cab drivers cheated our participants (for instance, by demanding more money than the cost of the trip when they paid by cash), they noted how the availability of customer support (through the Uber mobile app and phone support in Ola) meant that they could complain and get their money back. On the other hand, when auto-rickshaw drivers cheated them (for instance, when auto-rickshaw drivers did not hand back the right change, which they realized much after the conclusion of the trip) they were helpless.

In spite of these advantages, ride-hailing was not without its challenges. Participants struggled with many aspects, including finding the precise location of the cab on arrival, choosing and negotiating the destinations, and dealing with some drivers.

15.7 Challenges with Ride-Hailing

15.7.1 Finding the Cab

Almost all participants mentioned that finding the cab on the arrival was the most challenging aspect of the Uber or Ola journey. Although ride-hailing is door to door, factors including inaccurate GPS and limited addressing of maps in the country often resulted in the driver parking a few meters away from the exact pick-up location. Unable to determine where the cab had parked, people frequently took help from sighted people in the vicinity, including family members and security guards, to whom they communicated details like the cab number—available on the mobile app—to obtain assistance. However, unlike in the case of urban mass-transit services like buses, where participants sought help to get to the bus stop (often located far away from the participants' homes), find the right bus, and board, participants required "less" help with Uber and Ola cabs, where on most occasions one had to traverse a few meters to find and board a cab. Needless to say, the strategy of seeking sighted help (like with buses and trains) was contingent on the availability of the right help at the right time, without which finding the cab was impossible. Others shifted the onus onto the driver, giving them descriptive clues about themselves like the color of the shirt or that they were carrying a white cane to enable the driver to find them. Upon finding the cab, participants on certain occasions also sought help with finding the unoccupied seat. One participant, P7, described his experience with both strategies:

> Here, she [his wife] is there so she helps me get onto the cab. Or I directly tell them [drivers], I am blind you have to come and pick me...I used to go and stand near Rajalakshmi wines [and tell the driver] I am wearing so and so color of shirt, holding a white cane in my hand you have to find out, I cannot see...Then he used to come and pick. (P7)

However, participants described how drivers occasionally attempted to cir-cumvent road challenges like U-turns or crowded streets and consequently asked them to cross roads and walk from narrow roads to the main roads to find the cabs. As stated previously, navigating inaccessible roads can be pretty challenging for people with visual impairments; in fact, this is one of the reasons they choose to use ride-hailing services in the first place. This is generally the time at which interviewees disclosed their disability to the drivers, if they had not done so earlier, in the hope that drivers will oblige and pick them up.

Most noted that disclosing their disability resulted in drivers being more accommodating and, in this case, traveling to their location to pick them up. While some had no concerns in disclosing their disability to the driver to receive assistance from them, others were wary of doing so, citing how disclosing their disability to someone unfamiliar to them posed safety concerns. Many described how the drivers' assistance was central to the accessibility of ride-hailing services, and how when they did not receive help, it resulted in a loss of independence. As P16 explained,

> Many days back—I had booked a cab and had to wait almost half an hour to make out where he is and finally I could do it with someone's help. We don't

want to depend on someone that's why we book cab but even after booking cab we have to seek someone's help to board. So when driver can help us I think we can definitely stop asking people and we can be on time. (P16)

Participants believed that taking help from sighted people to find the cab reduced their sense of independence, one of the primary motivators of using ride-hailing services in the first place. For a lot of them, using Uber and Ola cabs stretched their budget, but they were willing to do so as long as it afforded increased independence. Others found other modes of transportation, like cars driven by drivers, to be more convenient, as ride-hailing was never door to door or on-demand enough.

15.7.2 Choosing and Negotiating Destinations

While the mobile apps were in and by themselves accessible, choosing and negotiating destination addresses was challenging for people with visual impairments. This was primarily because many addresses in the country are similar, making the required address hard to identify. For instance, many roads in the country share the same names (6th Cross or 4th Main are common street names), and there is not enough detail in the app to distinguish between similar addresses. For instance, P10 recalled an incident where he picked the wrong address while booking a cab for his sister (see Figure 15.1):

My sister was admitted to a hospital. It was named Peerless hospital and that is situated in a place called Bagajyothi...I searched Peerless hospital...I booked that. My sister is just standing there and suddenly the driver called me [and asked] where are you. I said, 'Ya—my sister is standing there'...He was surprised—"Where, I am in Salt Lake." Salt Lake is far far away from Bagajyothi...then I came to know that there is a Peerless lab that is situated in Salt Lake and there is a Peerless hospital in Bagajyothi. (P10)

The insufficient addressing of maps in India is compounded by the inability of people with visual impairments to drag and drop a pin on an address in a map (in both Uber and Ola cabs), which also made picking precise residential locations very difficult. Our participants worked around this by identifying landmarks close to their final destinations and then navigating the final few meters on their own.

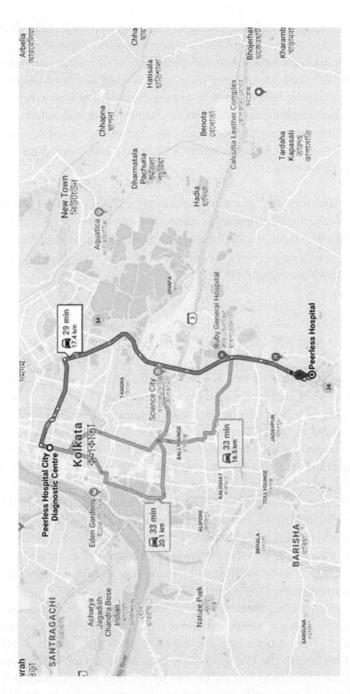

Figure 15.1 Peerless Hospital vs. Peerless Hospital City Diagnostics Center (Peerless Lab)

15.7.3 Driver Troubles

Finally, as mentioned earlier, passengers relied on drivers to help them with certain aspects of the ride-hailing journey. However, drivers were a mixed bunch—our participants encountered both drivers who were accommodating and helpful and those who cheated them and were less accommodating. Almost all participants noted instances where drivers did not know how to help or assist them, and had very little understanding of disability. Although riders could seek recourse from customer support, this was not always possible, and some participants noted how bad experiences with drivers had in fact dissuaded them from using ride-hailing altogether.

> A few drivers they don't give a damn about my vision impairment. Whatever I have given to them they just accept it and run away...Sometimes some drivers ask some shitty questions. And you are uncomfortable answering them, they begin from how do you study and go on and on...So sometimes you become uncomfortable answering them. (P26)

15.8 Conclusion

Thus, we see that in spite of the many benefits of ride-hailing, including the increased sense of independence, there are challenges to its use by people with visual impairments. There are opportunities to improve their overall ride-hailing experience, and the onus here is on the ride-hailing corporations to take the necessary measures, including improving access to technology features and sensitizing drivers to the needs of people with visual impairments. The state, too, can play a role here, and, by creating appropriate incentive and penalty mechanisms, can ensure that ride-hailing corporations accommodate the needs of people with disabilities. In a recent example from the Global North, the New York City government passed a mandate which necessitated that services like Uber and Lyft make at least 25% of their services wheelchair accessible by 2023 (Ottaway, 2018), a move which would ensure that wheelchair users are able to access and use the cabs to their benefit.

As suggested earlier, some aspects of the technology were inaccessible to people with visual impairments, who had to rely on sighted help and workarounds to use parts of the mobile app. For instance, many participants noted many unlabeled buttons in the Ola app, which included the interface to choose cab types and the interface to rate the driver—which they had to complete to book the next ride. Subsequently, they had to resort to seeking sighted help to circumvent these interface troubles. In this light, accessibility testing of the mobile apps prior to release, which should include checks for unlabeled buttons and making clear the changes to navigation and workflows on software updates, would represent a good first step. Overall, participants found Ola to be less accessible than Uber (which had fewer or no unlabeled buttons), resulting in a preference for the latter (over other factors like cost and availability). We attribute the accessibility of Uber to it being an international app—which has to adhere to accessibility standards in the United States (by conforming to the Americans with Disabilities Act). On the other hand, in India, the National Policy on Universal Electronic Accessibility (2019) only holds state-owned information and communications technology (ICT) bodies accountable to compliance measures, and not private players. While the Rights of PWDs Bill, introduced in 2016, does hold private players responsible for the non-implementation of compliance measures, the bill remains unimplemented in many states across the country (Narasimhan, 2016; Kameswaran & Hulikal Muralidhar, 2019) which in part explains why Ola, an Indian app, was inaccessible to our participants. Furthermore, we also noted how the visual components of both these apps, i.e., the maps, posed several challenges to people with visual impairments (who had trouble distinguishing between locations and selecting destinations) and that exploring enhancements that might make maps more screen reader friendly can increase the overall accessibility of the apps.

Needless to say, any means to improve the ride-hailing experience of people with disabilities cannot solely focus on technical features and has to be "sociotechnical." Drivers are central to the ride-hailing experience, and sensitizing, training, and informing them about the needs of people with disabilities, including those with visual impairments, and ways to assist and accommodate them will improve the accessibility of the services. This, combined with a feature on the app which would allow people with disabilities to disclose their disability on the mobile app and inform drivers, will be beneficial, as it could reduce the work associated with explicitly seeking help during the different junctures of a ride-hailing trip. Flexible policies by Uber and Ola cabs can complement these sociotechnical efforts. For instance, incentivizing drivers (including in the case of shared/pool rides where they received a fixed amount) for any accommodations/assistance they provide could also result in an improved ride-hailing experience for people with visual impairments and disabilities to an extent. Ride-hailing is a growing market in India and the Global South, as evidenced by the increasing market size of corporations like Uber and Ola cabs (Bhattacharya, 2019), and our research suggests that for people with visual impairments the benefits of these services far outweigh the challenges resulting from their use. By paying attention to their needs and improving the usability of ride-hailing, Uber and Ola cabs can pave the way for inclusivity, which is particularly relevant in a Global South context like India, where people with disabilities are otherwise marginalized and excluded from everyday life.

15.8.1 Ride-Hailing in India

This study is one of the few studies that examine ride-hailing outside the Global North context, and there are similarities and differences to what other studies have found in the United States and Europe. Among the similarities are the motivations for using the services in the first place—the added flexibility and convenience of these services which are on demand and door to door. However, in the United States, ride-hailing in many locations happens to be cheaper than traditional taxis and is also one of the reasons why riders favor them (Glöss et al., 2016). On the other hand, ride-hailing in India is relatively expensive, especially for our participants for whom mass-transit access was available at no cost or subsidized rates. While not replacing their use of buses and trains, we found that Uber and Ola cabs complemented them, and were more than an effective "plan b." Further, unlike the United States and Europe, door-to-door transport like auto-rickshaws, and in many places even mass transit, especially buses, operate under the informal economy. Here we see the benefits of a more formalized system for the likes of our participants: for instance, payment, which does away with the need to bargain and the associated emotional work. Likewise, addresses can pose challenges in India due to the insufficient addressing of digital maps in the country. Moreover, many addresses are also hard to distinguish, as they are similar sounding, and here we see that participants have to work around the system by identifying and picking landmarks, which can be quite challenging and is also likely to be different from ride-hailing experiences in the Global North. Finally, previous studies report on riders describing drivers as being "just like me" (Dillahunt et al., 2017; Kameswaran et al., 2018a), which we did not see in our study. Also, as explained earlier, unlike drivers in the Global North, Uber and Ola cab drivers in India work full-time, often for larger transportation corporations, and many come from the working class, often from villages. On the other hand, riders like our participants often happen to belong to the middle/upper-middle class. This class divide is a likely explanation for why our participants were comfortable seeking help, and in fact "expected" the drivers to assist them when required, as they felt that it was part of the Uber and Ola cabs' "service" offering (and seeking help from the driver did not impinge on their sense of independence). In contrast, studies examining the ride-hailing experiences of people with visual impairments in the United States find that riders carefully measured, and, in fact, limited, the help they sought from the driver (Brewer & Kameswaran, 2019).

Our qualitative study on the use of ride-hailing services by people with visual impairments in metropolitan India found that ride-hailing has several benefits for them, including increased independence, flexibility, and accessibility, and reduced emotional work. However, using these services was not without its challenges, which included finding the precise location of the cab on arrival, choosing destination addresses, and dealing with drivers. Although people at large, including those without disabilities, are likely to experience the benefits and challenges of ride-hailing, its relative advantages and disadvantages are amplified for people with visual impairments, primarily because traditional modes of transportation are so much more inaccessible. Although largely accessible, improvements through accessibility testing, driver training/incentivization, and flexible policies can additionally improve these services for the likes of our participants.

Acknowledgments

We would like to thank Jacki O'Neill for her mentorship and Jatin Gupta for his assistance with data collection. We would also like to thank Jonathan Lazar and Michael Ashley Stein for their helpful comments on earlier drafts of this work.

References

Accessible Mass Transit—American Foundation for the Blind. 2017. *Report on Accessible Mass Transit in the United States of America.*

Bhattacharya, A. (2019). Ola vs Uber: The latest score in the great Indian taxi-app game. *Quartz.*

Brewer, R. & Kameswaran, V. (2019). Understanding trust, transportation, and accessibility through ridesharing. In *Proceedings of the SIGCHI conference on human factors in computing systems* (pp. 1–11). ACM.

Dillahunt, T. R., Kameswaran, V., Li, L., & Rosenblat, T. (2017). Uncovering the values and constraints of real-time ridesharing for low-resource populations. In *Proceedings of the 2017 CHI conference on human factors in computing systems* (pp. 2757–2769). ACM.

Dixit, P. (2018). People are mad about Uber and Ola drivers canceling rides based on their destination. *BuzzFeed News.*

Glöss, M., McGregor, M., & Brown, B. (2016). Designing for labour: Uber and the on-demand mobile workforce. In *Proceedings of the 2016 CHI conference on human factors in computing systems* (pp. 1–12). ACM.

Griswold, A. (2018). Uber drivers are forcing riders to cancel trips when fares are too cheap. *Quartz.*

Kameswaran, V., Cameron, L., & Dillahunt, T. R. (2018a). Support for social and cultural capital development in real-time ridesharing services. In *Proceedings of the 2018 CHI conference on human factors in computing systems* (pp. 1–12). ACM.

Kameswaran, V, Gupta, J., Pal, J., O'Modhrain, S., Veinot, T., Brewer, R., Parameshwar, A., Vidhya, Y., & O'Neill, J. (2018b). "We can go anywhere": Understanding independence through a case study of ride-hailing use by people with visual impairments in metropolitan India. In *Proceedings of the ACM on human–computer interaction*, 2(CSCW), 85. ACM.

Kameswaran, V., Muralidhar, S. H. (2019). Cash, digital payments and accessibility: A case study from India. In *Proceedings of the ACM on human–computer interaction*, 3 (CSCW), 97. ACM.

Marston, J. R. & Golledge, R. G. (2003). The hidden demand for participation in activities and travel by persons who are visually impaired. *Journal of Visual Impairment and Blindness*, 97, 475–488.

Narasimhan, N. (2016). *Mobile apps are excluding millions of Indians who want to use them.*

National Policy on Universal Electronic Accessibility of 2019. Ministry of Law and Justice—Government of India (2019). https://www.meity.gov.in/writereaddata/ files/National%20Policy%20on%20Universal%20Electronics%281%29.pdf

Ottaway, A. (2018). *Uber, Lyft & Via fight wheelchair-access mandate in NYC.* Courthouse News Service. https://www.courthousenews.com/uber -lyft-via-fight-wheelchair-access-mandate-in-nyc/

Pal, J., Ammari, T., Mahalingam, R., Alfaro, A. M. H., & Lakshmanan, M. (2013). Marginality, aspiration and accessibility in ICTD. In *Proceedings of the sixth international conference on information and communication technologies and development: Full papers—Volume 1* (pp. 68–78). ACM.

Pal, J., Viswanathan, A., Chandra, P., Nazareth, A., Kameswaran, V., Subramonyam, H., Johri, A., Ackerman, M., & O'Modhrain, S. (2017). Agency in assistive technology adoption: Visual impairment and smartphone use in Bangalore. In *Proceedings of the 2017 CHI conference on human factors in computing systems* (pp. 5929–5940). ACM.

Pal, J., Youngman, M., O'Neill, T., Chandra, P., & Semushi, C. (2015). Gender and accessibility in Rwanda and Malawi. In *Proceedings of the seventh international conference on information and communication technologies and development*, 5 (pp. 1–9). ACM.

Pucher, J., Korattyswaroopam, N., & Ittyerah, N. (2004). The crisis of public transport in India: Overwhelming needs but limited resources. Journal of Public Transport, 7(4), 1–20.

The Rights of Persons with Disabilities Act of 2016. Ministry of Law and Justice— Government of India (2016). http://www.upfcindia.com/documents/rpwd_ 101017.pdf

United Nations. (2006). Convention on the Rights of Persons with Disabilities. United Nations Treaty Series. 2515:3.

World Bank. (2007). *People with disabilities in India: From commitments to outcomes.* Human Development Unit, South Asia Region. The World Bank.

UI, M. (2016). *A young man with visual impairment shares how Ola-Uber allows him to travel with dignity*. Youth Ki Awaaz. https://www.youthkiawaaz.com/2016/09/ola-uber-helps-the-visually-impaired/

Bangalore. (2020). In *Wikipedia*. https://en.wikipedia.org/wiki/Bangalore

16

The Role of Ugandan Public Universities in Promoting Accessible Information and Communications Technology

Patrick Ojok, PhD
Lecturer and Head of Department, Department of Community and
Disability Studies, Kyambogo University

16.1 Introduction

The use of technology has the potential to promote positive academic and career outcomes for persons with disabilities (PWDs; Burgstahler, 2003). Information and communications technologies (ICTs) have had an extraordinary impact on everything, from teaching and learning, institutional management, administration, and finance, to external relations, library services, research production and dissemination, and student life (Guri-Rosenblit, 2009). However, the diffusion of ICTs has been uneven, with most of Africa being left in a technological apartheid (Castells, 1999). Research evidence from three Sub-Saharan African countries, for instance, indicates that only 8% of households with PWDs can afford the internet, about half the percentage for households without persons with disabilities (PWDs) (United Nations (UN), 2018). As Drucker (1994) rightly asserted, in an era where the application of knowledge counts, developing countries can no longer expect to base their development on cheap industrial labor.

Driven by the centrality of technology to development, as noted in Chapter 2: Addressing the Drivers of Digital Technology for Disability, many international development organizations support the promotion of ICTs in Africa, assuming that equipping people with computers enables them to leapfrog into the technological world of economic opportunities (Alzouma, 2008). Some authors even argue that ICT is a suitable vehicle for development in Sub-Saharan Africa because the flexibility and mobility of ICT tools are unimpeded by language, location, and nationality (Yemke, 2012) and can be adopted and used regardless of poverty, illiteracy, or the absence of electricity (Vodafone, 2005). In contrast, this paper argues that access to ICTs in Africa is largely impeded by context-specific circumstances, such as high illiteracy levels, conflict, poverty, local languages, power outages, and uneven telephone network coverage, that are often

Accessible Technology and the Developing World. Michael Ashley Stein and Jonathan Lazar, Oxford University Press.

overlooked by technology providers from developed countries. Nonetheless, Sub-Saharan Africa largely relies on Global North models to bridge the digital divide, without paying close attention to its peculiar contexts; the region has likewise given little thought to how various technologies can effectively be integrated into peoples' lives in order to alleviate poverty and stimulate ICT uptake (Mutula, 2005; Kyem, 2012).

Sub-Saharan Africa is the least ICT-advanced region in the world and lies at the deepest end of the digital divide (Gymamfi, 2005); see Chapter 3: Global Trends for Accessible Technologies in the Developing World. For example, 80% of the world's 140 million out-of-school children are found in the African region (The African Child Policy Forum, 2011), and 9 in 10 extremely poor people are projected to live in Sub-Saharan Africa by 2030 (World Bank Blogs, 2018). There exists a higher percentage of illiterate adults in Sub-Saharan Africa (27%) than in Eastern and South-Eastern Asia (10%), Northern Africa and Western Asia (9%), and Latin America and the Caribbean (4%) (UNESCO Institute of Statistics, 2017: Fact Sheet No. 45). In terms of access to technology, numerous studies confirm the widening digital divide in Sub-Saharan Africa (e.g., UN, 2018; Barry et al., 2009; Fuchs & Horak, 2008; Chinn & Fairlie, 2007). Though not the only factor, income differential and disparity in telecommunications infrastructure contribute to the digital divide in Sub-Saharan Africa (Chinn & Fairlie, 2007). This has led to a new form of poverty called *information poverty* (Gebremichael & Jackson, 2006), or the so called "info-poor countries" (Kagan, 2000). Foley and Ferri (2012) are right in their assertion that the technology divide is less about access to technology or the distinction between "haves" and "have-nots" but more about the deeper underlying meanings of "access." Unfortunately, ICT interventions tend to overlook the sociocultural peculiarities of developing countries. See Chapter 2: Addressing the Drivers of Digital Technology for Disability and Chapter 7: Digital Inclusion in the Global South.

Furthermore, despite its much-touted usefulness, technology is not a panacea for development but a double-edged sword that can exacerbate socioeconomic disparities (Castells, 1999). For example, while 7% of adults without disabilities in Uganda have used a computer, only 2% of those with disabilities have ever used one. Also, only 36% of PWDs in Uganda own a mobile phone compared with 52% of persons without disabilities (Uganda Bureau of Statistics [UBOS], 2018). Globally, 51% of PWDs in developing countries who need assistive devices are unable to receive them, mainly because they are unavailable, inadequate, or unaffordable (UN, 2018). And even where technology is affordable, available, and accessible, their use in certain developing countries is impeded by unreliable and unaffordable energy sources. The use or lack of technology can also reinforce or create unexpected and often subtler forms of exclusion (Foley & Ferri, 2012). This can manifest in numerous ways in universities. For example, students required to take an online course rather than a conventional campus course risk being further isolated and denied social opportunities relative to students who take courses on campus (Foley & Ferri, 2012). Also, students who are print disabled are excluded when examinations and/or lecture materials are provided in inaccessible formats. To this end, advocacy for ICT access must shift toward notions of equity that qualify and contextualize technology-centered disparities within local and societal histories, values, languages, and perceptions of success and disproportion (Foley & Ferri, 2012).

Research in ICT in Uganda is growing (e.g., Hamasha et al., 2013; Kiza & Pederson, 2012), with relatively few focusing on disability (e.g., Uganda Communications Commission, 2018; Ojok, 2018; Baguma, 2017). The literature shows that only few higher-education students can access or are aware of library electronic resources and databases relevant to their programs (Okello-Obura & Magara, 2008). A study on access to ICT by students with visual impairments in public universities in Uganda revealed that 40% were intermediate ICT users and 37% were beginners, while the majority had never uploaded/downloaded any learning material to/from an e-learning platform (Ojok, 2018).

The present study explored the state of accessible ICT services and their use by teaching and non-teaching staff in Uganda's public universities. It found that most of the staff were aware of national ICT laws and policies, but that the provision of accessible ICT was low. The barriers to ICT were largely infrastructural rather than technical or managerial in nature. Although the findings revealed areas for investment and collaboration in technology in universities in Uganda, universities should also explore local avenues of investing in technology development. Consequently, this chapter urges ICT technology developers, providers, and users to consider a deeper meaning of access that goes beyond physical access or availability of ICT infrastructure and instead to focus on addressing disability-specific factors that preclude PWDs in universities from accessing and utilizing ICT.

16.2 Background

16.2.1 University Education in Uganda

University education in Uganda, like other Sub-Saharan Africa countries, is still picking up from decades of neglect in governmental funding. International development partners had misadvised many African governments to prioritize primary and secondary education on the assumption that university education was less crucial to economic growth and poverty mitigation (Bloom et al., 2014). The result was reduced investment in, and financing of, higher education (Otieno, 2013). In Uganda, this policy shift led to the introduction of a cost-sharing arrangement where government-sponsored students pay part of the tuition while government pays a part. Since the 1990s, government sponsorship to public universities plummeted and was capped at 4,000 students. This number is extremely low for a country where many households struggle to meet the basics of life. As a result, access to higher education in Uganda is uneven, with most of the students enrolling for university education coming from affluent families that can afford the cost of higher education (Kasozi, 2009). Evidently, higher education disparity is even wider for PWDs, as only 192 students with disabilities were enrolled in public universities in 2019 (National Council for Disability, 2019).

Since 2002, affirmative action measures have increased disability representation in universities. Under a university affirmative action scheme, 64 students with disabilities who fail to get acceptance under direct admission are annually admitted to public universities. Such a positive step, though it is in the right direction, cannot ensure equity in university education, since the focus is on admission only (Onsongo, 2009), and therefore captures very few students. To illustrate, a recent study revealed that many of the students with disabilities admitted under this affirmative action scheme are restricted to specific arts courses (e.g., education, guidance and counseling, community-based rehabilitation), and are only admitted to Makerere University and Kyambogo University (Ojok, 2018). It is unclear whether the students applied only for the courses they were admitted to or were denied their courses of interest.

Nevertheless, there have been some noteworthy developments in ICT and disability inclusion in public universities in Uganda:

- Kyambogo University and Makerere University have passed institutional policies on disability, while Mbarara University of Science and Technology is developing one.
- Kyambogo University and Makerere University Business School have established disability resource centers that are internally funded by the universities. The disability resource centers will help coordinate disability support services at the universities.
- With support from the Open Society Foundation, Makerere University has established a Disability Law and Rights Center, and integrated disability studies into its law courses. The center holds an annual disability rights conference that is becoming a recognized national platform for informed dialogue on disability rights and justice issues.
- With funding from the Norwegian Agency for Development Corporation, an accessible web portal for sensory impairments has been developed (http://enableffs.github.io/#/home). Contents of the portal can be read in both English and Kiswahili, while audio-visual contexts are interpreted into sign language and an audio description. The web portal is a product of a collaboration between Oslo University, Kyambogo University, University of Dar es Salaam, and the Kenya Institute of Special Education.

16.2.2 ICT Policy and Legal Framework

Uganda is a State Party to the Convention on the Rights of Persons with Disabilities (CRPD) and its Optional Protocol. It also ratified the Marrakesh Treaty to Facilitate Access to Published Works for Persons Who Are Blind, Visually Impaired or Otherwise Print Disabled (Marrakesh Treaty) which entered into force with respect to Uganda (WIPO, 2021). As a State Party to the Convention, Uganda is obliged to take appropriate measures to ensure accessibility of ICT to PWDs and that private sector service providers provide information and services, including through the internet, in accessible and usable formats for PWDs (Article 9). Likewise, Uganda is bound by the Marrakesh Treaty requirement to ensure access to published information for persons who are print disabled. These standards must be reflected both in practice and in relevant domestic ICT policies.

Uganda's national ICT policy framework is largely disability-inclusive. For example, one of the statutory functions of the Uganda Communications Commission is to stimulate research into the development and use of new communications techniques and technologies, including those which promote accessible communications services (Uganda Communications Commission [UCC] Act, Section 5, [I]). Generally, the national ICT policy envisions full and equal participation of PWDs in the information society, and requires special ICT training for PWDs, women, and youth. The PWD Act 2019 itself has no substantive ICT provision but requires Uganda to promote the use of assistive devices and technology and the provision in Braille of public information such as government documents and publications (UCC Act, Section 15 1 [b–c]). The Communications Act, Chapter 106, requires the Commission to promote research into the development and use of techniques and technologies including those which promote accessibility of hearing-impaired people to communication services.

The Ministry of ICT is in the process of enacting a standalone ICT policy on disability. The draft ICT for Disability Policy 2017 seeks to apply ICT in all government entities, including education at all levels. The main objective of this policy is to use ICT as a means to reduce the marginalization of PWDs and create equal opportunities for them (see https://ict.go.ug/wp-content/uploads/2018/06/ICTs-for-Disability-Policy-Draft.pdf). One of the promising initiatives in the policy is to include *accessible service delivery* as part of the terms of operators' license agreements or introduce other measures to ensure that affordable and accessible ICTs are provided to PWDs. Further, accessibility of ICTs and related services will be a condition for public procurement, while every ministry, department, and agency will be required to designate a senior officer to coordinate ICT and disability issues.

Nevertheless, there remain policy gaps and practical barriers. For example, there is no policy on web accessibility. Other barriers to access of ICTs by PWDs include weak ICT policy implementation, unavailability of software in local languages, low levels of accessible ICT literacy even among the academia, and the high cost of assistive technologies (Ministry of ICT, 2017).

16.3 Conceptual Framework

Universal access is a continuous and dynamic process impacted by socioeconomic and cultural peculiarities of different regions (Oyedemi, 2004). The Universal Wheel of Access (ibid.) offers a holistic approach to achieving universal access in the developing world's context, especially in Africa. It challenges the notion that providing physical access to ICT products to people in resource-constrained regions necessarily guarantees their use. Hence, beyond providing ICT products, the Universal Wheel of Access proposes six elements that constitute any holistic universal access interventions in the developing world. These elements are: policy and reforms; universal service funds; service platforms; other social utilities; literacy; content and language relevance; and technology and social relevance. Overlooking the elements can render the physical presence of ICT devices inconsequential for certain groups in different regions (ibid.).

The Universal Wheel of Access is cognizant of sociocultural peculiarities affecting ICT access and use by the developing regions of the world, hence its choice as the analytical framework for this paper. It emphasizes that universal access programs should take into consideration the needs of the people, the sociocultural nature of their livelihoods, and the realization that physical contact with ICT infrastructure does not automatically connote real access (ibid.). This assertion is particularly useful for understanding the ICT needs of PWDs, as they often experience barriers that are unique to their disabilities. Technological design that is responsive to disability concerns will ensure genuine universal service for all (Goggin & Newell, 2000, p. 127) (Figure 16.1).

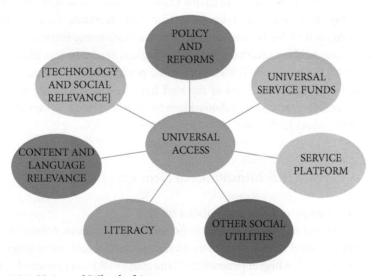

Figure 16.1 Universal Wheel of Access

16.4 Research Design and Methods

A cross-sectional survey design was used to ascertain the availability and use of accessible ICT services and products by the teaching and non-teaching staff of public universities in Uganda. Public universities were targeted because students with disabilities were more likely to be enrolled at these universities. A previous study had found that students with visual impairments admitted under the higher-education affirmative action scheme were studying only in Kyambogo and Makerere universities (Ojok, 2018). The other universities presumed they did not have the facilities to accommodate these students. A survey design was used despite the small number of participants, in order to explore a wider range of issues. Permission to collect data was obtained from the Vice Chancellors of the respective universities. 100 questionnaires were sent out to the public universities between February and June 2018. A total of 32 questionnaires were returned, representing a response rate of 32%. The survey questions focused on ICT infrastructure, existence of institutional policy on disability, provision of accessible ICT services, ICT monitoring, and barriers to accessible ICT.

16.5 Findings

16.5.1 Participants ($N = 32$)

A total of 32 staff from four public universities participated in the study; see Table 16.1. Slightly over half of the participants (53%) were from Kyambogo University, followed by Gulu University (23%), Uganda Management Institute (UMI) (13%), and Lira University (6%). The high response from Kyambogo University could have been because it is the researcher's home institution. Half of the participants held a master's degree as their highest academic qualification, followed by a bachelor's degree (28%), PhD (9%), postgraduate diploma (9%), and an ordinary diploma (3%). Most of the staff had been in university service for between 11 and 15 years (31%), followed by 6 to 10 years (25%), 5 years and below (22%), over 20 years (13%), and 16 to 20 years (9%).

16.5.2 Institutional Demographics

16.5.2.1 Awareness of Laws and Policies on ICT
Awareness of a law or policy is crucial to policy implementation. Most of the university staff surveyed in this study were aware of national and international laws on ICT. As expected, a higher proportion of the staff (84%) were aware of national

Table 16.1 Participants (N = 32)

Characteristic		n	%
University	Kyambogo	17	53
	Gulu	9	23
	Lira	2	6
	UMI	4	13
Period of university existence	5 years and below	3	10
	10 years and below	2	7
	15 years and below	18	58
	20 years and below	5	16
	Over 20 years	3	10
Highest academic qualification	PhD	3	9
	Master's Degree	16	50
	Bachelor's Degree	9	28
	Postgraduate Diploma	3	9
	Ordinary Diploma	1	3
Years in university service	5 years and below	7	22
	6–10 years	8	25
	11–15 years	10	31
	16–20 years	3	9
	Over 20 years	4	13

laws on ICT than of the CRPD and its mandates (52%). Hence, there is a need for more sensitization about the CRPD, as nearly half of the participants (48%) were not familiar with its provisions.

16.5.2.2 Existence of Institutional Policy on Disability
Passing an institutional policy is a significant first step toward addressing accessibility of ICT and other services. Most of the surveyed participants (45%) reported that their institutions did not have a policy on disability, 23% said their university was working on it, 16% were not sure, and 16% said their institutions had a policy on disability. Lack of institutional policies on disability and ICT can reinforce inequalities and lead to failure to address the access needs of PWDs.

16.5.3 Training on Accessible ICT for University Staff

Participants were asked whether their universities provided ICT accessibility training to teaching staff, students, and university managers. Overall, universities tended to focus ICT accessibility training on teaching staff (72%) and university managers (53%), with the least attention given to non-teaching staff (43%). The

Table 16.2 ICT Services Provided to University Staff (N = 32)

		Total	
Target for accessible ICT training	Response	n	%
Teaching staff	Yes	23	72
	No	9	28
Non-teaching staff	Yes	14	43
	No	18	56
University managers	Yes	17	53
	No	15	47
ICT needs assessment	Yes	4	13
	No	27	87

findings in Table 16.2 show that universities did not conduct ICT accessibility needs assessments to determine the appropriate and reasonable accommodations needed by students and staff with disabilities.

16.5.4 Availability of Accessible ICT Services and Products in Universities

Educational access and use of accessible ICT services and products depends on their availability in universities. Therefore, participants were asked: Which of the accessible ICT services, products, and systems are provided to students and staff with disabilities in their institutions? According to Table 16.3, none of the public universities provided real-time remote captioning, with the least available accessible ICT services being video and audio captioning (7%). The most available types of accessible ICT services in the four universities were: provision of separate accessible ICT labs (71%); accessible library environments (71%); accessible e-learning platforms (57%); accessible payment systems (52%); accessible admission process (48%); accessible student registrations (42%); assistive technology (37%); and accessible results management systems (33%). It is possible that certain accessible ICT services were missing for financial reasons or lack of demand by staff, or that the universities were unaware of them.

16.5.5 Institutional ICT Monitoring Mechanisms

Participants were asked whether their universities carried out ICT accessibility monitoring or had an accessible ICT monitoring committee. Overall, Figure 16.2

Table 16.3 Provision of Accessible ICT Services, Products and Systems University Students and Staff (N = 32)

		Total	
	Response	n	%
Accessible admission process	Yes	15	48.4
	No	16	51.6
Accessible student registrations	Yes	13	41.9
	No	18	58.1
Accessible e-learning platform	Yes	17	56.7
	No	13	43.3
Assistive technology	Yes	11	36.7
	No	19	63.3
Accessible results management system	Yes	10	33.3
	No	20	66.7
Accessible payment system	Yes	16	51.6
	No	15	48.4
Accessible library environment	Yes	22	71
	No	9	29
Accessible e-document conversion	Yes	10	33.3
	No	20	66.7
Separate accessible ICT labs	Yes	22	71
	No	9	29
Video captioning	Yes	2	6.9
	No	27	93.1
Audio description	Yes	2	6.9
	No	27	93.1
Real-time remote captioning	Yes	0	0
	No	29	100

shows there is a weak monitoring mechanism across all the universities. Only 21% of the participants reported that their universities monitored ICT accessibility on campus, while 39% said that their universities did not monitor ICT accessibility or were not sure. Less than half of the participants (41%) confirmed that their universities had an ICT monitoring committee, while others said there was no committee (25%), that they were not sure (25%), or that their universities were still working on forming a committee (8%). ICT monitoring enables universities not only to determine the ICT needs for different academic and administrative units, but also to better coordinate required support services for students and staff.

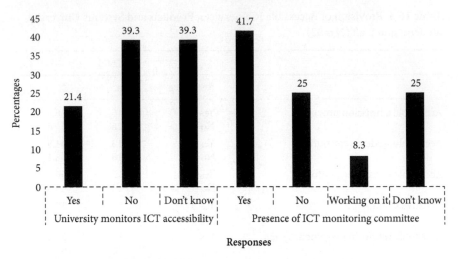

Figure 16.2 Availability of Institutional ICT Monitoring Mechanisms

16.5.6 Barriers to Using ICT in Teaching and Learning

Provision of adequate accessible ICT infrastructure can be a challenge for universities in the developing world. To gauge the barriers to ICT provisions, one of the survey questions asked: Is your university's capacity to provide accessible ICT for teaching and learning affected by a shortage or inadequacy in the following areas? Overall, the highest-ranked barriers to ICT across all universities were (see Table 16.4): insufficient number of desktop and laptops computers (56%); out of date/service computers (48%); insufficient number of internet-connected computers (41%); insufficient ICT technical support for staff (36%); lack of adequate ICT skills by lecturers (32%); and lack of, or inadequate, space (30%). It was apparent that ICT barriers in the universities were largely infrastructural rather than technical or managerial. This signifies a need to prioritize infrastructure development in the universities. In a bid to fill this gap, the African Development Bank recently supplied technology equipment to public universities, including computers for office and student use. The Faculty of Special Needs and Rehabilitation at Kyambogo University received ten desktops and ten laptops for its disability resource room. It also received Braille embossers, digital cameras, recorders, Braille scanners, and television sets for different academic units and the library.

THE ROLE OF UGANDAN PUBLIC UNIVERSITIES 333

Table 16.4 Barriers to Using ICT in Teaching and Learning (N = 32)

		Total	
	Responses	n	%
Insufficient number of computers	A lot	18	56
	A little	4	13
	Somewhat	8	25
	Not at all	2	6
Insufficient number of internet-connected computers	A lot	13	41
	A little	9	28
	Somewhat	6	19
	Not at all	4	13
Insufficient internet speeds	A lot	9	28
	A little	4	13
	Somewhat	12	38
	Not at all	7	22
Insufficient number of laptops	A lot	18	56
	A little	7	22
	Somewhat	3	9
	Not at all	4	13
Out of date/service computers	A lot	14	47
	A little	5	18
	Somewhat	6	20
	Not at all	5	17
Lack of adequate ICT lecturer skills	A lot	10	32
	A little	9	29
	Somewhat	8	26
	Not at all	4	13
Insufficient ICT technological support	A lot	11	36
	A little	5	16
	Somewhat	10	32
	Not at all	5	16
Inadequate space	A lot	9	30
	A little	6	20
	Somewhat	9	30
	Not at all	6	20
Pressure to prepare students for exams	A lot	6	19
	A little	6	19
	Somewhat	8	26
	Not at all	11	36
Lecturers not in favor of ICT use	A lot	5	16
	A little	2	7
	Somewhat	9	29
	Not at all	15	48
Unclear ICT benefit	A lot	3	10
	A little	2	7
	Somewhat	7	23
	Not at all	18	60
Using ICT is not a university goal	A lot	6	19
	A little	3	9
	Somewhat	4	13
	Not at all	19	59

16.6 Discussion and Emergent Issues

16.6.1 Policy Implementation

The promise of ICT for PWDs cannot be realized in developing countries without supportive legislation (Sarman et al., 2013). While the ICT legislative framework in Uganda is disability sensitive, the implementation is problematic (Ministry of ICT, 2017; Mugimba, 2008). For example, the PWDs Act 2006, which contained some ICT accessibility provisions for media houses and telephone companies (Section 22, para. 2 [a–b]), was repealed before implementation. It remains to be seen whether the draft National ICT for Disability Policy 2017 and the new PWDs Act 2019 will be implemented.

On the other hand, the recent social media tax called the Over the Top Tax has turned out to be a major barrier to internet access and social media use, especially by PWDs. Since it was introduced, 66% of PWDs reduced social media usage, 26% stopped using social media, and only 8% did not change their social media usage (UBOS, 2018). The tax will continue to adversely affect internet and social media usage among PWDs, since 93% of PWDs access the social media via mobile phones (UBOS, 2018). The cost of cellular data, the primary source of internet for many Ugandans, is expected to increase with the proposed policy to shift the social media tax to cellular internet data (Uganda Revenue Agency, 2020).

16.6.2 ICT Financing

Achieving universality in ICT access and usage is a capital-intensive endeavor for any low-income country, and calls for international collaboration. Due to lack of resources, even low-technology assistive devices can be too expensive for some individuals, families, and organizations, making the use of accessible technology an unrealistic goal (Grönlund et al., 2010). It is not a surprise, therefore, that in developing countries, assistive technology services and products are provided mostly by NGOs/DPOS (58%), families (48%), and governments (38%) (UNICEF, n.d.). Although universality, affordability, and availability are key principles of ICTs, the scarcity of resources makes these a far-fetched reality for PWDs in resource-constrained countries. A World Bank study on the digital divide reveals that not only are mobile phones expensive, but the median mobile phone owner in Africa spends over 13% of monthly income on phone calls and texting (World Bank, 2016). Technology penetration in higher education in Uganda is still low, but certainly an attractive investment priority among development partners such as the African Development Bank and the respective Swedish and Norwegian development agencies. In addition, ICT integration in education is a priority of the ministries of education and sports and the Ministry of ICT. Under the Rural Communications Development Fund program, the government of Uganda has connected several education institutions, including 9 universities and 43 tertiary institutions (UCC, 2014). Many universities charge a mandatory ICT fee, payable once by all first-year students. When used for the intended purpose, ICT fees are one of the ways universities can invest in accessible ICTs without depending on external aid.

16.6.3 Service Platforms

Access to ICT services can be gained through institutional access or direct home connections. In Uganda, because only a few can afford to have direct home connections, institutional access becomes the main gateway to ICT networks, often through places of work like education institutions. Fortunately, many universities now have fairly good internet connections and provide free WiFi connections to staff and students. Thus, the staff and students who own smart technology devices can access internet on campus with relative ease. It is, however, surprising that even though mobile phone ownership is relatively high among PWDs (69%), internet usage remains low at 16% (UBOS, 2019). This minimal internet use is attributable to possession of non-internet enabled mobile devices, the new social media tax, and the cost of cellular data. The situation appears similar in other developing countries, where only 19% of PWDs use the internet compared to 36% of persons without disabilities (UN, 2018).

16.6.4 Other Social Utilities

Lack of access to affordable and reliable utilities vital for ICT is one factor reinforcing the Ugandan digital divide. Only 18% of Ugandan households have an electricity connection, with an urban–rural electricity gap of 85% (Gillwald et al., 2019). The lack of electricity and underdeveloped ICT infrastructure are the primary causes of huge discrepancies in urban–rural internet use and mobile phone penetration. Further, in Uganda, places connected to the national grid experience frequent and prolonged power outages, commonly called *load shedding*, with some rural locations going without electricity for more than 96 hours, or even weeks. To illustrate, residents of Gulu district in Northern Uganda publicly burned a service vehicle for UMEME, Uganda's electricity provider, in protest over such outages (Uganda Radio Network, 2019). Due to unreliable electricity, smart phones with battery-consuming apps are less preferred in rural locations. Similarly, internet connectivity in upcountry universities is less reliable than in urban universities because of frequent power outages and uneven

distribution of telecommunication networks. Even in the capital city, power inter-
ruptions happen during teaching sessions and other time-bound tasks. Thus,
without attending to other utilities that affect ICT use, the presence of ICTs per se
will not single-handedly foster research and teaching in universities. Universities
can navigate this problem by investing in stand-by power generators. The other
solution is installing power-saving batteries that can save and provide power for
extra hours in the absence of electricity.

16.6.5 Digital Literacy

The availability of ICTs, while important, will not guarantee their use until literacy bottlenecks are addressed. In fact, technology products can lay redundant in
institutions unless the intended users possess the basic skills to use them. In
Uganda, over 36% of non-internet users are digitally illiterate, 23% stating that
they do not know how to use the internet (Gillwald et al., 2019). It may be possible to benefit from digital technology without or with low literacy, but one needs
basic or advanced literacy to navigate advanced technology functions. Some
scholars appear to underplay the role of literacy in technology, suggesting that
poverty, illiteracy, and lack of electricity do not prevent the adoption and use of
technology tools (Vodafone, 2005). On the contrary, lack of capacity is the
second-ranked barrier to ICT use by PWDs in Uganda, as 84% of PWDs do not
know how to use the internet (UBOS, 2018). This study's findings suggest that
lack of exposure to technology can contribute to digital illiteracy, even among literate people. For example, 9% of the public university staff in this study rated
their ICT knowledge at a beginner level, a clear sign that even literate elites can be
digitally illiterate without exposure to technology. It is therefore essential that
ICT interventions in universities should include basic computer skills training for
staff (Verhoest, 2000, p. 607).

16.7 Conclusion

Accessible technology research in Uganda is just emerging, and those focusing on
universities are sparse. The present study has explored the availability and use of
accessible ICT by teaching and non-teaching staff in Uganda's public universities.
The findings revealed a high awareness of national ICT policies, even though
some universities lacked institutional policies on disability. The most available
type of accessible ICT service in universities was providing a separate accessible
ICT facility, suggesting low awareness about ICT accessibility options. There was
a weak monitoring mechanism across all the universities, and none conducted
ICT accessibility needs assessments to determine the appropriate ICT products

for students and staff with disabilities. Overall, ICT barriers were largely infra-structural, implying a need to prioritize accessible ICT supplies to universities.

Accessible ICTs can provide PWDs with unprecedented levels of access to education, skills training and employment, and opportunities to participate in the economic, cultural, and social life of society. This is so because public universities in Uganda can serve functions that public libraries perform in the Global North. However, the lack of access to accessible ICTs in universities in Uganda is a major impediment to digital inclusion and the realization of disability-inclusive university education. To realize disability-inclusive university education, the starting point should be ensuring availability of accessible ICT products and software. To ensure widespread use of available ICTs, consideration should be given to peculiarities that can limit ICT, such as unreliable electricity, weak internet connections, and low digital literacy among intended users. Since universities generate internal revenue by charging an ICT fee, it is important they use such funds to procure accessible ICT products and services for staff and students. Likewise, students with disabilities on scholarships should receive extra support for assistive technology devices and products. Importantly, existing and proposed policies on ICT accessibility should be implemented at all levels of education.

References

Baguma, R. (2017, March). An audit of inclusive ICTs for education in Uganda. In *Proceedings of the 10th International Conference on Theory and Practice of Electronic Governance* (pp. 311–320). ACM. https://dl.acm.org/citation.cfm?id=3047339

Barry, B., Chukwuma, V., Bartons, C., Petitdidier, M., & Cottrell, L. (2009). *Digital divide in Sub-Saharan Africa universities: Recommendations and monitoring* (No. SLAC-PUB-13853). https://scholar.google.com/scholar?hl=en&as_sdt=0%2C5&q=Digital+Divide+in+SubSaharan+Africa+Universities%3A+Recommendations+and+Monitoring&btnG=

Bloom, D. E., Canning, D., Chan, K. J., & Luca, D. L. (2014). Higher education and economic growth in Africa. *International Journal of African Higher Education*, 1(1), 22–57.

Burgstahler, S. (2003). The role of technology in preparing youth with disabilities for postsecondary education and employment. *Journal of Special Education Technology*, 18(4), 7–19. http://journals.sagepub.com/doi/abs/10.1177/016264340301800401

Castells, M. (1999). Information technology, globalization and social development (No. 114). Geneva: UNRISD.

Chinn, M. D. & Fairlie, R. W. (2007). The determinants of the global digital divide: a cross-country analysis of computer and internet penetration. *Oxford Economic Papers*, 59(1), 16–44. http://www.jstor.org/stable/pdf/4500086.pdf?casa_token=gsNutTUQDI8AAAAA:ZVfsxJWTf3RsSiZL_4BT7j6SD66wuLmZpNf7Jg-QPLJ5Bn940

jgFoa1znjqe1mDzQzFuuq7AuXOkWy4iD4oM1ON9Magh5ScFGemTQ-ljaL_
saTKeDe4

Drucker, P. F. (1994). Trade lessons from the world economy. *Foreign Affairs*, 73(1), 99–108. http://www.jstor.org/stable/pdf/20045894.pdf?refreqid=excelsior%3A022 ba1f5b19b91b3ef161e80366bdb4b

Foley, A. & Ferri, B. A. (2012). Technology for people, not disabilities: Ensuring access and inclusion. *Journal of Research in Special Educational Needs*, 12(4), 192–200. https://onlinelibrary.wiley.com/doi/pdf/10.1111/j.1471–3802.2011.01230.x

Fuchs, C., & Horak, E. (2008). Africa and the digital divide. *Telematics and Informatics*, 25(2), 99–116.

Gebremichael, M. D. & Jackson, J. W. (2006). Bridging the gap in Sub-Saharan Africa: A holistic look at information poverty and the region's digital divide. *Government Information Quarterly*, 23(2), 267–280. https://pdfs.semanticscholar.org/ced1/8b1 1e6a2c4041125b4bdebed004db5bc43d9.pdf

Gillwald, A., Mothobi, O., Ndiwalana, A. N., & T. Tusubira (2019). *The State of ICT in Uganda*. Research ICT Africa. https://researchictafrica.net/wp/wp-content/ uploads/2019/05/2019_After-Access-The-State-of-ICT-in-Uganda.pdf

Goggin, G. & Newell, C. (2000). An end to disabling policies? Toward enlightened universal service. *The Information Society*, 16(2), 127–133. 10.1080/ 01972240050032889

Grönlund, Å., Lim, N., & Larsson, H. (2010). Effective use of assistive technologies for inclusive education in developing countries: Issues and challenges from two case studies. *International Journal of Education and Development Using ICT*, 6(4), 5–26.

Guri-Rosenblit, S. (2009). Distance education in the digital age: Common misconceptions and challenging tasks. International Journal of E-Learning & Distance Education/Revue internationale du e-learning et la formation à distance, 23(2), 105–122.

Gyamfi, A. (2005). Closing the Digital Divide in Sub-Saharan Africa: meeting the challenges of the information age. *Information Development*, 21(1), 22–30.

Hamasha, M. M., Rumbe, G., & Khasawneh, M. T. (2013). Optimisation of single-queue service delivery systems using a Markovian approach. International Journal of Industrial and Systems Engineering, 13(4), 424–441. https://doi.org/10.1504/ IJISE.2013.052608

Kagan, A. (2000). The growing gap between the information rich and the information poor both within countries and between countries: A composite policy paper. *IFLA Journal*, 26(1), 28–33.

Kasozi, A. B. K. (2009). *Access and equity to higher education in Uganda: Whose children attend university and are paid for by the state?* National Council for Higher Education. http://ahero.uwc.ac.za/index.php?module=cshe&action=downloadfile &fileid=36807145012339262082233

Kiiza, B., & Pederson, G. (2012). ICT-based market information and adoption of agricultural seed technologies: Insights from Uganda. *Telecommunications Policy*, *36*(4), 253–259. https://www.sciencedirect.com/science/article/abs/pii/S0308596112000110

Kyem, P. A. K. (2012). Is ICT the panacea to sub-Saharan Africa's development problems? Rethinking Africa's contentious engagement with the global information society. *Progress in Development Studies*, *12*(2–3), 231–244.

Ministry of ICT (2017). *ICT for Disability Policy*. Second Draft. https://ict.go.ug/wp-content/uploads/2018/06/ICTs-for-Disability-Policy-Draft.pdf

Mugimba, C. (2008). *ICT accessibility for persons with disabilities in Africa region: Uganda's country report.* https://afri-can.org/wp-content/uploads/2016/04/ICT-and-Disability-by-Mugimba-Christine.pdf

Mutula, S. M. (2005). Peculiarities of the digital divide in sub-Saharan Africa. *Program*, *39*(2), 122–138. https://www.emeraldinsight.com/doi/abs/10.1108/00330330510595706

National Council for Disability (2019). *Disability Status Report 2019*. https://afri-can.org/disability-status-report-uganda-2019/

Newell, G. G. C. (2000). An end to disabling policies? Toward enlightened universal service. *The Information Society*, *16*(2), 127–133.

Ojok, P. (2018). Access and utilization of information and communication technology by students with visual impairment in Uganda's public universities. *IJDS: Indonesian Journal of Disability Studies*, *5*(1), 65–80.

Okello-Obura, C. & Magara, E. (2008). Electronic information access and utilization by Makerere University students in Uganda. *Evidence Based Library and Information Practice*, *3*(3), 39–56. https://journals.library.ualberta.ca/eblip/index.php/EBLIP/article/view/935

Onsongo, J. (2009). Affirmative action, gender equity and university admissions–Kenya, Uganda and Tanzania. *London Review of Education*, *7*(1), 71–81. http://www.ingentaconnect.com/content/ioep/clre/2009/00000007/00000001/art00007

Oyedemi, T. (2004). Universal access wheel: Towards achieving universal access to ICT in Africa. *The African Journal of Information and Communication*, *2004*(5), 90–107.

Samant, D., Matter, R., & Harniss, M. (2013). Realizing the potential of accessible ICTs in developing countries. *Disability and Rehabilitation: Assistive Technology*, *8*(1), 11–20. https://www.tandfonline.com/doi/abs/10.3109/17483107.2012.669022

The African Child Policy Forum (2011). *The African report on child wellbeing 2011: Budgeting for children.* https://www.childwatch.uio.no/publications/research-reports/AfricanReport2011_english.pdf

Uganda Communications Commission (UCC). (2014). *Integrating ICT in education in Uganda: A report.* https://www.ucc.co.ug/files/downloads/ICT%20Integration%20into%20teaching%20and%20learning%20booklet%202014.pdf

Uganda Bureau of Statistics (UBOS). (2018). Disability Functional Limitations Survey 2017. https://www.ubos.org/wp-content/uploads/publications/072019Uganda_Functional_Difficulties_Survey_2017.pdf

Uganda Radio Network (2019, November 4). *Gulu residents burn UMEME truck.* https://ugandaradionetwork.net/story/gulu-residents-burns-umeme-truck-

Uganda Revenue Agency (URA). (2020). *URA wants OTT tax scrapped, levied on data instead.* https://www.softpower.ug/ura-wants-ott-tax-scrapped-levied-on-data-instead/

United Nations. (2018). *UN flagship report on disability and development.* https://www.un.org/development/desa/disabilities/publication-disability-sdgs.html

United Nations Educational, Scientific and Cultural Organization (UNESCO). (2017). *Fact sheet no. 45.* Institute of Statistics. http://uis.unesco.org/sites/default/files/documents/fs45-literacy-rates-continue-rise-generation-to-next-en-2017.pdf

Verhoest, P. (2000). The myth of universal service: Hermeneutic considerations and political recommendations. *Media, Culture & Society, 22*(5), 595–610.

WIPO. (2021). *Contracting parties. Marrakesh VIP treaty.* https://wipolex.wipo.int/en/treaties/ShowResults?search_what=C&treaty_id=843

World Bank Blogs (2018). *The number of extremely poor people continues to rise in Sub-Saharan Africa.* https://blogs.worldbank.org/opendata/number-extremely-poor-people-continues-rise-sub-sahara africa#:~:text=However%2C%20the%20number%20of%20people,live%20in%20Sub%2DSaharan%20Africa

17

Accessible Mobile Banking in India

Raja Kushalnagar
Gallaudet University

17.1 Introduction

For people with disabilities, the ability to access and use banking services is a vital aspect of their participation and inclusion in society, reducing inequality. Research has shown that access to banking reduces poverty, stimulates investment, and creates growth, particularly in rural areas. Advocates for people with disabilities have long approached issues of inclusiveness as a basic human right. Economic incentives alone do not assure accessible design of banking services, due to the existing inequality and lack of opportunities to participate in banking for people with disabilities. Policymakers and researchers have shown that access to formal banking reduces poverty, stimulates investment, and creates growth, particularly for the 70% of the population living in rural areas (Beck et al., 2005; Binswanger & Khandker, 1995; Morduch, 1994). In particular, the ability to access banking services and save securely would make a difference, as it increases the resilience of citizens with disabilities to accumulate and monitor capital and invest it, and shield against income shocks.

Bank services can be expensive to deploy, as they require years of innovation and iteration; e.g., automated teller machines. Global South countries such as India are relatively resource-constrained compared with their Global North peers, which means that banks have limited resources to meet their customers' needs, absent advances in affordability through alternative approaches such as internet or mobile phone-based banking. In resource-constrained environments, concern with economic efficiency and collective social wellbeing can override concern with individual rights and accessibility.

The introduction of new technologies in Global South countries such as India has enabled banks to offer cheaper, less resource-constrained services through new technologies such as online banking through the internet or mobile banking through mobile phone services. In other words, Global South banks do not have an incentive to invest in new technical infrastructure (e.g., bank branches and automated teller machines) when it is relatively expensive in comparison to

Accessible Technology and the Developing World. Michael Ashley Stein and Jonathan Lazar, Oxford University Press.
© Oxford University Press 2021. DOI: 10.1093/oso/9780198846413.003.0018

potential customer wealth. It can be cheaper, faster, and easier to provide services over the internet or over mobile networks.

The adoption of remote banking not only gives users access to financial services by way of their mobile phone account, but also functions as a stepping stone toward the adoption of a traditional bank account (United Nations (UN), 2019). For instance, due to the low financial literacy rates in India, which ranks near the bottom among Asia-Pacific countries (Mastercard International, 2015), 90% of transactions in India are still made using cash (Reserve Bank of India, 2018). India's number of bank branches per capita is lower than those in countries in the Global North, at 14.7 branches per 100,000 residents (World Bank, 2019a).

17.2 Branch Banking

Banks have historically set up staffed branches to offer required banking services such as depositing or withdrawing funds, and checking balances. However, increasing the number of bank branches increases operating and overhead costs, and even low-cost branches regularly lose money each year (Reddy & Kumar, 2006). Furthermore, when demand exceeds supply, individuals with more resources, such as wealthier farmers and other individuals, are able to monopolize services at the resource-constrained branches, which exacerbates inequality and inaccessibility (Reserve Bank of India, 2018). Finally, in terms of accessibility, even if staff are mandated to provide accessibility, customers may not be able to access branch services if the surrounding environment, often rural, is not accessible, for lack of paved roads, electricity, or internet connectivity.

17.3 Online Banking

In light of the resource constraints for providing branches, banks have turned to offering banking services through the internet. In doing so, they can eliminate some resource constraints, such as roads or buildings, and shift some of the resources that they are expected to have for everyday use—access to internet and electricity, for instance—to the customer. It has become increasingly popular with customers, due to its convenience and ease of use; it removes the necessity of physically going to a bank. Without appropriate guidelines or adequate resources, development and operations of online banking aim to provide service for a "standard" customer. The online banking services often ignore the wide diversity in the people who are trying to access the banks' websites. However, online banking is not a stand-alone service, and must be paired with branch banking (Reserve Bank of India, 2020). This may change if the country has infrastructure to secure and verify online identities, and to vet large transactions.

Table 17.1 Types of banking service

Provider	Characteristics	Functionality
Branch Banking	Branch bank services are run by a registered bank Customer travels to branch to interact with staff to use bank services Customer travels to bank branch to enroll	All banking services
Online Banking	Online bank services are run by a registered bank Customer interacts with bank's website or apps to use bank services Customer must visit branch to enroll or conduct major transactions	Balances, alerts, transfers
Mobile Banking	Banking service provided by phone provider through phone billing Customer uses mobile provider's apps or data service to add to/debit the phone bill balance Customer travels to phone provider agent or middleman to do enrollment or transactions	All banking services

17.4 Mobile Banking

Mobile banking is a new banking service approach through which banks and non-banks such as phone companies can leverage the ubiquity of customer-owned mobiles by offering specialized applications to access financial services access anywhere and anytime (Table 17.1). Furthermore, the companies can increase banking service capability by leveraging the fact that customers have already verified themselves to the mobile company when they purchased mobile service. The financial companies are able to leverage the ubiquity, convenience, and accessibility of mobile networks, which can greatly increase access for people with disabilities. Mobile application accessibility is usually greater than bank branch accessibility, due to the fact that mobile banking requires fewer resources than brank branches and the infrastructure is easier to modify. For instance, mobile banking has also become more popular and more necessary during the COVID-19 pandemic.

17.5 Disability Rights in India

Disability is a complex, culturally, and environmentally relative term about what a person can and cannot do. It is culturally relative, as cultures define their norms of being and doing differently. It is environmentally relative, as a person may be able to do a daily activity in some environments but not others. The norms for daily activities vary, and impairments considered to be disabling in one environment may not be in another. For instance, severe asthma may be disabling in an agrarian economy, but not in an office-based city economy (Elwan, 1999; Groce et al., 2011). Furthermore, with the increased survival rates from disabling accidents and disease, as well as population aging, the percentage of people with disabilities increases. For instance, an estimated 35% of people aged over 65 in the United States have a disabling condition (Administration for Community Living, 2017).

17.6 Disabilities in India

India is a large Global South country with a population of 1.35 billion, a per capita income of $2,015 (World Bank, 2019a), and a reported disability rate of 2.21%. For comparison, the population of the United States is 327 million, its per capita income is $62,641 (World Bank, 2019b), and its reported disability rate is 13.2% (University of New Hampshire Institute of Disability, 2018).

The reported disability rate in India of 2.21% is far lower than in the United States due to differences in definition of disability, a higher age structure in developed countries (older people report more disabilities), and more accurate information collection in the Global North (Elwan, 1999; Groce et al., 2011). For instance, the Indian Census's disabilities definition included only those with visual, hearing, speech, locomotor, and cognitive disabilities, while the American Census Survey used a definition of disabilities that, in addition to the above, included mental and emotional disabilities. The likelihood of disability undercount in India is further reinforced by statistics collected by the World Health Organization, which estimates a global prevalence of 15%, which is on the rise due to population aging and chronic diseases (World Health Organization, 2020). From solely clinical evidence, it is currently thought that at least 3.7% of the total population are affected by locomotor, visual, or communication-related disabilities, or by intellectual disabilities (Kumar et al., 2012), and that the reported disability rate in India of 2.21% is vastly underestimated.

In India, people with disabilities and their families are more likely than the rest of the population to live in poverty (Government of India, 2006). It is a two-way relationship—disability adds to the risk of poverty, and conditions of poverty increase the risk of disability. Poor households do not have adequate food, basic sanitation, and access to preventative healthcare. They live in lower-quality housing and work in more dangerous occupations. While disability causes poverty, it is also possible that poverty causes disability (Rao, 1990). The mechanisms are malnutrition, exposure to disabling disease, inadequate access to preventative and curative health care, and an enhanced risk of occupation-related accidents among the poor.

In India, around 36% of citizens with disabilities are gainfully employed, compared with around 66% of their peers, and 57% of citizens with disabilities are illiterate, compared with 27% of their peers without disabilities. A significant percentage of citizens are poor in absolute terms as well: 21.2% are classified as poor with an income of $1.90 per day or less (World Bank, 2019a) and have too little to invest in banking services without government assistance. Furthermore, the government has not sufficiently invested in human capital such as schooling: 27% of citizens are illiterate (Data For All, 2011), i.e., do not have the ability to read and write with understanding in any language.

17.7 Accessibility Laws and Regulations in India

India has several legal mandates to regulate banking services for people with disabilities: the Indian constitution, legislation, treaties and conventions, and regulatory rules. Article 15 of the Constitution (Government of India, 1951) includes people with disabilities and provides the right to gain equal and accessible access to public services, including banking. The government has also passed several legislative acts over the years to cover services open to the public by both government and private agencies: The National Trust Act (Government of India, 2018) and PWDs Act (Government of India, 2016) aim to promote accessibility of services open to the public, to increase the independence of people with disabilities. However, the efficacy of these laws was blunted, as the laws up to, but not including, the PWDs Act 2016 lacked enforcement mechanisms to require monitoring groups to develop metrics, to assess accessibility, and to monitor compliance.

For instance, the earlier laws specified that the company or agency shall, within the limits of economic capacity and development, for the benefit of PWDs, take special measures to provide accessibility. The laws did not specify consequences if they are not followed. Most companies therefore did not have personnel knowledgeable or invested in developing and providing accessibility.

Accessibility metrics and compliance is often highly sector-specific, and companies or agencies lacked experienced personnel to coordinate and provide accessible guidelines. Furthermore, companies and agencies often abused the economic hardship exemption in the law by stating that the accessibility work and guidelines were beyond the limits of their economic capacity, which is similar to the exemptions in other countries, such as the "undue burden" exception in the Americans with Disabilities Act in the United States.

The laws also did not include mandates or funding for dissemination or inclusion of people with disabilities in the oversight process. If mandated or advocacy groups were involved from the start, then the voice of people with disabilities could be heard and acted upon. However, compliance became much easier when the regulatory agencies in major sectors such as banking passed regulations that required individual banks to designate groups or individuals focused on accessibility.

For banking services, the Reserve Bank of India (RBI) is the primary bank regulator and has issued several notifications that mandate banks as designated groups to ensure that banking facilities are available in a non-discriminatory manner to all customers. The RBI has been conferred wide powers under the Banking Regulation Act (Government of India, 1949) under which it can supervise and control the various banking companies, and they are bound to follow its directions.

Section 35A of the Act specifies that in public interest or in the interest of banking policy, the RBI can issue such directions as it deems fit, and the banking companies or the banking company, as the case may be, shall be bound to comply with such directions. For example, it requires member banks to require that all branches offer accessible banking facilities including checkbook facilities, automated teller machines (ATMs), and lockers (safe deposit boxes).

17.8 Accessible Banking in India

As a Global South country with resource-constrained infrastructure, it can be difficult for people with disabilities to use banking services in India. People with disabilities face many accessibility issues when they consider banking and financial services.

For instance, if a customer who has a disability walks into a branch for a home loan, the branch often does not have the resources or knowledge of who to contact to accommodate the customer. Without help in the form of helpers or guarantors who are fully capable, the chances of obtaining finance from the banks are low, because banks are likely to give a person with disability a much lower credit rating based on their own criteria. These credit rating determinations put the customer with a disability at a disadvantage. Banks often offer complex products— for example, multiple varieties of checks or savings—which can be difficult to understand for people with cognitive disabilities. A customer with, e.g., dyslexia will likely have severe difficulty in filling out an application form, and banks are often not welcoming to people with disabilities in terms of the attitude of the staff toward such difficulties. Similarly, finding bank branches or ATMs which are accessible can be a challenge.

The Government of India passed the PWDs Act of 1995, that attempted to address physical access to banks for people with disabilities. However, the act assumed that the customer with a disability has a care giver, which does not foster independence. Twenty years later, RBI passed the National Trust Act, which required branches to accept guardianship certificates, issued either by the district court under the Mental Health Act or by the local-level committees under the National Trust Act—when bank accounts were opened or operated by legal guardians on behalf of people with disabilities (RBI, 2009a). The RBI regulations also directed that information about the opening of such bank accounts be displayed conspicuously, in both English and the regional language (RBI, 2009b). This act enabled PWDs like autism and cerebral palsy to open and operate accounts. In practice, it meant that customers had to have their designated guardian with them at all times and could not work with professional assistants. A few years later, the PWDs Act of 2016 addressed this, through which the emphasis on medical rehabilitation was replaced by an emphasis on social rehabilitation; for example, banks would be expected to provide customers with direct interaction with trained bank representatives.

The RBI also issued a regulation (RBI, 2008) that stipulated that all banking facilities, including internet banking facilities, be "invariably offered to the visually challenged without any discrimination," eliminating the need for a guardian by providing direct services. The circular also included specifics when necessary. For instance, it directed that all banks with ATMs must take necessary steps to provide ramps for all existing or future ATMs so that wheelchair users or PWDs can easily access them and also to arrange them in such a way that the height of the ATM does not block its use by a wheelchair user. It also required branches to add ramps at bank entrances so that PWDs or wheelchair users can enter the bank branches and conduct business without much difficulty. Furthermore, banks are expected to make at least one-third of new ATMs installed "talking ATMs" with Braille keypads and place them strategically in consultation with other banks to ensure that at least one talking ATM with Braille keypad is generally available in each locality, to cater to the needs of visually impaired persons.

17.9 Accessible Mobile Banking in India

Mobile banking is a new banking approach through which banks and their customers can leverage the ubiquity of mobile services by using mobile services to access bank services such as bill payment or check their balance anywhere and anytime. This approach leverages the ubiquity, convenience, and accessibility of mobile networks, which can greatly increase access for people with disabilities.

Mobile banking has been found to be very helpful in many parts of the world with little or no infrastructure, especially remote and rural areas

(Bureau of International Settlements, 2016). This aspect of mobile commerce is popular in India and other Global South countries, where most citizens do not have bank accounts. In most of these places, banks can only be found in big cities, and customers have to travel hundreds of miles to the nearest bank. For people with disabilities, this is a big step forward, as it means they do not have to deal with the hassle and inconvenience of going to a bank, where they may not find the assistance that they need. In other words, customers with bank accounts may find it easier to access their details and complete transactions through their mobile devices—and in India, 37% do (Bain & Company, 2012).

In developing countries like India, where a significant percentage of the population is living below the poverty line, cost is one of the most important obstacles in accessing banking services. However, while economically poor people may not be able to afford a phone and associated service by themselves, they often pool resources by sharing a phone to access public services such as mobile banking. This accessibility means that a lot of people with disabilities who live in rural areas, who were previously unable to access banks, can now do so using their mobile phones. Mobile banking also makes it much easier for customers with bank accounts to access their details and do transactions. Since physical banks are often difficult for people with disabilities to navigate, mobile banking could provide the best solution. However, there is a built-in assumption of a robust mobile infrastructure, which is not always present in all parts of India.

17.10 Recommendations

Without clear legal or regulatory mandates, banks tend not to value accessibility as a stand-alone starting point for investment. As a result, accessibility initiatives are funded through discretionary budgets. The benefits and costs of investment in accessibility infrastructure to make it universally accessible are not easily quantified. This may be because there is currently no accepted method to accurately measure the benefits of accessibility improvements. As a result, banks may ignore universal accessibility investments in favor of investments based on "local knowledge" or as part of larger projects justified using economic appraisal for outcomes associated with safety or efficiency. Banks may also base their budgets for mobile banking investments on unsolicited or prompted feedback from local communities. While these processes usually result in incremental improvements to the system, there is no way of monitoring the investment's true effectiveness for all people, let alone the particular benefit to people who have a particular need for universally accessible environments.

Without mandates, banks may not have the interest to collect and use customers' disability information and status for project appraisal and investment decision making, particularly when compared to the copious quantities of data they collect about daily banking activities. Efficiency and deployment speed can readily outweigh vague "accessibility" objectives in the absence of data about beneficiaries of this investment. Despite best practice guidelines, the absence of data about people with disabilities means that their needs cannot be transparently prioritized when trade-offs are made in new design. The lack of effective, convenient, and attributable ways to measure mobile banking investment in accessibility tends to result in ad hoc solutions driven by local complaints, which can improve accessibility locally, but not globally.

Making banking accessible is not just in the commercial interest of the bank but is also in line with its commitments under applicable legislative and regulatory acts. Accessibility should be treated not just as a corporate social responsibility measure, but as a responsibility to be fulfilled. Further, public banks have the biggest responsibility to implement these measures, as they are required to ensure that at least 4% of all hires are people with disabilities (Lok Sabha, 2018). If the banks did not comply, their services would not be accessible to a significant fraction of their own employees. For instance, the banks need to ensure that the internal banking software which is used is accessible for people with disabilities and can be accessed by them using the appropriate assistive technology like screen readers. Ultimately, making financial services more accessible will only mean that their customer base will grow.

State and national governments should encourage opening of bank accounts by people with disabilities so that any funds or scholarships can be directly transferred into their account, as opposed to being given to organizations which may not transfer it to the beneficiaries—helping to curb malpractices. Information on how people with disabilities can open an account—whether joint or single—and the formalities they need to fulfill should be made easily and readily available. This will encourage more people to open accounts for/with people with disabilities.

17.11 Banking Mandates for Accessibility

It is important to engage people with disabilities on their own terms and let them take the lead, and recognize that technology powerfully shapes human outcomes. It is fairly common for developers to focus on features that are relatively unimportant to individuals with disabilities, even when explicitly designing accessibility into communications programs. Investigation of a specific issue in collaboration with a person with disability (a user-centered design process) is important for understanding the balancing act between the technology, the interface, and individual user needs. In addition, banks should collaborate with broader disability organizations to produce and distribute banking forms and evaluate banking technologies.

To remedy the ad hoc accessibility design driven locally by banks, the National Informatics Centre mandated all banks to follow a global, best practice accessible design focus. The department disseminated best practices for online accessibility in 2009 (National Informatics Centre, 2009); these are required for all government sites and recommended for public accommodations such as banking. These guidelines are based on International Standards, including ISO 23026, the World Wide Web Consortium's Web Content Accessibility Guidelines, and the Disability Act of India, as well as the Information Technology Act of India. The guidelines were classified into three categories: mandatory, advisory, and voluntary, that can be used by organizations to assess their compliance with the National Informatics Centre guidelines, which mirrors the Web Content Accessibility Guidelines' matrix of A, AA, and AAA.

Accessibility also helps address broader access issues. For instance, for banking customers who are not comfortable with English, bank websites should be provided in each of the fourteen regional languages in India, such as Hindi, Bengali, Marathi, Tamil, and Kannada. Bank websites were previously available mostly in either Hindi or English. They all have their own scripts, which require use of Unicode (UTF-8), a computing industry standard for the consistent encoding, representation, and handling of text expressed in most of the world's writing systems. If Unicode is not used to display the fonts, the fonts can become garbled and a person using a screen reader will not be able to access the written material at all. Of course, for accessibility, the visual information should also be coupled with equivalent audio information and capable of being configured and controlled by the user. As in any case, the World Wide Web Consortium's (W3C) Web Content Accessibility Guidelines (WCAG) 2.1, with its four stated principles of web content—i.e., that it must be perceivable, operable, understandable, and robust—is the relevant international document (WCAG, 2018).

One of the biggest barriers for customers with disabilities at banks is the requirement for them to satisfy the mandatory requirement to validate their identity to the bank. This government mandate, "Know Your Customer," often requires a series of steps—a call, visit, and summons to the office or branch, which can be daunting. One potential fix for this would be to set up one or more regulatory requirements, such as a minimum percentage of customers with disabilities, to incentivize banks to come up with schemes to attract people with disabilities as customers and assure them of good and competent service without discrimination as well as incentives to invest or save. It is important to repeatedly emphasize that people with disabilities are potential customers, and they should be viewed that way.

17.12 Alternative Accessible Banking Models

Alternative models in developing countries have sprung up and been successful elsewhere. For instance, in Kenya, in 2007, major telecom company Vodafone deployed a branchless, mobile-centric financial payment platform called M-Pesa, which has since been deployed in many countries, mainly in the Global South, and particularly in Africa and Asia. M-Pesa was able to scale up the platform by collaborating with other national telecom operators to charge customers for transactions and allied activities. One major advantage of M-Pesa over traditional banks was that it was accessible to virtually all citizens, since it could be used on cheap, affordable flip phones that could be bought by anyone immediately without credit. Another advantage was that it relied on agents on commission, instead of branches. This combination of larger customer bases and the lower overheads of agents on commission meant that far fewer bricks-and-mortar offices were needed. In 2015, the average distance to the closest agent was nearly one-tenth that of the average distance to the closest bank (Logan, 2016). In turn, this has allowed mobile money users to diversify their informal risk-sharing networks and draw on a wider network of social support.

In Kenya, to comply with national accessibility laws and regulations, Vodafone invested resources to make the M-Pesa platform accessible. For instance, the interface is screen-readable on phones that support it, and providers have deployed interpreters for customer service. Safaricom introduced a voice biometric feature called Jitambulishe that enables customers, including those with visual disabilities, to unlock and access their accounts through voice. Safaricom also integrated it with an interactive voice response to provide enhanced access to blind customers. Blind customers were able to check balances in their mobile wallets using their voices, therefore reducing reliance on others and safeguarding them against risks associated with sharing of personal information, such as fraud. Before the introduction of interactive voice response and voice biometrics, M-Pesa customers with visual disabilities had to rely on the assistance of a trusted second party to conduct transactions over M-Pesa. When they did not have a trusted confidant close by, they would have to settle for help from strangers.

Other countries have tried to duplicate Vodafone's M-Pesa platform through adoption of open-standards mobile platforms, such as PalmPay in Nigeria and Ghana, or MoneyOnMobile in India. However, providers have had trouble scaling up using these platforms, because each provider has an incentive to extend and create different implementations of the open-standards platform. This has led to a fragmented market with competing business rules, standards, and protocols, which means customer reach is limited, and providers have difficulties in complying with national banking standards, especially security and accessibility. Furthermore, smaller providers have found it extremely difficult to set up agency networks, as these have been very labor intensive and expensive to set up and can take years to become effective.

It takes significant resources to foster market coordination and incentivize adoption of rules, standards, or protocols to promote accessibility. While Safaricom's Jitambulishe greatly increased accessibility for customers, especially those who are illiterate or have visual disabilities, its adoption by customers was slow, due to lower levels of smartphone usage (Kamau & Sunday, 2019). However, because the implementation cost for Jitambulishe was low, Safaricom was able to continue support indefinitely. Accessibility works best when the costs are either spread over the entire customer base, or when costs are low enough to support people with disabilities, so as to serve poor users or support small transactions.

Global South countries typically have more mobile service subscribers than bank customers, unlike the Global North, where there are about equal numbers of mobile subscribers and bank customers. For instance, in India, a Global South country, nearly all adults (95%) are mobile subscribers (Statista, 2019), while nearly 80% are bank customers (Bank, 2017). In contrast, in the United States, a Global North country, nearly all adults have mobile subscriptions and bank accounts—96% for the former (Pew Research Center, 2019) and 93% for the latter (Federal Deposit Insurance Corporation, 2017).

It would be helpful if there was a monitoring or evaluating mechanism to see how far traditional banks or alternative bank services are complying with regulatory requirements. Many online bank services do not have a feedback option, which results in people with disabilities not being able to provide their feedback to the developers and maintainers of the services. The combination of lack of accessibility, usability, and feedback has a significant effect on the degree of trust and satisfaction by people with disabilities (Armstrong et al., 2016; Debevc et al., 2015).

17.13 Conclusion

Traditional banking services in the developing world have often been slow to expand accessibility due to their focus on serving through physical branches. Branches are expensive to set up; in turn this makes accounts expensive to maintain, and many are old and expensive to update and make fully accessible. Although public services like banks are required to follow both legal and regulatory mandates, courts tend to be slow and it takes a long time to enforce legal mandates. On the other hand, banks have to regularly report their conformance to regulatory mandates and as a result are more responsive to regulatory requirements. As such, people with disabilities may be able to push for effective accessibility steps by focusing not only on national and local government but also on regulatory and bank stakeholders to ensure that the banks consider accessibility when designing or rolling out services.

For Global South countries like India, legal enforcement tends to be less well developed than regulatory enforcement. Therefore, it may be advantageous for national governments to require regulators to encourage accessibility of public services and to focus on incorporating technological advances to scale up services like mobile banking. Technological advances provide new avenues to increase accessibility of public services such as banking by altering the economic calculus of providing accessible public services. Furthermore, regulatory agencies should not lock into current technology, as the technology used today may not be relevant in the future. Instead, regulations ought to be technology neutral, so that the technological advances that can be used for full accessibility can be utilized to their fullest potential to enable people to use public services.

Top down, regulators can incorporate best practices worldwide into country-specific and sector-specific regulations, followed by mandates for accessibility training and awareness. It would be helpful if there was a monitoring or evaluating mechanism to see how far banks are complying with the standards or guidelines that have been set forth before them. Bottom up, local banks by virtue of constantly interfacing with their customers, glean information from their customers, including people with disabilities, on their economic and social banking needs. Furthermore, it can be helpful to explicitly require inclusion as a local metric so that banks can measure accessibility gaps for banking services.

References

Administration for Community Living. (2017). *2017 Profile of older Americans*. https://acl.gov/sites/default/files/Aging%20and%20Disability%20in%20America/2017OlderAmericansProfile.pdf

Armstrong, M., Brown, A., Crabb, M., Hughes, C. J., Jones, R., & Sandford, J. (2016). Understanding the diverse needs of subtitle users in a rapidly evolving media landscape. *SMPTE Motion Imaging Journal, 125*(9), 33–41. https://doi.org/10.5594/JMI.2016.2614919

Bain & Company. (2012, December 4). *Banking loyalty profile: Global edition*. https://www.bain.com/insights/customer-loyalty-in-retail-banking-2012/

Banking Regulation Act—Government of India (1949). https://rbidocs.rbi.org.in/rdocs/Publications/PDFs/BANKI15122014.pdf

Beck, T., Kunt-Demirguc, A., & Levine, R. (2005). SMEs, growth, and poverty: Cross-country evidence. *Journal of Economic Growth, 10*(3), 199–229.

Binswanger, H. P., & Khandker, S. R. (1995). The impact of formal finance on the rural economy of India. *Journal of Development Studies, 32*(2), 234–262. https://doi.org/10.1080/00220389508422413

Bureau of International Settlements. (2016). *Payment aspects of financial inclusion.* CPMI Papers, Bureau of International Settlements. https://www.bis.org/cpmi/publ/d144.htm

Data For All. (2011). *CensusInfo India 2011.* http://www.dataforall.org/dashboard/censusinfoindia_pca/

Debevc, M., Milošević, D., & Kožuh, I. (2015). A comparison of comprehension processes in sign language interpreter videos with or without captions. *PLOS ONE, 10*(5), e0127577. https://doi.org/10.1371/journal.pone.0127577

Elwan, A. (1999). *Disability and poverty: A survey of the literature.* Social Protection Department. World Bank.

Federal Deposit Insurance Corporation (FDIC). (2017). *FDIC national survey of unbanked and underbanked households.* https://www.fdic.gov/analysis/household-survey/2017/index.html

Groce, N., Kembhavi, G., Wirz, S., Lang, R., Trani, J.-F., & Kett, M. (2011). Poverty and disability—A critical review of the literature in low and middle-income countries. *SSRN Electronic Journal.* https://doi.org/10.2139/ssrn.3398431

Indian Constitution—Government of India (1951).

Kamau, M., & Sunday, F. (2019). *Hits and misses at Safaricom amid swelling billions on profit.* The Standard.

Kumar, S. G., Roy, G., & Kar, S. S. (2012). Disability and rehabilitation services in India: Issues and challenges. *Journal of Family Medicine and Primary Care, 1*(1), 69–73. https://doi.org/10.4103/2249–4863.94458

Lok, S. (2018). *Policy of Reservation to Persons with Disabilities.*, Pub. L. No. 36035/02/2017.

Mastercard International. (2015, April 14). *Asia-Pacific financial index literacy rates.* https://www1.mastercard.com/content/intelligence/en/research/reports/2015/mastercard-financial-literacy-index-report-2014h1.html

Morduch, J. (1994). Poverty and vulnerability. *The American Economic Review, 84*(2), 221–225. https://www.jstor.org/stable/2117833

National Policy for People with Disabilities—Government of India (2006).

Pew Research Center. (2019). *Internet and technology: Mobile fact sheet.* https://www.pewresearch.org/internet/fact-sheet/mobile/

National Informatics Centre. (2009). *Guidelines for Indian government websites.* http://egovstandards.gov.in/sites/default/files/GOI%20Web%20Guidelines.pdf

Rao, N. (1990). Integrating the disabled—A reality? *Indian Journal of Social Work, 51*(1), 149–157.

Reddy, A. A., & Kumar, P. (2006). Occupational structure of workers in rural Andhra Pradesh. *Journal of Indian School of Political Economy, 5*(1), 77–91. http://ispepune.org.in/PDF%20ISSUE/2006/JISPE1206/2006-1&2-5REDDY.PDF

Reserve Bank of India. (2008). *Banking facilities to the visually challenged.* https://www.rbi.org.in/Scripts/NotificationUser.aspx?Id=4226&Mode=0

Reserve Bank of India. (2009a). *Master circular on customer service*. DBOD.No.Leg. BC.9/09.07.006/2009-1022. https://rbidocs.rbi.org.in/rdocs/notification/PDFs/57 MCATT010709.pdf

Reserve Bank of India. (2009b). *Directions for opening and operation of accounts and settlement of payments for electronic payment transactions involving intermediaries*. https://www.rbi.org.in/scripts/NotificationUser.aspx?Mode=0&Id=5379

Reserve Bank of India. (2018). *Bank-wise ATM/POS/card statistics*. https://m.rbi.org. in/Scripts/ATMView.aspx

Reserve Bank of India. (2020). *Master direction—Know Your Customer (KYC) Direction, 2016 (Updated as on April 20, 2020)*.

Statista. (2019). *Forecast of mobile phone users in India*. https://www.statista.com/statistics/274658/forecast-of-mobile-phone-users-in-india/

The National Trust for Welfare of Persons with Autism, Cerebral Palsy, Mental Retardation and Multiple Disabilities (Amendment) Bill, 2018—Government of India (2018).

The Rights of Persons with Disabilities Act—Government of India (2016).

United Nations. (2019). Sustainable Development Goals. Retrieved from https://sustainabledevelopment.un.org/?menu=1300

University of New Hampshire Institute of Disability. (2018). *2018 annual compendium of disability statistics*. University of New Hampshire, Institute on Disability.

Web Content Accessibility Guidelines (WCAG) 2.1. (2018). https://www.w3.org/TR/WCAG21/

World Bank. (2017). *The global Findex database 2017*. https://globalfindex.worldbank.org/sites/globalfindex/files/2018-04/2017%20Findex%20full%20report_0.pdf

World Bank. (2019a). *World Bank data—India*. https://data.worldbank.org/country/india

World Bank. (2019b). *World Bank data—USA*.

World Health Organization. (2020, January). *Disability*. https://www.who.int/health-topics/disability#tab=tab_1

Subject Index

Notes

Tables and figures are indicated by an italic *t* or *f* following the page number.
Abbreviation used in this index

    CRPD - Convention on the Rights of Persons with Disabilities
    DARE Index - Digital Accessibility Rights Evaluation Index
    LMICs - low- and middle-income countries
    WAI - Web Accessibility Initiative